PROGRESS IN
MATHEMATICS

Volume 12

Algebra and Geometry

PROGRESS IN MATHEMATICS
Translations of Itogi Nauki — Seriya Matematika

1968: Volume 1 — Mathematical Analysis
      Volume 2 — Mathematical Analysis
1969: Volume 3 — Probability Theory, Mathematical
            Statistics, and Theoretical
            Cybernetics
      Volume 4 — Mathematical Analysis
      Volume 5 — Algebra
1970: Volume 6 — Topology and Geometry
      Volume 7 — Probability Theory, Mathematical
            Statistics, and Theoretical
            Cybernetics
      Volume 8 — Mathematical Analysis
1971: Volume 9 — Algebra and Geometry
      Volume 10 — Mathematical Analysis
      Volume 11 — Probability Theory, Mathematical
            Statistics, and Theoretical
            Cybernetics
      Volume 12 — Algebra and Geometry

In preparation:
      Volume 13 — Probability Theory, Mathematical
            Statistics, and Theoretical
            Cybernetics
      Volume 14 — Algebra, Topology, and Geometry
      Volume 15 — Mathematical Analysis

PROGRESS IN
MATHEMATICS

Volume 12

Algebra and Geometry

Edited by
R. V. Gamkrelidze

V. A. Steklov Mathematics Institute
Academy of Sciences of the USSR, Moscow

Translated from Russian by Nasli H. Choksy

ℙPLENUM PRESS • NEW YORK–LONDON • 1972

The original Russian text was published for the All-Union Institute of
Scientific and Technical Information in Moscow in 1970 as a volume of
Itogi Nauki — Seriya Matematika

Library of Congress Catalog Card Number 67-27902

ISBN 978-1-4757-0509-6   ISBN 978-1-4757-0507-2 (eBook)

DOI 10.1007/978-1-4757-0507-2

The present translation is published under an agreement with
Mezhdunarodnaya Kniga, the Soviet book export agency

© 1972 Plenum Press, New York

Softcover reprint of the hardcover 1st edition 1972

A Division of Plenum Publishing Corporation

227 West 17th Street, New York, N.Y. 10011

United Kingdom edition published by Plenum Press, London

A Division of Plenum Publishing Company, Ltd.

Davis House (4th Floor), 8 Scrubs Lane, Harlesden, NW10 6SE, England

PREFACE

This volume contains five review articles, three in the Algebra part and two in the Geometry part, surveying the fields of ring theory, modules, and lattice theory in the former, and those of integral geometry and differential-geometric methods in the calculus of variations in the latter. The literature covered is primarily that published in 1965-1968.

CONTENTS

ALGEBRA

GEOMETRY

INTEGRAL GEOMETRY

G. I. Drinfel'd

DIFFERENTIAL-GEOMETRIC METHODS IN THE CALCULUS OF VARIATIONS

N. I. Kabanov

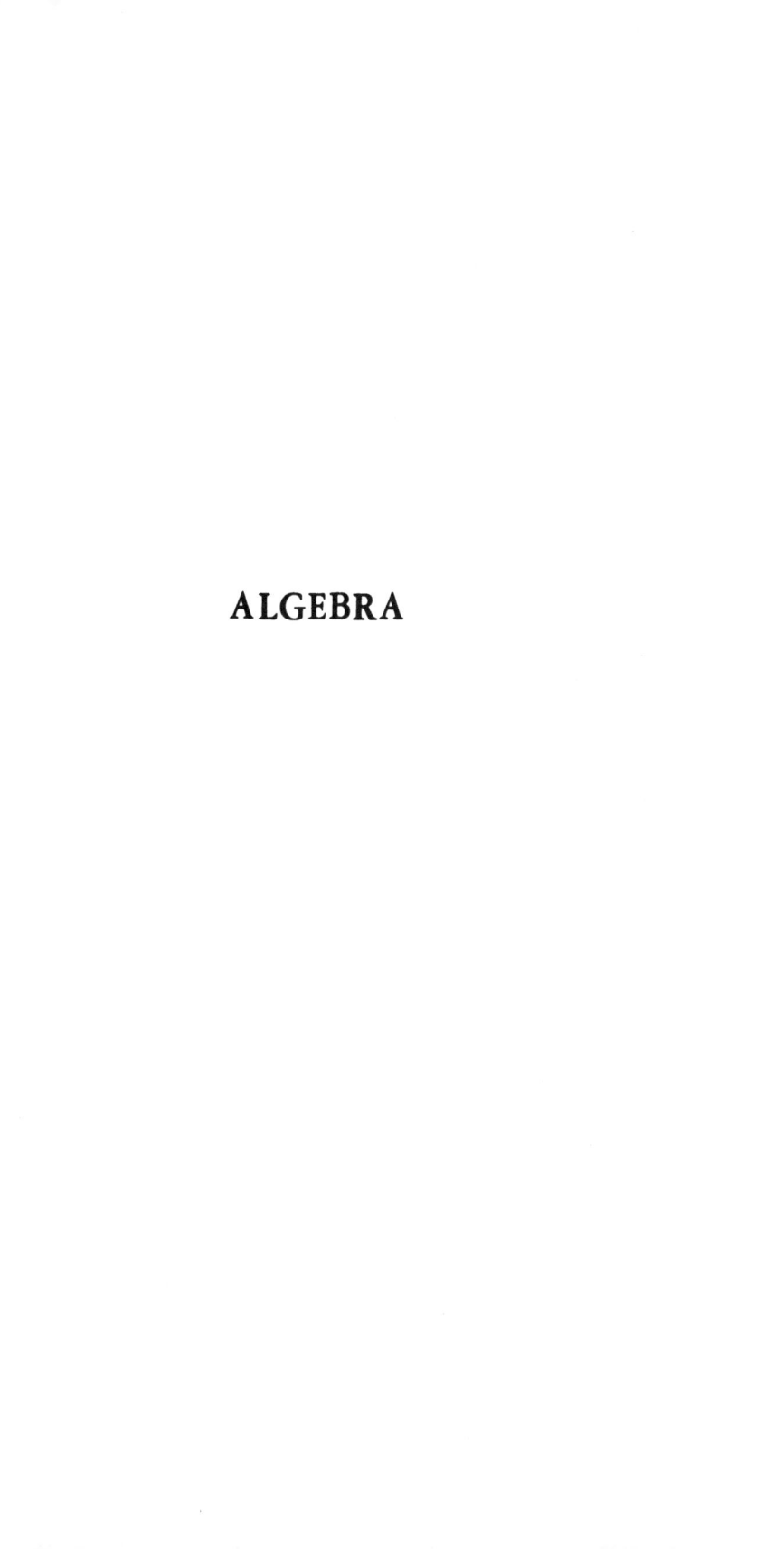

ALGEBRA

Ring Theory

L. A. Bokut', K. A. Zhevlakov, and E. N. Kuz'min

In this survey we cover the papers on ring theory reviewed in the Mathematics section of Referativnyi Zhurnal during 1966–1968, with the exception of Nos. 11 and 12 of 1968, with which the authors had no opportunity to acquaint themselves. In individual cases earlier results are mentioned, as well as articles which had not been reviewed upto the moment of writing.

§ 1. Associative Rings

The rings being considered in this section are associative. Unless we say otherwise the terminology is referred to the left (modules are left modules, an Artinian ring is a left-Artinian ring, etc.). The following branches of the theory of associative rings do not appear in this survey: commutative rings, extensions of rings, ordered rings, and generalizations of rings. Unless we specify otherwise, primality and radicality are to be understood in the sense of Jacobson.

1. Rings with Chain Conditions and Quotient Rings.

We call a ring R the (two-sided, left, right) order in a ring Q if Q is the classical full (two-sided, left, right) quotient ring for R. A ring R is called a left (right) Goldie ring if R satisfies the maximality condition for the left (right) annihilator ideals and if every direct sum of nonzero left (right) ideals contains only a finite number of terms. A left and right Goldie ring is called simply a Goldie ring. The following two theorems of Goldie [305, 306] have evoked large response: A ring R is a two-sided order in a prime ring with minimality condition if and only if R is a prime Goldie ring. A ring R is a right order in a semiprime ring with minimality condition if and only if R is a semiprime right Goldie ring.

The right orders in right-Artinian rings have been defined in [648, 649] and the one-sided orders in quasi-Frobenius rings in [376]. Recall that a ring R is called quasi-Frobenius if it is left- and right-Artinian and if for any left ideal A and any right ideal B of ring R there hold the equalities: $A = l(r(A))$ and $B = r(l(B))$, where r is the symbol for the right annihilator and l for the left.

Orders in prime rings have been studied also in [513]. A ring which is right- and left-Noetherian but does not have either a left or a right classical quotient ring, has been constructed in [647]. Classical quotient rings for rings not necessarily satisfying a chain condition, were analyzed in [515, 52, 226], while generalized classical quotient rings were analyzed in [117, 50].

Nonclassical quotient rings (in the sense of Jacobson and Utumi) were studied in [256, 276, 384, 407], and for Noetherian rings in [449, 681]. The construction of quotient rings in the sense of Jacobson for a class of rings wider than prime Goldie rings, has been described in [385].

Let us dwell on other aspects of rings with chain conditions. A ring R is called quasi-Frobenius if and only if every projective R-module is injective [274]. If S is a finite inverse semigroup and F is a field, then the semigroup algebra F(S) is quasi-Frobenius [731]. Every nilsubsemigroup of an Artinian or a Noetherian ring is nilpotent [257]. The condition when every proper ideal I of a Noetherian ring R will be generalized nilpotent ($\bigcap_{n \geqslant 1} I^n = 0$), was given in [495]. There exists a Noetherian ring whose Jacobson radical is not generalized nilpotent [339]. The lower nilradical of a Noetherian ring R is the intersection of all maximal nilpotent subrings of ring R [512]. If R is a nilring, being an algebra over a commutative integral domain Ω and satisfying a maximality condition for left annihilators, and if an identity over Ω, one of whose coefficients is 1, is fulfilled in R, then R is nilpotent [343].

A nonzero right ideal I of ring R is called uniform if I does not contain direct sums of right ideals of ring R. A prime ring R with unity possesses a uniform right ideal if and only if the lengths of the direct sums of right ideals in R are bounded [328]. Prime rings with maximal annihilator right ideals were considered in [422]. Conditions for the existence of a unity in a Noetherian ring have been given in [606, 401].

New results on the additive theory of ideals occur in [7-10, 12, 13, 36, 37, 105, 174, 493]. A large number of these papers were surveyed in the article "Rings" in Progress in Mathematics, Vol. 5: Algebra.

2. PI-Rings and Related Questions. The concept of a rational identity was introduced in [146]. Let D be an algebra with unity over an infinite field F, let $X = \{x_1, x_2, ...\}$ and $Y = \{y_1, y_2, ..., y_n\}$ be two systems of unknowns (infinite and finite), let F [X, Y] be a free associative algebra over F of the set $X \cup Y$, and let $p_i(x, y) = p_i(x, y_1, ..., y_{i-1})$ be fixed elements of the algebra F [X, y]. An element $\varphi(x, y)$ of algebra F [X, Y] is called a rational identity of algebra D if D satisfies the following quasi-identity:

$$\bigwedge_{i=1}^{n} (y_i p_i(x, y) - 1 = 0) \rightarrow \varphi(x, y) = 0.$$

The basic result of [146] is the following theorem: A rational identity fulfilled in all the matrix algebras F_n, $n \geq 1$, is fulfilled in all bodies over field F. Applications of the results of the paper to various aspects of algebra and projective geometry have been obtained. In particular, it has been proven that the multiplicative group of a noncommutative division ring* with an infinite center does not satisfy any identity whatsoever.

In [150] the theory of commutative Hilbert algebras and of Jacobson rings, developed by Curtis, Goldman, and Krull, is carried over to the noncommutative case. A ring R is called a Jacobson ring if any prime ideal I of ring R (R/I is a prime ring) is the intersection of maximal ideals. An algebra R over a field F is called a Hilbert algebra if for any maximal ideal $P \subset R$ the factor algebra R/P is finite dimensional. Let Ω be a commutative Jacobson ring, let $R = \Omega[x_1, ..., x_n]$ be a finitely-generated Ω-algebra, and let R satisfy some identity (with coefficients from Ω). Then:

---

*Translator's Note: For many years now the term skew field or sfield has been used to refer sometimes to not necessarily commutative fields and sometimes to definitely noncommutative fields. To make things more confusing the term division ring has been treated as a synonym. The Russian language has two distinct words: "telo" to mean a (general) field and "pole" to mean a commutative field. In this translation I have used "division ring" for the Russian word "telo" because this is what appears in the titles in the bibliography and in the usage as reflected in the reviews in Mathematical Reviews of these papers.

1) R is a Jacobson ring, 2) if M is a maximal ideal in R, then $\Omega \cap M$ is a maximal ideal in Ω and R/M is a finite-dimensional algebra over the field $\Omega / \Omega \cap M$. Hence it follows that a finitely-generated algebra over a Jacobson ring and over a Hilbert algebra is a Hilbert algebra. Jacobson rings have been treated also in [584].

Let $\mathfrak{M}$ and $\mathfrak{N}$ be two manifolds of associative algebras over a field F. The following class of algebras is called the product $\mathfrak{M}\mathfrak{N}$ of these manifolds: $\mathfrak{M}\mathfrak{N} = \{A \mid \exists B \lhd A, B \in \mathfrak{M}, A/B \in \mathfrak{N}\}$ (B $\lhd$ A means B is an ideal of A). It is not difficult to show that $\mathfrak{M}\mathfrak{N}$ is once again a manifold. The products of a commutative manifold of algebras over F by nilpotent manifolds of algebras over F (KD-algebra) were studied in [134]. It was shown that any KD-algebra is locally left-(right-)Noetherian. The sum of any finite number of KD-ideals is a KD-ideal. An algebra over a field of characteristic zero with the identity relation

$$x_1^{t_1}x_2^{t_2}x_3^{t_3} + x_2^{t_2}x_1^{t_1}x_3^{t_3} - x_2^{t_2}x_3^{t_3}x_1^{t_1} - x_3^{t_3}x_2^{t_2}x_1^{t_1} = 0,$$

$t_i \geq 1$, is a KD-algebra.

The following generalization of Hilbert's theorem on a basis has been obtained in [90]: Let R be an associative algebra with a finite number of generators over a field F of characteristic zero and let the identity

$$xy^m = \sum_{r=1}^{m} \alpha_r y^r x y^{m-r}, \quad \alpha_i \in F.$$

be fulfilled in R. Then R is left-Noetherian. The preceding identity follows, for example, from the identity $v(x_1, \ldots, x_m) = 0$, where v is a (commutator) right-normalized Lie word in the variables.

The identities of a full matrix algebra were studied in [31]. It was shown that the identities of the algebra F_n, n = 2, 3, 4, 5, 6, F is a field, do not follow from the standard identity $S_{2n} = 0$ (this was known for n = 2 (Latyshev, P. Cohn)). Finitely-generated PI-algebras over a field were considered in [583]. Certain ideas from noncommutative algebraic geometry (manifolds, dimensionality of manifolds, etc.) were developed. Results which generalized the commutative case were obtained.

If the Jacobson radical of ring R is nilpotent, then the polynomial ring R[x] possesses a strongly pivotal monomial if and only
if a certain identity is fulfilled in it [373]. Let R be an algebra
over a field, satisfying an identity of degree d, and let S be a multiplicative subsemigroup (with the zero of the algebra) of R. Then
1) If S is nilpotent of index n, then the ideal of algebra R, generated
by $S^{[d/2]}$ is nilpotent of index $(d + 1)^{2n}$, 2) If S is a nilsemigroup,
then S is nilpotent [147]. In this same paper [147] significantly
more general results were obtained. Let R be a prime ring satisfying an identity with coefficients from Ω, where Ω is some collection of endomorphisms of the additive group of ring R. Then: 1)
All the divisors of zero of ring R are two-sided, 2) R has a two-
sided classical quotient ring Q, Q is a central prime algebra of
dimension $n^2 < \infty$ over its own center, 3) $R \subseteq H_n$, H is a field [149].
Every T-ideal containing a right-normalized commutator $[x_1, x_2,
x_3, x_4]$ is finitely-generated (as a T-ideal) [91].

Rings in which the ideal generated by commutators is a nilring, were studied in [403, 404]. A finite-dimensional algebra
(over an arbitrary field) possessing a nilbasis is nilpotent [138].
A ring which is the sum of two nilpotent subrings is nilpotent [400].
A ring R is called a radical extension of its own subring A if R
is a nilring modulo A. If R is a radical extension of a commutative
semiprime subring and R does not contain nilideals ($\neq 0$), then R
is commutative [159]. The structure of nilpotent finite-dimensional algebras was analyzed in [562].

3. Radical of Rings. Paper [633] presents an exam-
ple of a simple radical ring, in the sense of Jacobson (this ring is
an algebra over a field F which can be taken arbitrarily). It has
been proven that in the class of associative rings the Kurosh chain
for constructing the lower radicals can be stabilized at the spot
ω_0, where ω_0 is the first infinite ordinal [668]. In the class of all
rings the Kurosh chain cannot be stabilized at any ordinal whatsoever [108]. If P_1 is a hereditary class of rings, closed relative to
homomorphisms, then the Kurosh chain breaks off at the third
step [160].

The radicals of rings with operators were considered in
[252]. Let S be a Kurosh-Amitsur radical defined on the class of
associative rings. For every ideal I of ring R the radical S(I) is
an ideal in R. This is true even for alternative rings [157]. The

preradicals of rings (the functor $f \colon R \to fR$, where fR is an ideal, and $(fR)^\mu \subseteq fR^\mu$ for any epimorphism μ) were treated in [511]. We remark that preradicals of modules had been previously considered in [6]. Papers [106, 107] dealt with the general theory of radicals of rings (see the article "Rings" in Progress in Mathematics, Vol. 5). One class of radicals of associative rings was defined in [101]. These radicals turned out to be special in the sense of Andrunakievich. Special radicals were considered also in [180].

A radical of an Artinian ring is the intersection of maximal nilpotent subrings [171]. The projective limit of radical (semiprime) rings will be a radical (semiprime) ring [364]. Let F (G) be a group algebra of a finite group over an algebraically closed field of characteristic p > 2, and Let N be a radical of this algebra. All groups G for which N is commutative have been described [714]. A new characterization of the Levitzki radical was given in [719]. The well-known results of Amitsur and Bovdi on semiprime semigroup rings were generalized in [639]. The semisimplicity of the Bovdi crossed product of a right-ordered group G and of a ring R of characteristic $\neq 2$, whose radical does not contain zero divisors, was proven in [63].

4. Questions Connected with the Commutativity Property. The center of a universal enveloping algebra U of a finite-dimensional Lie algebra L over a field of characteristic zero was studied in [254]. In particular, if the radical of algebra L is nilpotent, then the center is not trivial. Wedderburn's theorem on the commutativity of a finite division ring was generalized in [559]. A ring R is said to be generalized commutative if for any x, y $\in$ R there exist positive numbers n (x, y) and m (x, y) such that $(xy)^{m(x, y)} = (yx)^{n(x, y)}$. A semiprime generalized-commutative ring is commutative [191]. A noncommutative algebra all of whose proper subalgebras are commutative is called a single-step algebra. Finite-dimensional single-step algebras without unity were described in [27]. A ring R is called subcommutative if for any a, b $\in$ R there exist c, d $\in$ R such that ab = bc and ab = da. Subcommutative rings were examined in [54, 296, 591]. The following properties of a ring, always satisfied in the commutative case, were considered in [51]: (α) Every principal ideal (a) generated by nonunitary elements a (not having one-sided inverses)

differs from the ring R; (β) In the ring R every maximal ideal A is completely prime (R/A has no zero divisors). A ring of infinite series $R[[x_1, \ldots, x_n]]$ is an α-ring (a β-ring) if and only if R is an α-ring (a β-ring). In [460, 478] it was proved again that a ring R in which for every x there exists n(x) > 1 such that $x^{n(x)} = x$, is commutative.

5. Matrix Rings. The book [120] is devoted to the study of maximal commutative subalgebras of a complete matrix algebra F_n, where F is a field, $n \geq 2$. It is proven that for any k, $2 \leq k \leq n$, in F_n there also exist subalgebras of nilpotency index k. In the extreme case (k = 2, k = n) the number of classes of adjoint subalgebras of the type indicated is finite (it equals (n−1) and 1, respectively). The case of the complex number field is delineated separately. Problems still exist. In [49] it was proven that in the algebra F_6, where F is an algebraically closed field of characteristic zero, the number of classes of adjoint maximal commutative subalgebras is finite. The conjecture (see [300]) that maximal commutative subalgebras F_n have a dimension $\geq$ n was refuted in [242] (a subalgebra of the type indicated of dimension 13 was constructed in F_{14}).

Let R be a ring, $M_\rho(R)$ be the ring of finite-rowed infinite matrices over R, $M_\rho^*(R)$ be the ring of bounded-rowed matrices over R (for every matrix A there exists N such that $a_{ik} = 0$ when k > N), M_ρ' be the ring of homogeneous finite-rowed matrices (for every matrix A there exists n such that if n(i) is the number of nonzero elements in the i-th row, then $n(i) \leq n$). In [570] it was proven that $\Gamma(M_\rho^*(R)) = M_\rho^*(\Gamma(R))$, where Γ is the Jacobson radical. Further, $\Gamma(M_\rho'(R)) = M_\rho'(\Gamma(R))$ if and only if $\Gamma(R)$ is nilpotent. In [632] it was proven that R is primitive if and only if $M_\rho(R)$ is primitive.

The rings R and S are said to be Morita-equivalent if the category of right R- and S-modules are equivalent (see [525]). In [238] it was proven that the following properties of a ring are equivalent: 1) $R = K_n$, K is the right domain of free ideals (i.e., K is a noncommutative domain with invariant basis number and all right ideals of K are free, as R-modules); 2) R is Morita-equivalent to the right domain of free ideals. Furthermore, also equivalent are the conditions: 1) $R = K_n$, K is the domain of principal right ideals; 2) R is Morita-equivalent to the domain of principal right ideals.

6. Multiplicative Semigroups of Rings. The following problem of Mal'tsev, arising at the end of the Thirties, was solved in [208,406,21]: Does there exist an associative ring R without zero divisors, not embeddable in a division ring, whose multiplicative semigroup is embeddable in a group? It turned out that such rings do exist. It was shown in [21] that they exist even in the class of semigroup algebras. The existence of rings with the above-mentioned properties in the class of group rings remains an open problem. Besides, we do not know as yet even an example of a group algebra without zero divisors, not embeddable in a division ring.

One class of rings without zero divisors has been described (in terms of generators and ratios) in [23, 24]. The results in these papers were essentially used in [21]. In [121] it was proven that every irreducible element without a free term of a free algebra remains irreducible in the algebra of infinite series. [76] is on expansions into irreducible elements in weak Bezout rings.

7. Topological Rings. Division Ring. An analog of the Nagata-Higman theorem on the nilpotency of associative nilrings of bounded index for topological rings was formulated in [94]. A number of papers [733-736] have studied linearly compact rings. Here it was proven, in particular, that the following conditions are equivalent for a linearly compact ring R:

1) R is regular in the sense of von Neumann;

2) the closure of any principal left ideal contains a right-sided unity;

3) the Jacobson radical of ring R equals zero.

An example was presented in [97] of a nontrivial topologization of the rational number field, i.e., a topologization, different from the discrete, the p-adic, the Archimedean ones or their combinations, while [98] presented an example of a complete locally-bounded nonnormalizable extension of the field of p-adic numbers. Commutative locally-bicompact rings satisfying the boundedness condition and the maximality condition for ideals containing a given open ideal, were studied in [548]. A meaningful supplement to

Jacobson's density theorem was obtained in [722]. Here it was proven that a connected locally-bicompact primitive ring with a minimal left ideal is isomorphic to a complete matrix ring over the real or complex number field or over the quaternion division ring. A nondiscrete locally-bicompact prime ring with a minimal left ideal is isomorphic to a complete matrix ring over a locally-bicompact division ring. If A is a nondiscrete locally-bicompact primitive algebra over a topological field F which either is nondiscrete or is of characteristic zero or is noncountable, then A is finite dimensional over its own center and contains a unity. A pseudonormalizability criterion for topological rings was obtained in [17].

Let Δ be a division ring, $\{\Delta_n\}$ a sequence of matrix rings over Δ, whose orders range over the number sequence (n) $=\{n_i \mid n_0 = 1, n_{i+1} = n_i q_{i+1}, q_{i+1} > 1, i = 0, 1, 2, \ldots\}$. A rank-function is defined in the ring $R = \lim_{\rightarrow} R_n$. The complement $\hat{R} = CR(\Delta)$ of the ring R in the rank-metric is a continuous regular prime ring and $\hat{R}$ does not depend upon the choice of the sequence (n) [127]. These assertions were stated (without proof) by von Neumann. A characterization of the class of rings $\{CR(\Delta)\}$, where Δ is a field, was obtained in [128]. In [245, 316-318] the above-mentioned von Neumann constructions were studied for the case when Δ is a regular ring. The results here partially overlap those in [127]. In particular, if Δ if a regular ring with a rank-function, then $CR(\Delta)$ is a regular ring with a rank-function prolonging the rank-function of ring Δ [316] and $CR(\Delta)$ does not depend upon the choice of the sequence (n) [318].

If Q is the rational number field, $A_1, \ldots, A_n$ are nth-order square matrices with elements algebraically independent over Q, K is a purely transcendental extension of field Q, generated by the elements of matrices A_i, K_n is a complete matrix algebra over K, then, as was established in [67, 68], the matrices A_i generate a a certain division subring D in K_n, having a dimension n^2 over its own center Z, and the degree of transcendentality of Z over Q equals $(m-1) n^2 + 1$. These assertions were proven in [583] from a more general position.

It is well known that finite-dimensional algebras with a divisor over the real number field R exist if and only if the dimension of the algebra equals 1, 2, 4, or 8. The class of eight-dimensional algebras with a divisor over R, close to the Cayley-Dickson algebra, has been analyzed in [78,79] — these are quadratic algebras in which any two elements generate a subalgebra of not more than four dimensions. It turns out that also in the case of an arbitrary base field an algebra with analogous properties may be described completely [83]. We note the following assertions. Let F be a field of characteristic $\neq 2$, and D be a quadratic algebra with a divisor over F, in which any elements generate a subalgebra of no more than four dimensions. Then the algebra D is finite dimensional over F and the dimension of D equals 1, 2, 4, or 8.

The quotient division ring of a universal enveloping algebra U for a finite-dimensional Lie algebra L over a field of characteristic zero was studied in [193, 656]. If L is a semisimple algebra, then the center of the quotient division ring coincides with the quotient field for the center of algebra U [656]. The same thing is true if L is a nilpotent algebra, whereas, in general, this property is violated for solvable Lie algebras.

Of interest from the systematic point of view is paper [503] wherein is given a new proof of Albert's theorem on finite division rings with associative powers.

Examples have been constructed in [237] of extensions K/T, where K and T are division rings, such that $[K:T]_R = n < \infty$ and $[K:T]_L = \infty$ (here $[K:T]_R$ is the right dimension of K over T, $[K:T]_L$ is the left dimension).

8. Other Aspects of Ring Theory. Algebras in which every subalgebra is an ideal were described in [465]. The considerably more difficult question of describing rings all of whose subrings are ideals (the Hamiltonian rings) was examined in [1-4]. Rings in which every additive subgroup is a subring were described in [489].

The right (left) domains of free ideals were studied in [115] (in this paper such rings are called right and, respectively, left Cohn rings; see Paragraph 5 for the definition). In particular, it is proven that the condition of the absence of a zero divisor is a consequence of the remaining conditions in the definition of right (left) domains of free ideals. A right domain of free ideals exists which is not a left domain of free ideals.

The definition of a free n-nilpotent sum of associative algebras, where n is an integer, $n \geq 2$, was introduced in [64]. It was proven that if an algebra is decomposed into a variety of nilpotent sums (n, m-sums, $n \neq m$), then one of these sums is a direct sum. Ultraproducts of associative rings were considered in [283]. It was proven that the ultraproduct of primitive rings is primitive. The following question was studied in [170]: Let A and B be associative algebras with isomorphic subalgebra lattices. What is the relation between A and B? It was proven that if A is the ring of matrices of order ≥ 3 over a division ring, finite-dimensional over the center, then there exists a semilinear isomorphism or an anti-isomorphism of A onto B. Let R be a ring and A and B be isomorphic subrings of R. The isomorphism $\varphi: A \to B$ is called continuable if the ring R is imbeddable into a ring R' in which φ is continued upto an automorphism of ring R'. There exist a ring R, two subrings A and B, and two isomorphisms φ_1, $\varphi_2: A \to B$ such that one of these isomorphisms is continuable while the other is not [25]. An associative ring without nilelements is a subdirect sum of rings without zero divisors [14]. The analog of Magnus' theorem on freedom for associative algebras with one defining relation was obtained in [461].

§2.   Lie Algebras and Their Generalizations

__1.   Lie Algebras.__ The theory of infinite-dimensional representations of nilpotent Lie algebras over the complex number field, paralleling the theory of representations of Lie groups [69], was constructed in [255]. If U is a universal enveloping algebra of a nilpotent Lie algebra L, then the ideal $I \subset U$ is the kernel of some irreducible representation of L when one of the following equivalent conditions is fulfilled: 1) I is the maximal ideal in U; 2) the center of the algebra U/I consists only of scalars; 3) the algebra U/I is isomorphic to one of the algebras A_n ($n \geq 0$) with generators p_1, p_2, ..., p_n; q_1, q_2, ..., q_n and with defining relations $[p_i, p_j] = [q_i, q_j] = 0$, $[p_i, q_j] = \delta_{ij} \cdot 1$. The representation of algebra L is not defined on its kernel. Analogous questions for solvable Lie algebras were treated in [253]. An interesting addition to Ado's theorem was obtained by Hochschild [346]. It turned out that for any finite-dimensional Lie algebra L over an arbitrary field there exists an exact linear representation ρ such that the operator $\rho(x)$ is nilpotent each time that the operator ad x, $x \in L$, is nilpotent.

Of great interest is the problem of classifying finite-dimensional simple Lie algebras of characteristic $p > 0$ (it is assumed additionally that the algebras being considered are restricted). This problem becomes a problem of classifying finite prime groups, however, in contrast to group theory, where numerous series of prime groups are known and where from time to time there appear new prime groups not occurring in any one of these series, there are comparatively few known classes of simple Lie p-algebras. As Kostrikin and Shafarevich [75] have shown, all these classes are connected in a defined manner with simple Lie Algebras over a field of characteristic zero. In the case of finite-dimensional prime Lie algebras over a field of characteristic zero the Chevalley procedure (reduction mod p, complemented, possibly, by a factorization with respect to the center) leads to the so-called algebras of classical type. In particular, any simple p-algebra with a nonsingular Killing form is an algebra of classical type (Seligman). An analog of an infinite-dimensional simple Cartan algebra over a field of characteristic zero is a simple Lie p-algebra of Cartan type, which decomposes into the four series: 1) W_n, $\dim W_n = np^n > 2$; 2) S_n, $\dim S_n = (n-1)(p^n-1)$, $n > 2$; 3) H_n, $n = 2m$, $\dim H_n = p^n - 2 > 2$;

4) K_n, $n = 2m + 1$, $p > 2$, $\dim K_n = \left\{ \begin{array}{l} p^n, \; n + 3 \not\equiv 0 \pmod{p}, \\ p^n - 1, \; n + 3 \equiv 0 \pmod{p}. \end{array} \right\}$

The algebras W_n, S_n, H_n, and K_n are called, general, special, Hamiltonian, and contact, in accordance with the names of the corresponding series of infinite Cartan algebras. Each of them is realized as a certain algebra of differentiations of a truncated polynomial ring $k[x_1, \ldots, x_n]$ where k is a base field, $x_i^p = 0$. A peculiarity of Cartan-type algebras is the presence in them of the so-called long filtration. We say that the subalgebra chain $\mathfrak{A}_0 \supset \mathfrak{A}_1 \supset \cdots \supset \mathfrak{A}_i \supset 0$ in the Lie algebra $\mathfrak{A}$ defines a filtration of length r if $\mathfrak{A}_i = \{x \in \mathfrak{A}_{i-1}, x\mathfrak{A} \subset \mathfrak{A}_{i-1}\}$ ($i \geq 1$). From the definition it follows at once that $\mathfrak{A}\mathfrak{A} \subseteq \mathfrak{A}_{i-1}$ ($i > 0$), $\mathfrak{A}_i \mathfrak{A}_j \subseteq \mathfrak{A}_{i+j}$ ($i, j \geq 0$) and each of the subalgebras $\mathfrak{A}_i$ ($i > 0$) is an ideal in $\mathfrak{A}_0$. In the factor space $\mathfrak{A}/\mathfrak{A}_0 = V$ there acts the representation Γ of the algebra $\mathfrak{A}_0/\mathfrak{A}_1 = \mathfrak{A}_0$. A filtration of length $r \geq 2$ exists in a simple Lie p-algebra $\mathfrak{A}$ ($p > 5$) if and only if the set $C^* = \{x \in \mathfrak{A}, (\mathrm{ad}\,x)^2 = 0\}$ is nonempty. The set C^* is empty in algebras of classical type, while it is nonempty in Cartan-type algebras (when $p > 5$). The classification hypothesis, stated in [75], assumes that all simple Lie p-algebras with $p > 7$ are exhausted by algebras of classical and Cartan types. The following assertion has been proved in this direction. Let $\mathfrak{Q}$ be a simple Lie p-algebra over an algebraically closed field k of characteristic p > 7, possessing a proper subalgebra $\mathfrak{Q}_0$ such that dim V < p, dim L_0 < p, and the representation Γ is irreducible. Then $\mathfrak{Q}$ is either a classical algebra or, when r ≥ 2, a Cartan-type algebra from among the series 1)-3).

A proper subalgebra in $\mathfrak{Q}$ is said to be invariant if it is invariant relative to all automorphisms of $\mathfrak{Q}$. There are no such subalgebras in simple Lie p-algebras of classical type. If $\mathfrak{Q}$ is a Cartan-type algebra, then, for example, the nilpotent subalgebra C generated by the set C^* will be invariant.

Any Cartan-type algebra possesses a unique maximal invariant subalgebra $\mathfrak{Q}_0$ which coincides with the normalizer of the subalgebra C and defines a filtration of length p = 2 [75]. In connection with the classification hypothesis stated above there arises the question: will every simple Lie p-algebra $\mathfrak{Q}$ over an algebraically closed field of characteristic p > 5 be an algebra of classical type if $C^* = \varnothing$? An affirmative answer to this question is given in [72], contiguous to [75], under the additional assumption that $\mathfrak{Q}$ admits of the Cartan expansion $\mathfrak{Q} = H + \Sigma A^\alpha$ such that $(adx)^{p-1} = 0$ for at least one element x ≠ 0 from A^α or from H. Simple Lie algebras over a field of characteristic two were considered in [212]. Rigid constraints on the expansion of $\mathfrak{Q}$ with respect to the Abelian subalgebra H were found under which $\mathfrak{Q}$ turns out to be an algebra of classical type. A classification to within isomorphism of simple Lie algebras of type E_6 over finite fields, and also over real-closed and p-adic fields, has been carried out in [284]. A non-trivial classification of irreducible representations of a three-dimensional simple Lie algebra over an algebraically closed field of characteristic p > 0 was obtained in [104]. In dimension p these representations are parametrized by a three-dimensional algebraic manifold.

A number of assertions on semisimple Lie algebras over fields of finite characteristic have been obtained. Let $\mathfrak{Q}$ be a Lie algebra over a field of characteristic p > 5, let $\mathfrak{Q}^2 = \mathfrak{Q}$ the center of $\mathfrak{Q}$ be trivial, and let $\mathfrak{Q}$ admit of an expansion relative to the Cartan subalgebra H such that the root subspaces $\mathfrak{Q}_\alpha$ are one-dimensional and $\alpha(\mathfrak{Q}_\alpha\mathfrak{Q}_{-\alpha}) \neq 0$. Then $\mathfrak{Q}$ is a direct sum of simple algebras satisfying the same conditions as $\mathfrak{Q}$ and, moreover, each simple term is either an algebra of classical type or a Cartan-type algebra of rank one [198]. For any semisimple Lie algebra $\mathfrak{Q}$ over a field F of characteristic p > 3 with a nonsingular Killing form there exists a separable extension K of field F such that $\mathfrak{G}_F \otimes K$ is a direct sum of simple p-algebras of classical type [172]. If a semisimple Lie p-algebra $\mathfrak{G}$ over an algebraically closed field of characteristic p > 0 admits of an exact p-representation of power n < p−1, then $\mathfrak{G}$ decomposes into a direct sum of simple Lie p-algebras of classical type with a nonsingular Killing form [74].

A description of Lie algebras of characteristic p > 3, admitting of an exact linear representation with a nonsingular form from the left, was obtained in [197] and, in particular, the class of non-simple Lie algebras not expandable into a direct sum was indicated. If in a restricted Lie p-algebra the p-mapping has a trivial kernel, then such an algebra is Abelian [229]. A number of interesting results on Lie p-algebras were obtained in [642]. Properties of a special form of subalgebras in Lie algebras of rank one over an algebraically closed field of characteristic p > 3 were considered in [41, 42]. A new construction of a free resolvent for computing the cohomologies of Hopf algebras and of restricted Lie algebras was proposed in [490]. The interpretation of group algebras as Hopf algebras of special form was given in [678].

Finite-dimensional Lie algebras of characteristic zero with different conditions on its subalgebra lattice were examined in [431, 432]. An automorphism φ of a Lie algebra is called regular if $\varphi(x) = x$ implies x = 0. Any finite-dimensional Lie algebra possessing a regular automorphism is solvable [77]. By analogy with group theory the Frattini subalgebra Φ of a Lie algebra $\mathfrak{G}$ is defined as the intersection of all maximal subalgebras of $\mathfrak{G}$. If $\mathfrak{G}$ is nilpotent, then $\Phi = \mathfrak{G}^2$. In the general case Φ lies in the nilradical of $\mathfrak{G}$ and if the base field is algebraically closed, then Φ is an ideal [484]. The first cohomology group $H'(\mathfrak{G}, \mathfrak{G})$ of a Lie algebra $\mathfrak{G}$ has been computed for the case when $\mathfrak{G}$ is the algebra of all triangular nth-order matrices over a field F or is a subalgebra of this algebra, of matrices with zero trace [687]. The manifolds of Lie algebras form a multiplication semigroup; Mal'tsev posed the question: Is this semigroup free? The answer turned out to be in the affirmative for Lie algebras over a field of characteristic zero [102].

It is very well known that in any Lie algebra $\mathfrak{G}$ there exists a unique maximal locally-nilpotent ideal N (the Plotkin radical). A simple example shows that the ideal N cannot be characteristic (that N is characteristic means that N is closed relative to all differentiations of Ω) even in the case of algebras of characteristic zero [329]. This answers one of Plotkin's questions. The question of the existence of the Brown-McCoy radical in Lie algebras is taken up in [9]. It happens that such a radical, roughly speaking, does not exist except for finite-dimensional algebras over a field of characteristic zero, where there has been defined, for example, the classical radical, while the modular classes coincide with the classes of simple perfect Lie algebras. Finally, we note [66] wherein certain infinite-dimensional simple Lie algebras are constructed.

Let Γ be the algebra of an addition group and Γ_0 its subgroup of index two. Let $\Gamma_1 = \Gamma/\Gamma_0$; the elements of Γ_0 and Γ_1 are called even and odd, respectively. The direct sum of the modules L_λ ($\lambda \in \Gamma$) with the following multiplication rules for homogeneous components is called a Γ-graded Lie algebra: 1) $L_\lambda L_\mu \subseteq L_{\lambda+\mu}$; 2) $x_\lambda x_\mu = -(-1)^{\lambda\mu}x_\mu x_\lambda$; 3) $(-1)^{\lambda\nu}x_\lambda(x_\mu x_\nu) + (-1)^{\lambda\mu}x_\mu(x_\nu x_\lambda) + (-1)^{\nu\mu}x_\nu(x_\lambda x_\mu)$; 4) $(x_\lambda x_\lambda)x_\lambda = 0$. As the domain of the operators serves either a field F of characteristic $\neq 2$ or, more commonly, an associative-commutative ring K with unity and with divisors by 2 [617]. For Γ-graded algebras (GLA) a representation theory has been developed and analogs of the Birkhoff-Witt theorem and of the theorem on the existence of an exact finite-dimensional representation for a finite-dimensional GLA have been proved. We remark that the space V of representation ρ also should be Γ-graded, $\rho(x_\lambda)$ maps V_μ into $V_{\lambda+\mu}$ and $\rho(x_\lambda x_\mu) = \rho(x_\lambda)\rho(x_\mu) - (-1)^{\lambda\mu}\rho(x_\mu) \cdot \rho(x_\lambda)$. The cohomologies and the deformations of graded Lie algebras were considered in [541, 542].

2. Mal'tsev Algebras and Other Generalizations of Lie algebras. Considerable progress has been achieved in the theory of Mal'tsev algebras which are a very natural generalization of a Lie algebra. Criteria for the solvability and the semisimplicity of finite-dimensional Mal'tsev algebras over a field of characteristic zero, analogous to Cartan's criteria for Lie algebras, were proven in [467]. Engel's theorem, which plays an important role in the lattice theory of Lie algebras, has been carried over to Mal'tsev algebras [60]. Furthermore, this theorem turns out to be valid also for binary Lie algebras of arbitrary characteristics [80, 81]. The nilradical (the maximal nilpotent ideal coinciding with the sum of all nilpotent ideals of the algebra) has been correctly defined in finite-dimensional binary Lie al-

gebras, in particular, in finite-dimensional Mal'tsev algebras. Certain results in the theory of Lie algebras have been carried over to Mal'tsev algebras and to the infinite-dimensional case. Thus, a solvable Mal'tsev algebra of characteristic $\neq 2$, satisfying the n-th Engel condition, is locally nilpotent [82]. A locally nilpotent radical has been correctly defined in the class of Mal'tsev algebras satisfying the n-th Engel condition. In [85] it was proven that if A is an arbitrary Mal'tsev algebra of characteristic $\neq 2$, then the sum of locally finite ideals of A is a locally finite ideal, and the sum of locally nilpotent ideals is a locally nilpotent ideal. Hence it is clear that it follows that in such an algebra there exists a unique maximal locally-finite ideal and a unique maximal locally-nilpotent ideal, results which are very well known in the theory of Lie algebras. Finally, we note the classification of non-Lie central simple Mal'tsev algebras of finite dimension over fields of characteristic $\neq 2,3$, obtained in [88].

To date it is not known whether there exist simple binary Lie algebras other than Lie algebras and Mal'tsev algebras. It was proven in [624] that if A is a finite-dimensional simple binary Lie algebra over an algebraically closed field of characteristic zero and if the Killing form of algebra A is invariant and nonsingular, then A is a Lie algebra or a non-Lie Mal'tsev algebra. Binary Lie algebras over a field of characteristic zero, in which there are no divisors of zero in the sense of [89], were considered in [86]. Finite-dimensional anticommutative algebras over a field of characteristic zero, on which there is an additional structure of a general Lie triple system in the sense of Yamaguti, were examined in [626, 627]. Such an algebra A is either a Lie algebra or a Mal'tsev algebra or satisfies the identity

$$I\,(x,\,y,\,z)\,w = I\,(w,\,x,\,yz) + I\,(w,\,y,\,zx) + I\,(w,\,z,\,xy),$$

where $I(x, y, z) = (xy)z + (yz)x + (zx)y$. In the latter case A is once again a Lie algebra if in A there exists an element $u \neq 0$ such that the operator $R_u: x \rightarrow xu$ acts diagonally in A.

§3. Alternative and Jordan Rings

1. **Alternative Rings.** It was proven in [59] that in an alternative ring A a quasiregular ideal S without elements of order two in the additive group is nilpotent if A satisfies the mini-

mality condition for the right ideals contained in S. Every alternative ring A satisfying a minimality condition for right ideals is an extension of a nilpotent ring with the aid of a finite direct sum of rings each of which has the structure either of a full matrix algebra over some associative skew field or of a Cayley-Dickson algebra over its own center [56]. If A is an alternative ring with a minimality condition for right ideals without elements of order two in the additive group, B is a subring of it, C is a right ideal, B contains C and is nilring modulo C, then B is nilpotent modulo C [56]. In particular, every nilsubring of an alternative Artinian ring is nilpotent.

Let N (R) be an associative center and let Z (R) be a commutative center of ring R, i.e., N (R) = $\{n \in R | (n, R, R) = 0\}$, Z (r) = $\{z \in R | [z, R] = 0\}$, where (a, b, c) = (ab) c − a(bc), [a, b] = ab − ba. The ring R is called semiprime if it does not contain trivial ideals, i.e., nonzero ideals T such that T^2 = 0. Now, let R be a semiprime alternative ring, A its one-sided ideal, then N (A) = N (R) ∩ A, Z (A) = Z (R) ∩ A. An ideal A of an alternative ring R is semiprime as a ring if and only if it does not contain trivial ideals of the whole ring R. If R is a semiprime alternative ring without elements of order three in the additive group, then N (R) and Z (R) are nonzero [645].

An example has been constructed of an alternative algebra with four generators of rank 23, in which the square of the commutator does not lie in the associative center, i.e., $([a, b]^2, c, d) \neq 0$ for some a, b, c, d, [355]. Hence it follows that in a free alternative ring with four or more generators there is a trivial ideal because not only the Kleinfeld identity $([a, b]^2, c, d) = 0$ but also the Dorofeev identity ([a, b] [c, d] + [c, d] [a, b], e, f) = 0 are not fulfilled in every semiprime alternative ring [58]. The presence of a zero divisor in a free alternative ring with three generators follows from the identity $[[a, b]^4, a]$ (a, b, c) = 0. In every free alternative ring R there is an ideal contained in the associative center of R [646].

Among other results we note the following. Let A be an alternative algebra over F, B be a subalgebra of it, e be the unity in B. If e = $\alpha [x, y]^4$ for some x, y ∈ B and $\alpha \in$ F, then e ∈ N (A), and if, moreover, the ideal of algebra B, generated by all its associators, equals B, then e ∈ Z (A) [354]. Every alternative commutative prime ring is a field [61]. The known inductive chain for the construction of a lower radical is stabilized in alternative rings on the second infinite ordinal [668].

The following result has been obtained [353] in the theory of right-alternative rings. Let R be a right-alternative ring without nilpotent ideals and without elements of order two in the additive group, possessing an idempotent e, and let $R = R_0(e) + R_{1/2}(e) + R_1(e)$ be the Albert decomposition of ring R into the sum of three modules in the idempotent e. If there are no nilpotent elements in R_0 and R_1, then R is alternative.

Rings satisfying the identity $\alpha(y, x, x) - (\alpha + 1)(x, y, x) + (x, x, y) = 0$, where $\alpha \neq -1, 1, -1/2, -2$, were analyzed in [663]. It turned out that if such a ring possesses a nontrivial idempotent, is prime, and does not have elements of order two and three in the additive group, then it is alternative.

Other generalizations of alternative rings are discussed in [213, 396].

2. Jordan Rings. The question of polynomials which would equal zero identically in all special Jordan algebras but would be nonzero in free Jordan algebras has remained open for a long time. It has been ascertained [135] that there are no such polynomials of two variables and that such polynomials of three variables do exist [480], but neither their form nor even their minimal possible degree is known. Much progress on this question was made in [304]. Let A be a free associative algebra from the free generators c, y, z; J_0 is a subalgebra of algebra $A^{(+)}$, generated by x, y, and z, J is a free Jordan algebra whose free generators are the same x, y, z, and φ is the natural homomorphism of J onto J_0. It has been proven that in J there do not exist polynomials of degree < 8 belonging to Ker φ. If $\{a, b, c\} = a(bc) - b(ca) + c(ab)$, then, for example, the following eighth-degree polynomial

$$4\{\{z\{xyx\}z\}y(zx)\} - 2\{z\{x\{y(xz)y\}x\}z\} - 4\{(xz)y\{x\{zyz\}x\}\} + 2\{x\{z\{y(xz)y\}z\}x\}$$ belongs to Ker φ.

Now, let x^{-1}, y^{-1} also be in J and J_0. It was proven in [501] that Ker φ does not contain elements which have degree 0 or 1 in z and that every Jordan algebra generated by two elements and their inverses is special (cf. [135] and [480]).

A Jordan algebra J is called regular if for every $a \in J$ there

exists an element $b \in J$ such that $2\,(ab)\,a - a^2 b = a$ (if $J = A^{(+)}$,
where A is an associative algebra, then this is precisely equivalent to the regularity, in the sense of von Neumann, of algebra A).
It has been proven [696] that in the class of Jordan algebras (of
characteristic $\neq 2$) the extension of a regular algebra by means of
a regular algebra is once again a regular algebra and, consequently,
a regular radical exists in the class of Jordan algebras. The
existence of a locally nilpotent radical in arbitrary operator Jordan
rings and the coincidence of this radical with a locally solvable one
were proven in [57]. Therein also was proven that a locally solvable Jordan ring with a minimality condition for ideals is nilpotent
and an example was constructed showing that nilpotency and solvability are not equivalent even on special Jordan rings: there
exists a special nonnilpotent Jordan ring A such that $(A^2)^2 = 0$.

The study of Jordan algebras satisfying a certain finiteness
condition was initiated in [369]. Let $U_a = 2R_a^2 - R_{a^2}$ (here and
below R_a is the operator of right multiplication, L_a of left). In an
algebra A the subspace B is called a quadratic ideal if $AU_b \subseteq B$
for any $b \in B$. A Jordan algebra A satisfying the following conditions is examined: (I) $A \ni 1$, (II) $\forall a \in A$, $U_a \neq 0$, (III) A satisfies
a minimality condition for the principal quadratic ideals generated
by idempotents, and each such ideal contains a minimal quadratic
ideal. Then A is a finite direct sum of simple Jordan algebras S_i
satisfying (I) – (III). And every S_i is one of the following algebras:
1) a Jordan division ring, 2) $S_i \ni 1$, $1 = e_1 + e_2$, where e_i are orthogonal idempotents and $S_i U_{e_k}$ is a Jordan division ring, 3) the Jordan
algebra of Hermitian matrices over one of the following algebras:
a) a Cayley algebra (the involution is the standard one), b) the
quaternion algebra (the involution is the standard one), c) a direct
sum of two anti-isomorphic associative division rings (the involution transposes the direct summands), d) and associative division
ring with involution.

It was proven in [505] that if we set $U_a = R_a(R_a + L_a) - R_{a^2}$
in a noncommutative Jordan algebra A, then thanks to the fact that
$A^{(+)}$ is a commutative Jordan algebra, all the results of [369] just
cited may be reformulated in corresponding fashion for noncommutative Jordan algebras.

Let A be a simple Jordan algebra of characteristic $\neq 2$ with
unity and with orthogonal idempotents e_1 and e_2 such that $e_1 + e_2 = 1$,

and let $A_i = \{x \in A \mid xe_i = x\}$ be a Jordan division ring. Further, let Q be either an associative division ring with involution σ or a direct sum of two division rings transposable by the involution σ, let γ be an invertible diagonal matrix from Q_2, let $H(Q_2, \gamma)$ be the Jordan algebra of all matrices A from Q_2 such that $A^*\gamma = \gamma A$ ($a_{ij}^* = a_{ji}\sigma$). It turned out [554] that either $A \cong H(Q_2, \gamma)$ or A is a Jordan algebra defined by a nonsingular bilinear form on the space over the extension of the base field.

Among the articles devoted to finite-dimensional Jordan algebras we should first of all take note of [370]. In this paper an analog of the classical concept of the Cartan subalgebra of a Lie algebra was defined for Jordan algebras. The notion of associator nilpotency of a Jordan algebra was used in making this definition. Let $\mathfrak{A}$ be a Jordan algebra, A its multiplication algebra, R_a the operator of multiplication by an element a, and let $R_{a,b} = R_a R_b - R_{ab}$. The element a is said to be associator nilpotent if the operator R_{a^i,a^j} is nilpotent for any i and j. The algebra $\mathfrak{A}$ is called associator nilpotent of index n if every nth-order associator formed from its elements equals zero. An analog of the well-known Engel theorem was proved: A finite-dimensional Jordan algebra $\mathfrak{A}$ (with unity) over an infinite-dimensional field is associator nilpotent if and only if every element of $\mathfrak{A}$ is associator nilpotent. If $\mathfrak{B}$ is an associator-nilpotent subalgebra of a finite-dimensional Jordan algebra $\mathfrak{A}$, then the Lie algebra $\mathfrak{L}_{\mathfrak{A}}(\mathfrak{B})$ of linear transformations of the space $\mathfrak{A}$, generated by linear transformations of the form $R_{b,c}$, where b, c, $\in \mathfrak{B}$, is nilpotent. The Cartan subalgebra of algebra $\mathfrak{A}$ is defined thus: it is the associator-nilpotent subalgebra $\mathfrak{B}$ such that the Fitting null component $\mathfrak{A}_0$ of algebra $\mathfrak{A}$ relative to the nilpotent Lie algebra $\mathfrak{L}_{\mathfrak{A}}(\mathfrak{B})$ of linear transformations coincides with $\mathfrak{B}$. If $\mathfrak{B}$ equals n, then for a $\in \mathfrak{A}$ we set $\mathfrak{Z}_a = \{z \in \mathfrak{A} \mid z R^n_{a_m,a_z^i} = 0,\ i, j = 0,\ 1, 2, \ldots\}$. An element a is said to be associator regular if $\mathfrak{Z}_a$ has minimal dimension. If the base field is infinite, then for every associator-regular element a, $\mathfrak{Z}_a$ is a Cartan subalgebra. If the base field is algebraically closed and is of characteristic zero, then any two Cartan subalgebras are conjugate relative to inner automorphism.

Now let A be a finite-dimensional Jordan or alternative algebra of characteristic zero, G be some automorphism group of it, S be any maximal G-invariant semisimple subalgebra and T be

an arbitrary G-invariant semisimple subalgebra of A. It has been
proven in [675] that then there exists an automorphism u of algebra A carrying T into S and u = exp d, where d is a nilpotent
inner derivation of A, belonging to radical of the enveloping associative algebra of A and commuting with G. Recall that a derivation D of a Jordan algebra is called inner if $D = \sum [R_{b_i}, R_{c_i}]$,
where R_x is the operator of right multiplication by element x. In
a finite-dimensional central simple Jordan algebra A (dim A $\neq$ 3)
every derivation is inner [509], and the Lie algebra of its derivations is semisimple [509, 334]. If A is an arbitrary Jordan algebra, then the set I of all its inner derivations is an ideal in the Lie
algebra $\mathfrak{Q}$ of all derivations; if A is of characteristic p $\neq$ 0,2 and
d is any derivation of A and $d^p \in I$, then d $\in$ I [278].

Let A be a Jordan algebra with unity over a field of characteristic $\neq$ 2. The connection between algebra A and a specified
Lie algebra $\mathfrak{Q}$ (A) was established in [416]. The algebra A is imbedded in $\mathfrak{Q}$ (A) in the following sense: A is a subspace of $\mathfrak{Q}$ (A) and
multiplication in A is expressed in terms of multiplication in $\mathfrak{Q}$ (A).
It turns out here that a simple algebra is imbedded in a simple
one. In case algebra A is finite dimensional the radicals of algebras A and $\mathfrak{Q}$ (A) prove to be very intimately interrelated; this is
made use of to re-prove the existence of the Wedderburn decomposition for Jordan algebras.

Among the papers devoted to the connection between Jordan
and associative algebras we note the following elegant result [371].
Let R be a finite-dimensional central simple associative algebra
with involution σ, of degree $\geq$ 3, H(R, σ) be the Jordan algebra of
all σ-symmetric elements of R relative to Jordan multiplication.
It turns out that every finite-dimensional special simple Jordan
algebra of degree $\geq$ 3 is isomorphic with some H(R, σ).

Now let R be a simple associative ring with involution. In
[341] it was conjectured that every Jordan homomorphism of
H(R, σ) into H(R', σ') can in unique fashion be continued upto the
usual homomorphism of R into R' (the validity of the conjecture was
proven in [372] for the case when R is a complete matrix ring over
a division ring). The validity of this conjecture was proved in [486]
under certain additional constraints (which complete matrix rings
satisfy automatically). Here, however, it was proven that the Jor-

dan derivation $H(R, \sigma)$ can be uniquely continued upto the usual derivation if R is a prime ring.

As before, let $U_a = 2R_a^2 - R_{a^2}$ and $P(x, y) = \frac{1}{2}(U_{x+y} - U_x - U_y)$. A subalgebra B in A is called a P-ideal if $AP(u, v) \subseteq B$ for all $u, v \in B$. If a is a fixed element of A, then we can give a new operation $x^\perp y = aP(x, y)$, relative to which A turns into a Jordan algebra, called the mutation of algebra A. The study of mutations in [468] permitted the classification of minimal P-ideals. The connections between mutations and the structure group of algebra A were studied in [508, 335].

Among the papers devoted to noncommutative Jordan algebras we should take note of [500]. Let D_n be a complete matrix algebra over some associative algebra D. In D_n we introduce the new multiplication: $x \cdot y = \lambda xy + (1-\lambda)yx$, where λ is a fixed element of the base field. We obtain an algebra $D_n(\lambda)$ relative to this multiplication. It was proven in [500] that if A is a noncommutative Jordan algebra of characteristic $\neq 2$, containing ≥ 3 connected orthogonal idempotents, then either $A \cong D_n(\lambda)$ or $A \cong H(D_n, \sigma)$, where D is associative and σ is a canonical involution in D_n when $n > 3$, while D is alternative and σ is an involution relative to which the elements of the associative center are self-adjoint, when $n \leq 3$. Here it was also proven that if A is a finite-dimensional noncommutative Jordan algebra of characteristic $\neq 2$ and R is a radical of it (a maximal nilideal), then A/R is semisimple, contains a unity, and is the direct sum of simple algebras; finite-dimensional central simple noncommutative Jordan algebras were classified.

Conditions under which certain noncommutative Jordan algebras have a multiplication table with specified properties were analyzed in [308].

Other generalizations of Jordan algebras were considered in [623]. A generalization of Jordan algebras are commutative algebras in which the left multiplication operators L_x form a Lie triple system relative to the composition $[[L_x, L_y], L_z] = L_{(x, y, z)}$, where $[L_x, L_y] = L_x L_y - L_y L_x$ is a commutator and $(x, y, z) = (xy)z - x(yz)$ is an associator. Such algebras are called LT-algebras and may be characterized by the identity

$$2x[x(xy)] + x^3y = 3x(x^2y).$$

The study of LT-algebras was begun in [552]. The notion of a
radical of an LT-algebra was defined in [576] analogous to the
Jacobson radical in the associative case (the intersection of all
maximal modular ideals). This radical possesses the usual properties; in the case of finite-dimensional Jordan algebras it coincides with the nilradical. Any finite-dimensional semisimple LT-
algebra over a field of characteristic zero is Jordan. It was
shown in [577] that a radical of a finite-dimensional LT-algebra
does not necessarily split into semidirect summands and conditions were found as to when such a splitting does exist.

3. Algebras Arising in Genetics. A finite-
dimensional nonassociative algebra over a field F is called a baric
algebra if a nontrivial homomorphism φ: $x\varphi = \omega(x) \in F$ exists;
$\omega(x)$ is called the weight of x. A baric algebra in which the set N
of elements of zero weight is an ideal and all powers of N are
ideals, is called a special train algebra. Such algebras were
studied in [310]. Special train algebras corresponding to one genetic object turn out to stand in a certain relation to each other,
namely, a special isotopy; at the same time, algebras corresponding to different genetic objects are not isotopic to each other in
this sense. Commutative special train algebras with an idempotent
were considered in [348]. The direct product of special train algebras is once again a special train algebra [349]. Gamete algebras
are studied in [350]; it is proved that they are Jordan; from this
there are rapidly derived certain consequences for genetics (for
example, in the absence of selection the gamete ratios do not
change from generation to generation, etc.).

BIBLIOGRAPHY

1. V. I. Andriyanov, "Periodic Hamiltonian rings," Mat. Sb., 74(2):241-261 (1967).
2. V. I. Andriyanov, "Composite Γ-rings," Sverdl. Gos. Ped. Inst., Vol. 54, 12-21 (1967).
3. V. I. Andriyanov, "Composite Hamiltonian rings," Mat. Zap. Ural'skii Univ., 5(3):15-30 (1966).
4. V. I. Andriyanov and P. A. Freidman, "Hamiltonian rings," Uch. Zap. Sverdl. Gos. Ped. Inst., Vol. 31, 3-23 (1965).
5. V. A. Andrunakievich and V. I. Arnautov, "Invertibility in topological rings," Dokl. Akad. Nauk SSSR, 170(4):755-758 (1966).
6. V. A. Andrunakievich, V. I. Arnautov, and Yu. M. Ryabukhin, "Rings," in:

R. V. Gamkrelidze (Ed.), Progress in Mathematics, Vol. 5, Plenum Press, New York (1969), pp. 127-177.

7. V. A. Andrunakievich and Yu. M. Ryabukhin, "One-sided primarity in arbitrary rings," in: Mathematical Investigations, Vol. 1, Part 1 [in Russian], Kishinev (1966), pp. 3-31.

8. V. A. Andrunakievich and Yu. M. Ryabukhin, "Primarity in rings," in: Mathematical Investigations, Vol. 1, Part 2 [in Russian], Kishinev (1966), pp. 65-97.

9. V. A. Andrunakievich and Yu. M. Ryabukhin, "The existence of the Brown-McCoy radical in Lie algebras," Dokl. Akad. Nauk SSSR, 179(3):503-506 (1968).

10. V. A. Andrunakievich and Yu. M. Ryabukhin, "Prime ideals in noncommutative rings," Dokl. Akad. Nauk SSSR, 165(1):13-16 (1965).

11. V. A. Andrunakievich and Yu. M. Ryabukhin, "Connected rings," in: Investigations in General Algebra [in Russian], Kishinev (1965), pp. 3-24.

12. V. A. Andrunakievich and Yu. M. Ryabukhin, "On the additive theory of ideals in rings, modules, groupoids," Dokl. Akad. Nauk SSSR, 168(3):495-498 (1966).

13. V. A. Andrunakievich and Yu. M. Ryabukhin, "Additive theory of ideals in systems with quotients," Izd. Akad. Nauk SSSR, Ser. Mat., 31(5):1057-1090 (1967).

14. V. A. Andrunakievich and Yu. M. Ryabukhin, "Rings without nilpotent elements and completely prime ideals," Dokl. Akad. Nauk SSSR, 180(1):9-11 (1968).

15. V. I. Arnautov, "Topologically weakly regular rings," in: Investigations in Algebra and Mathematical Analysis [in Russian], Kartya Moldovenyaske, Kishinev (1965). pp. 3-10.

16. V. I. Arnautov, "Topological rings with a given local weight," in: Investigations in General Algebra [in Russian], Kishinev (1965), pp. 25-36.

17. V. I. Arnautov, "Pseudonormalizability criterion for topological rings," Algebra Logika, Seminar, 4(4):3-24 (1965).

18. V. I. Arnautov, "Topologization of the ring of integers," Bul. Akad. Shtiintse RSSMold., No. 1, 3-15 (1968).

19. A. Blokh, "One generalization of the notion of a Lie algebra," Dokl. Akad. Nauk SSSR, 165(3):471-473 (1965).

20. A. A. Bovdi, "Group rings of torsion-free groups," Sibirsk. Mat. Zh., 1(4):555-558 (1960).

21. L. A. Bokut', "Embedding of rings in a division ring," Dokl. Akad. Nauk SSSR, 175(4):755-758 (1967).

22. L. A. Bokut', "Embedding theorems in the theory of algebras," Colloq. Math., 14:349-353 (1964).

23. L. A. Bokut', "Factorization theorems for certain classes of rings without zero divisors, II," Algebra Logika, Seminar, 4(5):17-46 (1965).

24. L. A. Bokut', "Factorization theorems for certain classes of rings without zero divisors, I," Algebra Logika, Seminar, 4(4):25-52 (1965).

25. L. A. Bokut', "The continuations of isomorphisms of rings," Algebra Logika, Seminar, 7(1):33-45 (1968).

26. L. P. Busarkina, "Some set-theoretic decompositions of rings," Mat. Zap. Ural'skii Inst., 5(1):15-27 (1965).

27. L. P. Busarkina and N. F. Sesekin, "First power algebras," Mat. Zap. Ural'skii Inst., 5(1):28-34 (1965).

28. A. I. Veksler, "Lattice orderability of algebras and rings," Dokl. Akad. Nauk SSSR, 164(2):259-262 (1965).

29. B. B. Venkov, "Homologies of identity groups in algebras with a divisor," Tr. Mat. Inst. Akad. Nauk SSSR, 80:66-89 (1965).

30. M. I. Vodinchar, "Pseudonormalizability criterion for topological algebras," in: Mathematical Investigations, Vol. 2, Part 1 [in Russian], Kishinev (1967), pp. 133-140.

31. M. Gavrilov, "Identity relations in a full matrix algebra," Godishnik Sofiisk. Univ., Mat. Fak., 59:45-48 (1964-1965 (1966)).

32. V. I. Gemintern, "Semiprime rings with atomic structure of left ideals," Sibirsk. Mat. Zh., 8(4):947-951 (1967).

33. E. S. Golod, "Certain problems of the Burnside type," Report Abstracts, Internat. Congr. Mathematicians [in Russian], Moscow (1966), p. 142.

34. E. L. Gorbachuk, "Splittability of torsion and pretorsion in the category of right Λ-modules," Mat. Zametki, 2(6):681-688 (1967).

35. I. M. Goyan, "The Baer radical of nearrings," Bul. Akad. Shtiintse RSSMold., No. 4, 32-38 (1966).

36. I. M. Goyan, "Driprimarity in rings," in: Mathematical Investigations, Vol. 1, Part 2 [in Russian], Kishinev (1966), pp. 151-171.

37. I. M. Goyan and Yu. M. Ryabukhin, "On the axiomatic additive theory of Riley ideals," in: Mathematical Investigations, Vol. 2, Part 1 [in Russian], Kishinev (1967), pp. 14-25.

38. L. Ya. Gringlaz, "Locally nilpotent quasirings," Mat. Zap. Ural'skii Univ., 5(1):35-42 (1965).

39. R. A. Danilov, "Composition algebra of entire functions," Collection of Articles on Mathematics, Chelyabinsk State Ped. Inst., No. 1, Part 2 [in Russian], (1966), pp. 102-110.

40. Hai-ch'uan Jang, "The proof of one of the properties of quadratic alternative rings," Jilin Shida Xuebao, No. 1, 1-3 (1965).

41. A. Kh. Dolotkazin, "Lie algebras of rank one with nonzero inner product. I," Izv. Vyssh. Uchebnyk Zavedenii, Matematika, No. 4, 36-43 (1966).

42. A. Kh. Dolotkazin, "Lie algebras of rank one with nonzero inner product. II," Izv. Vyssh. Uchebnykh Zavedenii, Matematika, No. 5, 70-77 (1966).

43. A. Kh. Dolotkazin, "The Lie algebras $U_0(x_1, \ldots, x_n)$," Summaries of the Scientific Aspirants Conference of 1965 at the Kazan Univ. Math. Mech., Report Abstracts [in Russian], Kazan (1967), pp. 13-14.

44. G. I. Domracheva, "Some remarks on lattice-ordered algebras and their ideals," Uch. Zap. Novgorodsk. Golovn. Ped. Inst., 7:3-16 (1966).

45. Yu. A. Drozd, "Representations of cubic Z-rings," Dokl. Akad. Nauk SSSR, 174(1):16-18 (1967).

46. Yu. A. Drozd and V. V. Kirichenko, "Representations of rings lying in a second-order matrix algebra," Ukrainsk. Mat. Zh., 19(3):107-112 (1967).

47. Yu. A. Drozd and V. V. Kirichenko, "Hereditary orders," Ukrainsk. Mat. Zh., 20(2):246-248 (1968).

48. Yu. A. Drozd and V. M. Turchin, "The number of modules of representations in genus for second-order integral matrix rings," Mat. Zametki, 2(2):133-138 (1967).

49.   Z. M. Dyment, "Maximal commutative nilpotent subalgebra of a sixth-order matrix algebra," Vestsi Akad. Nauk BSSR, Ser. Fiz.-Mat. Nauk, No. 3, 53-68 (1966).

50.   V. P. Elizarov, "Plane extensions of rings," Dokl. Akad. Nauk SSSR, 175(4):759-761 (1967).

51.   V. P. Elizarov, "Two properties of associative rings," Mat. Zametki, 2(3):225-232 (1967).

52.   V. P. Elizarov, "Pseudo division rings and quotient rings," Sibirsk. Mat. Zh., 8(4):782-787 (1967).

53.   V. P. Elizarov, "Nonsingularly dimensional rings," Sibirsk. Mat. Zh., 6(5):1181-1184 (1965).

54.   V. P. Elizarov, "Subcommutative 2-rings," Mat. Zametki, 2(6):639-694 (1967).

55.   V. P. Elizarov, "Correction to my paper 'Quotient rings of associative rings'," Izv. Akad. Nauk SSSR, Ser. Mat., 31(3):727 (1967).

56.   K. A. Zhevlakov, "Alternative Artinian rings," Algebra Logika, Seminar, 5(3): 11-36 (1966).

57.   K. A. Zhevlakov, "Solvability and nilpotency of Jordan rings," Algebra Logika, Seminar, 5(3):37-58 (1966).

58.   K. A. Zhevlakov, "Lower nilradical of alternative rings, Algebra Logika, Seminar, 6(4):11-17 (1967).

59.   K. A. Zhevlakov, "Radical ideals of an alternative ring," Algebra Logika, Seminar, 4(4):87-102 (1965).

60.   K. A. Zhevlakov, "The nilradical of Mal'tsev algebras," Algebra Logika, Seminar, 4(5):67-78 (1965).

61.   K. A. Zhevlakov, "Remarks on prime alternative rings," Algebra Logika, Seminar, 6(2):21-33 (1967).

62.   A. E. Zalesskii, "Solvable groups and crossed products," Mat. Sb., 67(1):154-160 (1965).

63.   A. E. Zalesskii, "The semisimplicity of a crossed product," Sibirsk. Mat. Zh., 6(5):1009-1013 (1965).

64.   O. A. Ivanova, "Nilpotent decomposition of associative algebras," Mat. Sb., 71(3):423-432 (1966).

65.   P. S. Kazimirs'kii, "One remark on the theory of domains of principal ideals, Vestnik L'vov Politekhn. Inst., No. 16, 96-97 (1967). [in Ukrainian]

66.   I. L. Kantor, "Infinite-dimensional prime graded Lie algebras," Dokl. Akad. Nauk SSSR, 179(3):534-537 (1968).

67.   A. A. Kirillov, "Certain algebras with divisor over the field of rational functions," Funkts. Analiz i Ego Prilozh., 1(1):101-102 (1967).

68.   A. A. Kirillov, "Division rings generated by matrices with independent elements," Uspekhi Mat. Nauk, 23(1):227-228 (1968).

69.   A. A. Kirillov, "Unitary representations of nilpotent Lie groups," Uspekhi Mat. Nauk, 17(4):57-110 (1962).

70.   V. V. Kirichenko, "Orders all of whose representations are completely decomposable," Mat. Zametki, 2(2):139-144 (1967).

71.   A. U. Klimyk, "Decomposition of a direct product of irreducible representations of semisimple Lie algebras over irreducible representations," Ukrainsk. Mat. Zh., 18(5):19-27 (1966).

72. A. I. Kostrikin, "Squares of adjoint endomorphisms in simple Lie p-algebras," Izv. Akad. Nauk SSSR, Ser. Mat., 31(2):445-487 (1967).

73. A. I. Kostrikin, "Simple Lie algebras of finite characteristic," Uspekhi Mat. Nauk, 23(2):204-205 (1968).

74. A. I. Kostrikin, "A theorem on semisimple Lie p-algebras," Mat. Zametki, 2(5):465-474 (1967).

75. A. I. Kostrikin and I. R. Shafarevich, "Cartan pseudogroups and Lie p-algebras," Dokl. Nauk SSSR, 168(4):740-742 (1966).

76. É. G. Koshevoi, "Multiplicative semigroups of one class of rings without zero divisors," Algebra Logika, Seminar, 5(5):49-54 (1966).

77. V. A. Kreknin, "Solvability of a Lie algebra with a regular automorphism," Sibirsk. Mat. Zh., 8(3):715-716 (1967).

78. E. N. Kuz'min, "Certain classes of algebras with divisor," Algebra Logika, Seminar, 5(2):57-102 (1966).

79. E. N. Kuz'min, "Algebras with divisor over the real number field," Dokl. Akad. Nauk SSSR, 172(5):1014-1017 (1967).

80. E. N. Kuz'min, "Engel's theorem for binary Lie algebras," Dokl. Akad. Nauk SSR, 176(4):771-773 (1967).

81. E. N. Kuz'min, "Anticommutative algebras satisfying the Engel condition," Sibirsk. Mat. Zh., 8(5):1026-1034 (1967).

82. E. N. Kuz'min, "Locally nilpotent radical of Mal'tsev algebras satisfying the n-th Engel condition," Dokl. Akad. Nauk SSSR, 177(3):508-510 (1967).

83. E. N. Kuz'min, "One class of anticommutative algebras," Algebra Logika Seminar, 6(4):31-50 (1967).

84. E. N. Kuz'min, "Locally finite Mal'tsev algebras," Algebra Logika, Seminar, 6(4):27-30 (1967).

85. E. N. Kuz'min, "Algebraic sets in Mal'tsev algebras," Algebra Logika 7(2):42-47 (1968).

86. E. N. Kuz'min, "Zero divisors in binary Lie algebras," Sibirsk. Mat. Zh., 9(1):97-103 (1968).

87. E. N. Kuz'min, "Simple Mal'tsev algebras over a field of characteristic zero," Dokl. Akad. Nauk SSSR, 181(6):1324-1326 (1968).

88. E. N. Kuz'min, "Mal'tsev algebras and their representations," Algebra Logika, 7(4):48-69 (1968).

89. V. N. Latyshev, "Zero divisors and nilelements in a Lie algebra," Sibirsk. Mat. Zh., 4(4):830-836 (1963).

90. V. N. Latyshev, "Generalization of Hilbert's theorem on the finiteness of bases," Sibirsk. Mat. Zh., 7(6):1422-1424 (1966).

91. V. N. Latyshev, "Finite generation of a T-ideal with the element $[x_1, x_2, x_3, x_4]$," Sibirsk. Mat. Zh., 6(6):1432-1434 (1965).

92. E. M. Levich, "Prime and strictly prime rings," Izv. Akad. Nauk LatvSSR, Ser. Fiz. i Tekhn. Nauk, No. 6, 53-58 (1965).

93. Z. S. Lipkina, "Pseudonormability of topological rings," Sibirsk. Mat. Zh., 6(5):1046-1052 (1965).

94. L. M. Makharadze, "Nilpotency of topological rings," Soobshch. Akad. Nauk GruzSSR, 40(2):257-261 (1965).

95. V. I. Monastyrskii, "Periodic subgroups of the multiplicative group of a division

ring," IZv. Akad. Nauk BSSR, Ser. Fiz.-Mat. Nauk, No. 1, 29-31 (1965).

96. G. M. Mubarakzyanov, "Some theorems on solvable Lie algebras," Izv. Vyssh. Uchebn. Zavedenii, Matematika, No. 6, 95-98 (1966).

97. A. F. Mutylin, "Example of a nontrivial topologization of the rational number field. Complete locally restricted fields," Izv. Akad. Nauk SSSR, Ser. Mat., 30(4):873-890 (1966).

98. A. F. Mutylin, "Nonnormable extension of the p-adic field," Mat. Zametki, 2(1):11-14 (1967).

99. A. F. Mutylin, "Correction to the paper 'Example of a nontrivial topologization of the rational number field. Complete locally restricted fields'," Izv. Akad. Nauk SSSR, Ser. Mat., 32(1):245-246 (1968).

100. M. E. Novodvorskii, "Nondecomposable finite-dimensional representations of Lie algebras," Vestnik Mosk. Univ., Mat., Mekh., No. 6, 52-55 (1966).

101. Fêng-wên Nie, "The F-potent property of rings," Acta Scient. Natur. Univ. Jilinensis, No. 2, 25-33 (1964).

102. V. A. Parfenov, "Manifolds of Lie algebras," Algebra Logika, Seminar, 6(4):61-73 (1967).

103. A. V. Roiter, "On the theory of integral representations of rings," Mat. Zametki, 3(3):361-366 (1968).

104. A. N. Rukakov and I. R. Shafarevich, "Irreducible representations of a simple three-dimensional Lie algebra over a field of finite characteristic," Mat. Zametki, 2(5):439-454 (1967).

105. Yu. M. Ryabukhin, "Primarities," in: Mathematical Investigations, Vol. 2, Part 1 [in Russian], Kishinev (1967), pp. 67-82.

106. Yu. M. Ryabukhin, "Semistrictly hereditary radicals in primitive classes of rings," in: Research on General Algebra [in Russian], Kishinev (1965), pp. 111-122.

107. Yu. M. Ryabukhin, "Hypernilpotent and special radicals," in: Research on Algebra and Mathematical Analysis [in Russian], Kartya Moldovenyaske, Kishinev (1965), pp. 65-72.

108. Yu. M. Ryabukhin, "Lower radicals of rings," Mat. Zametki, 2(3):239-244 (1967).

109. V. N. Salii, "Rings with commuting idempotents," in: Theory of Semigroups and Its Application [in Russian], Saratov Univ., Saratov (1965), pp. 279-285.

110. Wang-chieh Hsieh, "Seminimality conditions and certain generalizations of the Wedderburn-Artin lattice theorem. I," Acta Scient. Natur. Scholar. Super. Sinens., P. Math. Mech. et Astron., 1(3):209-223 (1965).

111. Wang-chieh Hsieh, "Semiminimality conditions and certain generalizations of the Wedderburn-Artin lattice theorem. II," Acta Scient. Natur. Scholar. Super. Sinens., P. Math. Mech. et Astron., 1(4):299-317 (1965).

112. Wang-chieh Hsieh, "Structure of rings with nuclei and the nucleus of rings with nuclei," Acta Scient. Natur. Univ. Jilinensis, No. 2, 65-84 (1964).

113. V. M. Seliverstov, "The *-regularity of a matrix ring over a given ring," Sibirsk. Mat. Zh., 9(1):229-233 (1968).

114. L. A. Simonyan, "Certain radicals of Lie algebras," Sibirsk. Mat. Zh., 6(5):1101-1107 (1965).

115. L. A. Skornyakov, "Cohn rings," Algebra Logika, Seminar, 4(3):5-30 (1965).

116. L. A. Skornyakov, "Locally bicompact biregular rings (supplement)," Mat. Sb., 69(4):663 (1966).

117. L. A. Skornyakov, "The Elizarov quotient ring and the localization principle," Mat. Zametki, 1(3):263-268 (1967).

118. L. A. Skornyakov, "Left-chain rings," Izv. Vyssh. Uchebn. Zavedenii, Matematika, No. 4, 114-117 (1966).

119. D. A. Suprunenko and V. I. Monastyrnyi, "Silov subgroups of the multiplicative group of a division ring," Dokl. Akad. Nauk BSSR, 9(4):217-218 (1965).

120. D. A. Suprunenko and R. I. Tyshkevich, Commutative Matrices [in Russian], Nauka i Tekhnika, Minsk (1966), 104 pp.

121. B. V. Tarasov, "Free associative algebras," Algebra Logika, Seminar, 6(5):93-105 (1967).

122. P. A. Freidman, "Rings associative modulo their own radicals," Mat. Zap. Ural'skii Univ., 5(1):67-72 (1965).

123. P. A. Freidman, "Rings with a distributive subring lattice," Mat. Sb., 73(4):513-534 (1967).

124. G. M. Tsukerman, "Rings of endomorphisms of free modules," Sibirsk. Mat. Zh., 7(5):1161-1167 (1966).

125. A. I. Cheremisin, "Completely orderable rings," Algebra Logika, Seminar, 4(6):29-46 (1965).

126. A. I. Cheremisin, "Completely orderable rings," Algebra Logika, Seminar, 4(2):67-85 (1965).

127. B. Chernyshov, "CR-rings and their isomorphisms," Sibirsk. Mat. Zh., 7(5):1168-1193 (1966).

128. B. Chernyshov, "Finitely-approximable prime continuous regular rings," Sibirsk. Mat. Zh., 8(3):695-704 (1967).

129. B. M. Shain, "O-rings and LA-rings," Izv. Vyssh. Uchebn. Zavedenii, Matematika, No. 2, 111-112 (1966).

130. M. A. Shatalova, "l_A- and l_I-rings," Sibirsk. Mat. Zh., 7(6):1383-1399 (1966).

131. M. A. Shatalova, "Certain aspects of the theory of lattice-ordered rings," Uspekhi Mat. Nauk, 21(5):267-268 (1966).

132. M. A. Shatalova, "Irreducible extensions of lattice-ordered rings," Sibirsk. Mat. Zh., 8(2):406-414 (1967).

133. M. A. Shatalova, "One class of lattice-ordered rings," Uch. Zap. Mosk. Obl. Ped. Inst., 185:125-134 (1967).

134. A. I. Shirshov, "Certain identity relations in algebras," Sibirsk. Mat. Zh., 7(4):963-966 (1966).

135. A. I. Shirshov, "Special J-rings," Mat. Sb., 38(2):149-166 (1956).

136. É. G. Shutov, "Certain immersions of rings," Uch. Zap. Leningrad. Gos. Ped. Inst. im. Gertsena, 328:280 (1967).

137. I. K. Shcherbatskii, "Upper and lower isolated M-components," Bul. Akad. Shtiintse RSSMold., No. 1, 89-93 (1968).

138. A. Abian, On the nilpotency of algebras. Amer. Math. Monthly, 74(1, part I):33-34 (1967).

139. A. A. Albert, New results on associative division algebras. J. Algebra, 5(1):110-132 (1967).

140. A. A. Albert, On associative division algebras of prime degree. Proc. Amer. Math. Soc., 16(4):799-802 (1965).

141. A. A. Albert, On exceptional Jordan division algebras. Pacif. J. Math., 15(2): 377-404 (1965).

142. H. P. Allen, Jordan algebras and Lie algebras of type D_4. Bull. Amer. Math. Soc., 72(1, part I):65-67 (1966).

143. H. P. Allen, Jordan algebras and Lie algebras of type D_4. Diessert. Abstrs, B27(5):1533 (1966).

144. H. P. Allen, Jordan algebras and Lie algebras of type D_4. J. Algebra, 5(2):250-265 (1967).

145. H. P. Allen, Rings with involution. Israel J. Math., 8(2):99-106 (1968).

146. S. A. Amitsur, Rational identities and applications to algebra and geometry. J. Algebra, 3(3):304-359 (1966).

147. S. A. Amitsur, Nil semi-groups of rings with a polynomial identity. Nagoya Math. J., 27(1):103-111 (1966).

148. S. A. Amitsur, An embedding of PI-rings. Proc. Amer. Math. Soc., 3:3-9 (1952).

149. S. A. Amitsur, Prime rings having polynomial identities with arbitrary coefficients. Proc. London Math. Soc., 17(3):470-486 (1967).

150. S. A. Amitsur and C. Procesi, Jacobson-rings and Hilbert algebras with polynomial identities. Ann. mat. pura ed appl., 71:61-72 (1966).

151. G. Ancochea, Algèbres associatives unitaires de dimension finie sur corps commutatif. J. reine und angew. Math., 222(1/2):75-78 (1966).

152. G. Ancochea, Algèbres associatives unitaires sur un corps commutatif. J. Algebra, 9(2):176-180 (1968).

153. C. T. Anderson, On an identity common to Lie, Jorfan and quasiassociative algebras. Doct. diss. Ohio State Univ. (1964), 53 pp. Dissert Abstr., 25(8):4716 (1965).

154. F. W. Anderson, Lattice-ordered rings of quotients. Canad. J. Math., 17(3):434-448 (1965).

155. J. M. Anderson, Sobre derivaciones generalizades. Rev. colomb. mat., 1(3):29-32 (1967).

156. T. A. Anderson, A note on derivations of commutative algebras. Proc. Amer. Math. Soc., 17(5): 1199-1202 (1966).

157. T. Anderson, N. Divinsky, and A. Sulinski, Hereditary radicals in associative and alternative rings. Canad. J. Math., 17(4):594-603 (1965).

158. S. Andreadakis, On the derivations and automorphisms of Lie algebras. Arch. Math., 17(1):36-43 (1966).

159. E. P. Armendariz, On radical extensions of rings. J. Austral. Math. Soc., 7(4): 552-554 (1967).

160. E. P. Armendariz and W. G. Leavitt, The hereditary property in the lower radical construction. Canad. J. Math., 20(2):474-476 (1968).

161. A. Shigemoto, On the automorphism ring of division algebras. Kodai Math. Semin. Repts., 18(4):368-372 (1966).

162. A. Shigemoto, On invariant subspaces of division algebras. Kodai Math. Semin. Repts., 18(4):332-334 (1966).

163. S. Aumercier, Problèmes d'extensions dans les anneaux. I. C. r. Acad. sci., AB262(13):A733-A735 (1966).

164. S. Aumercier, Problèmes d'extensions dans les anneaux. II. C. R. Acad. sci., AB262(19):A1031-A1033 (1966).

165. S. Aumercier, Problèmes d'extensions dans les anneaux. III. C. r. Acad. sci., AB262(20):A1095-A1097 (1966).

166. V. K. Balachandran, Regular elements of L*-algebras. Math. Z., 93(2):161-163 (1966).

167. V. K. Balachandran and P. R. Parthasarathy, Cartan subalgebras of an L*-algebra. Math. Ann., 166(4):300-301 (1966).

168. B. Banaschewski, Maximal rings of quotients of semi-simple commutative rings. Arch. Math., 16(6):414-420 (1965).

169. B. A. Barnes, Modular annihilator algebras. Canad. J. Math., 18(3):566-578 (1966).

170. D. W. Barnes, Lattice isomorphisms of associative algebras. J. Austral. Math. Soc., 6(1):106-121 (1966).

171. D. W. Barnes, On the radical of a ring with minimum condition. J. Austral. Math. Soc., 5(2):234-236 (1965).

172. R. T. Barnes, On splitting fields for certain Lie algebras of prime characteristic. Proc. Amer. Math. Soc., 17(4):930-935 (1966).

173. W. E. Barnes, On the Γ-rings of Nobusawa. Pacif. J. Math., 18(3):411-422 (1966).

174. W. E. Barnes and W. M. Cunnea, On primary representations of ideals in non-commutative rings. Math. Ann., 173(3):233-237 (1967).

175. F. Bartolozzi, Su una classe di quasicorpi (sinistri) finiti. Rend. mat. e applic., 24(1):165-173 (1965).

176. K. Baumgartner, Bemerkungen zum Isomorphieproblem der Ringe. Monatsh. Math., 70(4):299-308 (1966).

177. W. E. Baxter, Concerning the commutator subgroup of a ring. Proc. Amer. Math. Soc., 16(4):803-805 (1965).

178. W. E. Baxter and E. F. Haeussler, Generating submodules of simple rings with involution. Duke Math. J., 33(3):595-603 (1966).

179. W. E. Baxter and W. S. Martindale, Rings with involution and polynomial identities. Canad. J. Math., 20(2):465-473 (1968).

180. M. Becheanu, Remarque sur les radicals speciaux. Rev. roumaine math. pures et appl., 10(3):357-360 (1965).

181. J. C. Beidleman, Nonsemi-simple distributively generated near-rings with minimum condition. Math. Ann., 170(3):206-213 (1967).

182. J. C. Beidleman, On the theory of radicals of distributively generated near-rings. I. The primitive radical. Math. Ann., 173(2):89-101 (1967).

183. J. C. Beidleman, On the theory of radicals of distributively generated near-rings. II. The nil-radical. Math. Ann., 173(3):200-218 (1967).

184. J. C. Beidleman, Quasi-regularity in near-rings. Math. Z., 89(3):224-229 (1965).

185. J. C. Beidleman, On near-rings and near-ring modules. Doct. diss. Penn. State Univ. (1964), 178 pp.; Dissert. Abstr., 25(8):4716-4717 (1965).

186. J. C. Beidleman, A radical for near-ring modules. Mich. Math. J., 12(3):377-383 (1965).

187. J. C. Beidleman, Distributively generated near-rings with descending chain condition. Mat. Z., 91(1):65-69 (1966).

188. J. C. Beidleman, Strictly prime distributively generated near-rings. Math. Z., 100(2):97-105 (1967).

189. J. C. Beidleman and R. H. Cox, Topological near-rings. Arch. Math., 18(5): 485-492 (1967).

190. L. P. Belluce and S. K. Jain, Prime rings with a one-sided ideal satisfying a polynomial identity. Pacif. J. Math., 24(3):421-424 (1968).

191. L. P. Belluce, I. N. Herstein, and S. K. Jain, Generalized commutative rings. Nagoya Math. J., 27(1):1-5 (1966).

192. A. Bergmann, Hauptnorm und Struktur von Algebren. J. reine und angew. Math., 222(3/4):160-194 (1966).

193. P. Bernat, "Sur le corps enveloppant d'une algèbre de Lie résoluble," Bull. Soc. Math. France, Mém. 7 (1966), 175 pp.

194. S. J. Bernau, On semi-normal lattice rings. Proc. Cambridge Philos. Soc., 61(3): 613-616 (1965).

195. M. Bertrand, Algèbres non associatives et algèbres génétique, Gauthier-Villars, Paris (1966), 103 pp.

196. A. R. Blass and C. V. Stanojevic, On certain classes of associative rings. Amer. Math. Monthly, 75(1):52-53 (1968).

197. R. E. Block, The Lie algebras with a quotient trace form. Ill. J. Math., 9(2): 277-285 (1965).

198. R. E. Block, On the Mills-Seligman axioms for Lie algebras of classical type. Trans. Amer. Math. Soc., 121(2):378-392 (1966).

199. D. Boccioni, Caratterizzatione di une classe di anelli generalizzati. Rend. Semin. mat. Univ. Padova, 35(1):116-127 (1965).

200. D. Boccioni, Struttura degli anelli generati dai loro elementi unita sinistry. Rend. Semin. mat. Univ. Padova, 39:86-101 (1967).

201. D. Boccioni, Immersione di un anello in anello avente un numero cardinale qualsiasi di elementi unita sinistry. Rend. Semin. mat. Univ. Padova, 39:102-122 (1967).

202. D. Boccioni, Somme dirette e rappresentazioni di anelli E_s-generati. Rend. Semin. mat. Univ. Padova, 39:123-135 (1967).

203. A. H. Boers, L'anneau à quatre elements. Proc. Koninkl. nederl. akad. wet., 28(1):14-21 (1966).

204. A. H. Boers, Les anneaux 4- et 5-associatifs et quelques generalisations. Proc. Koninkl. nederl. akad. wet., A69(1):6-13 (1966).

205. J. W. Bond, Weak minimal generating set reduction theorems for associative and Lie algebras. Ill. J. Math., 10(4):579-591 (1966).

206. P. R. Bongale, Filtred Frobenius algebras. Math. Z., 97(4):320-325 (1967).

207. W. Bos, Ein einfacher Beweis eines Satzes von Hurwitz über reelle Divisionsalgebren. Math.-phys. Semesterber., 12(1):79-86 (1965).

208. A. J. Bowtell, On a question of Mal'cev. J. Algebra, 7(1):126-139 (1967).

209. A. Braier, Exponential form of the elements of some algebras. Bull. Inst. politehn. Iasi, 13(1/2):115-122 (1967).

210. M.-P. Brameret, Treillis d'idéaux et structure d'anneaux. Sémin. Dubreil et Pisot Fac. Sci. Paris, 16(1):1/01-1/12 (1962-1963).

211. H. Braun and M. Koecher, Jordan-algebren. Berlin, Springer, (1966) 375 pp.

212. G. Brown Simple Lie algebras over fields of characteristic two. Math. Z., 95(3): 212-222 (1967).

213. R. B. Brown, On generalized Cayley-Dickson algebra. Pacif. J. Math., 20(3): 415-422 (1967).

214. R. H. Bruck, The algebra of dimension-linking operators. Canad. Math. Bull., 8(2):203-222 (1965).

215. P. Burmeister and J. Schmidt, On the completion of partial algebras. Colloq. math., 17(2):235-245 (1967).

216. R. G. Buschman, Quasi-inverses of sequences. Amer. Math. Monthly, 73(4):134-135 (1966).

217. D. Butuc and M. Scutaru, Relatti si lärgirea notiunii de numär. Gaz. mat. (RSR), A72(10):370-378 (1967).

218. L. J. Cada, A structure theory for linear associative nilpotent algebras. Doct. diss. Cathol Univ. America, (1964), 52 pp.; Dissert. Abstr., 26(2):1058 (1965).

219. W. H. Caldwell, Hypercyclic rings. Pacif. J. Math., 24(1)29-44 (1968).

220. R. E. Chandler and K. Koh, Applications of a function topology on rings with unit. Ill. J. Math., 11(4):580-585 (1967).

221. A. J. Chandy, Rings generated by the inner automorphisms of non-Abelian groups. Doct. diss. Boston Univ. Graduate School., Dissert. Abstrs., 26(6)3365 (1965).

222. Hai-chuan Chang, Alternative rings. Sci. Abstr. China, Math. and Phys. Sci. 2(3):2-3 (1964).

223. Chong-yun Chao, A nonimbedding theorem of nilpotent Lie algebras. Pacif., J. Math., 22(2):231-234 (1967).

224. Chong-yun Chao, Some characterizations of nilpotent Lie algebras. Math. Z., 103(1):40-42 (1968).

225. N. Chaptal, Anneaux dond le demi-groupe multiplicatif est inverse. C. r. Acad. sci., AB262(5):A274-A277 (1966).

226. S. U. Chase and C. Faith, Quotient rings and direct products of full linear rings. Math. Z., 88(3):250-264 (1965).

227. Kim-lin Chew, Extensions of rings and modules. Doct diss. Univ. Brit. Columbia, (1965), 129 pp.; Dissert. Abstr., 26(8):4681 (1866).

228. Wei-liang Chow, On unmixedness theorem. Amer. J. Math., 86(4):799-822 (1964).

229. Byoung-song Chwe, On the commutativity of restricted Lie algebra. Proc. Amer. Math. Soc., 16(3):547 (1965).

230. W. E. Clark, A note on semiprimary PP-rings. Osaka J. Math., 4(1):177-178 (1967).

231. J. R. Clay, Imbedding an arbitrary ring in nontrivial near-ring. Amer. Math. Monthly, 74(4):406-407 (1967).

232. J. R. Clay and J. J. Malone, Jr., The near-rings with identities on certain finite groups. Math. scan., 19(1):146-150 (1966).

233. A. Climescu, Asupra unei clase de algebre de ordinul patru. Bull. Inst. politechn. Iasi., 13(1-2):1-4 (1967).

234. A. Climescu, A nouă clasa de inele slabe. Bull. Inst. politehn. Iasi., 10(3-4):1-4 (1964).

235. J. A. Cohn and D. Livingstone, On the structure of group algebras. I. Canad. J. Math., 17(4):583-593 (1965).

236. P. M. Cohn, Hereditary local rings. Nagoya Math. J., 27(1):223-230 (1966).

237. P. M. Cohn, On a class of binomial extensions. Ill. J. Math., 10(3):418-424 (1966).

238. P. M. Cohn, A remark on matrix rings over free ideals rings. Proc. Cambridge Philos. Soc.,62(1):1-4 (1966).

239. P. M. Cohn, Sur une classe d'anneaux hereditaires. Semin. Dubreil et Pisot Fac. Sci. Paris, 19(1):7/01-7/05 (1967).

240. P. M. Cohn, On the free product of associative rings. Ill. J. Algebra, 8(3):376-383 (1968).

241. L. M. Court, The impossibility of rings with dual distributive laws. Riv. mat. Univ. Parma, 6:31-36 (1965).

242. R. C. Courter, The dimension of maximal commutative subalgebras of K_n. Duke Math. J., 32(2):225-232 (1965).

243. C. G. Cullen and C. A. Hall, Classes of functions on algebras. Canad. J. Math., 18(1):139-146 (1966).

244. M. Curzio,Una generalizzazionedegli spazi di Sperner. Ricerche. mat., 16(1): 145-153 (1967).

245. B. P. Dawkins and I. Halperin, The isomorphism of certain continuous rings. Canad. J. Math., 18(6):1333-1344 (1966).

246. R. De Marr, Partially ordered fields. Amer. Math. Monthly, 74(4):418-420 (1967).

247. F. R. De Meyer, Galois theory in rings and algebras. Doct. diss. Univ. Ore.,(1965), 59 pp.; Dissert. Abstr., 26(8):4683 (1966).

248. F. R. De Meyer, Galois theory in separable algebras over commutative rings. Ill. J. Math., 10(2):287-295 (1966).

249. F. R. De Meyer, Some notes on the general Galois theory of rings. Osaka J. Math., 2(1):117-127 (1965).

250. S. E. Dickson, Decomposition of modules. I. Classical rings. Math. Z., 90(1):9-13 (1965).

251. N. Divinsky, Rings and Radicals, Allen and Unwin, London (1965).

252. N. Divinsky and A. Sulinski, Kurosh radicals of rings with operators. Canad. J. Math., 17(2):278-280 (1965).

253. J. Dixmier, Représentations irréductibles des algèbres de Lie résolubles. J. math. pures et appl., 45(1):1-66 (1966).

254. J. Dixmier, Sur le centre de l'algèbre enveloppant d'une algèbre de Lie. C. r. Acad. sci., 265(15):A408-A410 (1967).

255. J. Dixmier, Représentations irreductibles des algèbres de Lie nilpotentes. Anais. Acad. brasil. ciênc., 35(4):491-519 (1963).

256. M. Djabali, Anneau de fractions d'un J-anneau. Semin. Dubreil et Pisot Fac. Sci. Paris, 18(1):8/01-8/12 (1967).

257. M. Djabali, Demi-groupes nilpotents et radical d'un anneau noethérien ou artinien.C.r. Acad. sci., 264(1):A493-A495 (1967).

258. F. Drollinger, Über die Struktur nilpotenter und auflösbarer Liescher Algebren. Comment. math. helv., 42(4):259-284 (1967).

259. D. Dubois, Modules of sequences of elements of a ring. J. London Math. Soc., 41(1):177-180 (1966).

260. D. Dubois, A note on David Harrison's theory of preprimes. Pacif. J. Math., 21(1):15-19 (1967).

261. F. H. Eckstein, "Some results in ring theory: I. An extension of the Wedderburn-Artim theorem. II. Semigroup methods in ring theory," Doctoral dissertation, Tulane Univ. (1967), 84 pp.; Dissert. Abstr., B28(7):2931-2932 (1968).

262. F. H. Eckstein, Complete semi-simple rings with ideal neighbourhoods of zero. Arch. Math., 18(6):587-590 (1967).

263. K. E. Eldridge, Descending chain condition rings with cyclic quasi-regular group. Doct. diss. Univ. Colo. (1965), 63 pp., Dissert. Abstr., 26(11):6736 (1966).

264. S. Elliger, Über das Rangproblem bei Körpererweitrungen. J. reine und angew. Math., 221:162-175 (1966).

265. S. Elliger, Über galoissche Körpererweiterungen von unendlichem Rang. Math. Ann., 163(4):359-361 (1966).

266. S. Elliger, Konstruktion einfacher Ringe über einem einfachen Ring. Math. Ann., 176(1):15-27 (1968).

267. Endliche Gruppen und Liesche Ringe, Ber. 12-18. 7.65. Math. Forschungsinst. Oberwolfach (1965), 9 pp.

268. M. Endo, On the topologies of the rational number field. Comment. math. Univ. St. Pauli, 16(1):11-20 (1967).

269. M. Endo, A note on locally compact division rings. Comment. math. Univ. St. Pauli, 14(1):57-64 (1966).

270. I. Enescu, Les algèbres d'ordre cinq associatives, commutatives et à unité principale, à scalaires réels. Nota I. Algèbres indécomposables. Bull. Inst. politechn. Iasi., 13(1):1-2, 13-22 (1967).

271. I. Enescu, Asurpa caracterului de algebre polinomiale al unor numere hipercomplexe utilizate in studiul unor oscilatii liniare. Bull. Inst. politehn. Iasi, 12(3/4): 23-28 (1966).

272. D. B. Erickson, Orders for finite noncommutative rings. Amer. Math. Monthly, 73(4):376-377 (1966).

273. W. T. van Est, On Ado's theorem. Proc. Koninkl. nederl. akad. wet., A69(2): 176-191 (1966).

274. C. Faith, Rings with ascending condition on annihilators. Nagoya Math. J., 27(1):179-191 (1966).

275. C. Faith, On Köthe rings. Math. Ann., 164(3):207-212 (1966).

276. C. Faith and Y. Utumi, Maximal quotient rings. Proc. Amer. Math. Soc., 16(5): 1084-1089 (1965).

277. G. Fardoux, Bases réduites d'idéaux de A (X). A anneau principal. C. r. Acad. sci., AB262(21):A1146-A1148 (1966).

278. J. R. Faulkner, The inner derivations of a Jordan algebra. Bull. Amer. Math. Soc., 73(2):208-210 (1967).

279. J. Fell, Algebras and fiber bundles. Pacif. J. Math., 16(3):497-503 (1966).

280. J. Fell and J. Mather, Barely faithful algebras. Amer. Math. Monthly, 72(9): 1001-1003 (1965).

281. H. Fell and A. J. Goldman, Realization of semi-multiplies as multipliers. Amer. Math. Monthly, 72(6):639-641 (1965).

282. E. Feller, A type of quasi-Frobenius ring. Canad. Math. Bull., 10(1):19-27 (1967).

283. M. Ferrandon, Ultraproduits et anneaux primitifs. Semin. Dubreil et Pisot. Fac. Sci. Paris, 19(2):16/01-16/09 (1967).

284. J. Ferrar, On Lie algebras of type E_6. Bull. Amer. Math. Soc., 73(1):151-155 (1967).

285. J. Ferrar, Generic splitting fields of composition algebras. Trans. Amer. Math. Soc., 128(3):506-514 (1967).

286. J. Ferrar, "On Lie algebras of type E_6," Doctoral dissertation, Yale Univ. (1966), 115 pp.; Dissert. Abstr., B27(8):2776 (1967).

287. P. A. Fillmore, An Archimedean property of cardinal algebras. Michigan Math. J., 11(4):365-367 (1967).

288. M. Flato and D. Sternheimer, Algèbres unifiantes; application à l'algèbre de Lie du groupe de Poincaré, C. r. Acad. sci., 260(13):3532-3534 (1965).

289. I. Fleischer, Ein Satz aus der abstracten Idealtheorie, Avhandl. utg. Norske, Vid-acad., Oslo. I Mat.-Naturvid. kl., 6 (1964), 12 pp.

290. I. Fleischer, Note sur les espaces à norme non-archimedienne. Proc. Koninkl. nederl. acad. wet., A68(4):630-631 (1965); Indagationes math., 27(4):630-631 (1965).

291. F. Forelli, Homomorphisms of ideals in group algebras. III. J. Math., 9(3):410-417 (1965).

292. E. Fried, Beiträge zur Theorie der Frobenius-Algebren. Math. Ann., 155(4):265-269 (1964).

293. L. Fuchs, Riesz rings. Math. Ann., 166(1):24-33 (1966).

294. L. Fuchs, On partially ordered algebra. I. Colloq. math., 14:115-130 (1964).

295. N. Funayama, Imbedding a regular ring in a regular ring with identity. Nagoya Math. J., 27(1):61-64 (1966).

296. G. Galbura and I. D. Ion, Asupra localizării descompunerilor primare in sub-comutativ. An. Univ. Bucuresti, Ser. stiint. natur. math.-mecan., 15(1):9-16 (1966).

297. N. Ganesan, Properties of rings with a finite number of zero divisors. II. Math. Ann., 161(4):241-246 (1965).

298. E. Gentile, A uniqueness theorem on rings of matrices. J. Algebra, 6(1):131-134 (1967).

299. C. George and M. Lévy-Nahas, Finite-dimensional representations of some non-semisimple Lie algebras. J. Math. Phys., 7(6):980-988 (1966).

300. M. Gerstenhaber, On dominance and varieties of commuting matrices. Ann. Math., 73(2):324-348 (1961).

301. M. Gerstenhaber, On the construction of division rings by the deformations of fields. Proc. Nat. Acad. Sci. U. S. A., 55(4):690-692 (1966).

302. R. W. Gilmer, Jr., If R(x) is Noetherian, R contains identity. Amer. Math. Monthly, 74(6):700 (1967).

303. K. Glazek, Algebry *-laczne i γ-algebry. Acta Univ. wratisl., No. 58, pp. 5-19 (1967).

304. C. M. Glennie, Some identities valid in special Jordan algebras. Pacif. J. Math., 16(1):47-59 (1966).

305. A. W. Goldie, Semi-prime rings with maximum condition. Proc. London Math. Soc., 10(38):201-220 (1960).

306. A. W. Goldie, The structure of prime rings under ascending chain conditions. Proc. London Math. Soc., 8(32):589-608 (1958).

307. A. W. Goldie, Localization in non-commutative noetherian rings. J. Algebra, 5(1):89-105 (1967).

308. J. I. Goldman, On a class of nodal noncommutative Jordan algebras. Trans. Amer. Math. Soc., 128(1):176-183 (1967).

309. O. Goldman and Chin-han Sah, On a special class of locally compact rings. J. Algebra, 4(1):71-95 (1966).

310. H. Gonshor, Special train algebras arising in genetics. II. Proc. Edinburgh Math. Soc., 14(4):333-338 (1965).

311. H. Gonshor, On abstract affine near-rings. Pacif. J. Math., 14(4):1237-1240 (1964).

312. B. Gordon and T. S. Motzkin, On the zéros of polynomials over division rings. Trans. Amer. Math. Soc., 116(4):218-226 (1965).

313. R. Gordon, Rings faithfully represented on their left socle. J. Algebra, 7(3):303-342 (1967).

314. R. Gordon, "Rings faithfully represented on their left socle," Doctoral. dissertation, Calif. Inst. Technol. (1966), 78 pp.; Dissert. Abstr. B27(5):1542 (1966).

315. E. Granirer and M. Rajagopalan, A note on the radical of the second conjugate algebra of a semigroup algebra. Math. scand., 15(2):163-166 (1965).

316. I. Halperin, Regular rank rings. Canad. J. Math., 17(5):709-719 (1965).

317. I. Halperin, Extension of the rank function. Studia math., 27(3):325-335 (1966).

318. I. Halperin, Von Neumann's manuscript on inductive limits of regular rings. Canad. J. Math., 20(2):477-483 (1968).

319. H. Halpern, The maximal GCR ideal in an AW-algebra. Proc. Amer. Math.Soc., 17(4):905-914 (1966).

320. A. Hanna, On the ring of integers. Amer. Math. Monthly, 74(10):1241-1243 (1967).

321. M. Harada, QF-3 and semi-primary PP-rings. I. Osaka J. Math., 2(2):357-368 (1965).

322. M. Harada, QF-3 and semi-primary PP-rings. II. Osaka J. Math., 3(1):21-27 (1966).

323. M. Harada, Hereditary orders which are dual. J. Math. Osaka City Univ., 14(2): 107-115 (1963).

324. M. Harada, On semi-primary PP-rings. Osaka J. Math., 2(1):153-161 (1965).

325. M. Harada, Multiplicative ideal theory in hereditary orders. J. Math. Osaka City Univ., 14(2):83-106 (1963).

326. M. Harada, Note on orders over which an hereditary order is projective. Osaka J. Math., 4(1):151-156 (1967).

327. M. Harada, Supplementary results on Galois extension. Osaka J. Math., 2(2): 343-350 (1965).

328. R. Hart, Simple rings with uniform right ideals. J. London Math. Soc., 42(4): 614-617 (1967).

329. B. Hartley, Locally nilpotent ideals of a Lie algebra. Proc. Cambridge Philos. Soc., 63(2):257-272 (1967).

330. R. G. Hathway, On non-commutative non-associative algebras. Doct. diss. Univ. Wisc. (1966), 76 pp.; Dissert. Abstr., B28(3):1002 (1967).

331. A. Hattori, Simple algebras over a commutative ring. Nagoya Math. J., 27(2): 611-616 (1966).

332. A. Hattori, On strongly separable algebras. Osaka J. Math., 2(2):369-372 (1965).

333. K. Hauschild, Über die Unentscheidbarkeit nicht assoziativer Schiefringe. Wiss. Z. Hunboldt-Univ. Berlin. Math.-naturwiss. Reihe, 15(5):681-683 (1966).

334. K.-H. Helwig, Über Automorphismen und Derivationen von Jordan-Algebren. Proc. Koninkl. nederl. acad. wet., 70:381-394 (1967); Indagationes math., 31(4): 381-394 (1967).

335. K.-H. Helwig, Über Mutationen von Jordan-Algebren. Math. Z., Vol. 103, Nos. 1-7 (1968).

336. K.-H. Helwig, Eine Verallgemeinerung der formal-reelen Jordan-Algebren. Invent. Math., 1(1):18-35 (1966).

337. R. L. Hemminger, On the Wedderburn principal theorem for commutative power-associative algebras. Trans. Amer. Math. Soc., 121(1):36-51 (1966).

338. I. N. Herstein, A theorem on left Noetherian rings. J. Math. Analysis and Applic., 15(1):91-96 (1966).

339. I. N. Herstein, A counterexample in Noetherian rings. Proc. Nat. Acad. Sci. U. S. A., 54(4):1036-1037 (1965).

340. I. N. Herstein, Anelli alternativi ed algebre composizione. Rend. Math. e applic., 23(3-4):364-393 (1964).

341. I. N. Herstein, Lie and Jordan structures in simple associative rings. Bull. Amer. Math. Soc., 67:517-531 (1961).

342. I. N. Herstein, Special simple rings with involution. J. Algebra, 6(3):369-375 (1967).

343. I. N. Herstein and L. Small, Nil rings satisfying certain chain conditions; an addendum. Canad. J. Math., 18(2):300-302 (1966).

344. K. Hirata and K. Sugano, On semisimple extensions and separable extensions over non-commutative rings. J. Math. Soc. Japan, 18(4):360-373 (1968).

345. R. D. Hirsch, A note on non-commutative polynomial rings subject to degree-preservation. J. London Math. Soc., 42(2):333-335 (1967).

346. G. Hochschild, An addition to Ado's theorem. Proc. Amer. Math. Soc., 17(2): 531-533 (1966).

347. P. Holgate, Sequences of powers in genetic algebras. J. London Math. Soc., 42(3): 489-496 (1967).

348. P. Holgate, Genetic algebras associated with polyploidy. Proc. Edinburgh Math. Soc., 15(1):1-9 (1966).

349. P. Holgate, The genetic algebra k linked loci. Proc. London Math. Soc., 18(3): 315-327 (1968).

350. P. Holgate, Jordan algebras arising in population genetics. Proc. Edinburgh Math. Soc., 15(4):291-294 (1967).

351. W. G. van Hoorn and B. van Rootselaar, Fundamental notions in the theory of seminearrings. Compositio math., 18(1-2):65-78 (1966).

352. Pang-chieh Hsieh, Rings with semi-minimum condition. Scientia sinica, 14(3): 343-362 (1965).

353. M. M. Humm, On a class of right alternative rings without nilpontent ideals. J. Algebra, 5(2):164-175 (1967).

354. M. M. Humm and E. Kleinfeld, On extensions of Cayley algebras. Proc. Amer. Math. Soc., 17(5):1203-1205 (1966).

355. M. M. Humm and E. Kleinfeld, On free alternative rings. J. Combin. Theory, 2(2):140-144 (1967).

356. K. Hunter, Nilpotence of nil subrings implies more general nilpotence. Arch. Math., 18(2):136-139 (1967).

357. M. S. Huzurbazar, The multiplicative group of a division ring. Math. Student, 32(1-2):7-10 (1964).

358. M. Ikeda, Über die einstufig nichtkommutativen Ringe. Nagoya Math. J., 27(1): 371-379 (1966).

359. E. C. Ingraham, On the existence of inertial subalgebras. Doct. diss. Univ. Ore. (1965), 52 pp.; Dissert. Abstr., 26(8):4690-4691 (1966).

360. E. C. Ingraham, Inertial subalgebras of algebras over commutative rings. Trans. Amer. Math. Soc., 124(1):77-93 (1966).

361. D. B. Ion, Asupra algebrelor Jordan de tip A. Studii si cercetări mat. Acad. RPR, 17(2):301-310 (1965).

362. D. B. Ion, Asupra unor proprietăți ale constantelor de structură ale algebrelor Jordan de tip A. Studii și cercetări mat. Acad. RPR, 19(7):1031-1038 (1967).

363. D. B. Ion, Asupra unor proprietăți ale rădăcinilor algebrelor Jordan de tip A. Studii și cercetări mat. Acad. RPR, 18(2):309-313 (1966).

364. I. D. Ion, Radicalul limitei projective de inele asociative. Studdi și cercetari. mat. Acad. RPR, 16(6):1141-1145 (1964).

365. R. Iordănesku, Sur des représentations des algèbres $A^1_{m, n}$. Rev. romaine mat. pures et appl., 10(9):1403-1421 (1965).

366. R. Iordănesku, Les représentations quasi-irréducibles des algèbres $A^2_{m, n}$ dans des algèbres $A^q_{m, n}$. Rev. romaine math. pures et appl., 10(10):1583-1591 (1965).

367. R. Iordănesku, Les représentations quasi-irréductibles des algèbres $A^3_{m, n}$. Rev. roumaine math. pures et appl., 11(7):843-845 (1966).

368. J. R. Isbell, Embedding two ordered rings in one ordered ring, Part I. J. Algebra, 4(3):341-364 (1966).

369. N. Jacobson, Structure theory for a class of Jordan algebras. Proc. Nat. Acad. Sci. U.S.A., 55(2):243-251 (1966).

370. N. Jacobson, Cartan subalgebras of Jordan algebras. Nagoya Math. J., 27(2):591-609 (1966).

371. N. Jacobson, Associative algebras with involution and Jordan algebras. Proc. Koninkl. nederl. akad. wet., A69(2):202-212 (1966); Indagationes math., 28(2):202-212 (1966).

372. N. R. Jacobson, Homomorphism of Jordan rings of self-adjoint elements. Trans. Amer. Math. Soc., 72:310-322 (1952).

373. S. K. Jain, Polynomial rings with a pivotal monomial. Proc. Amer. Math. Soc., 17(4):942-945 (1966).

374. S. K. Jain, Rings with constraints. Math. Student., 34(3-4):175-178 (1966).

375. S. K. Jain, On unitary of II-unitary rings. Publs. math., 11(1):241-244 (1964).

376. J. P. Jans, On orders in quasi-Frobenius rings. J. Algebra, 7(1):35-43 (1967).

377. G. J. Janusz, Primitive idempotents in group algebras. Proc. Amer. Math. Soc., 17(2):520-523 (1966).

378. T. L. Jenkins, Another characterization of the Jacobson radical. Amer. Math. Monthly, 74(10):1237 (1967).

379. T. L. Jenkins, The upper radical determined by regular rings. Amer. Math. Monthly, 74(10):1240 (1967).

380. W. E. Jenner, On regular automorphisms of algebras. Portug. math., 24(1-2): 47-48 (1965).

381. Ch. U. Jensen, A remark on semi-hereditary local rings. J. London Math. Soc., 41(3):479-482 (1966).

382. A. Johnson, Order in logic and integral domains. Amer. Math. Monthly, 72(4): 386-390 (1965).

383. D. G. Johnson, The completion of an archimedean ƒ-ring. J. London Math. Soc., 40(3):493-496 (1965).

384. R. E. Johnson, Rings with zero right and left singular ideals. Trans. Amer. Math. Soc., 118(6):150-157 (1965).

385. R. E. Johnson, Remarks on a paper of Procesi. J. Algebra, 2(1):38-41 (1965).

386. R. E. Johnson, Unique factorization in a principal right ideal domain. Proc. Amer. Math. Soc., 16(3):526-528 (1965).

387. R. E. Johnson, Prime matrix rings. Proc. Amer. Math. Soc., 16(5):1099-1105 (1965).

388. W. R. Johnstone, Generalized Hilbert functions and resolutions associated with maximal one-sided ideals. Math. Z., 98(3):243-258 (1967).

389. T. Kanzaki, On galois algebra over a commutative ring. Osaka J. Math., 2:309-317 (1965).

390. T. Kanzaki, On Galois extension of rings. Nagoya Math. J., 27(1):43-49 (1966).

391. T. Kanzaki, On commutator rings and Galois theory of separable algebras. Osaka J. Math., 1(1):103-115 (1964).

392. T. Kanzaki, Special type of separable algebra over a commutative ring. Proc. Japan Acad., 40(10):781-786 (1964).

393. I. Kaplansky, Lie Algebras, Lectures in Modern Mathematics, Vol. 1, Wiley, New York (1963), pp. 115-132.

394. H. Karzel, Unendliche Dicksonsche Fastkörper. Arch. Math., 16(4):247-256 (1965).

395. F. Kasch, Projective Frobenius-Erweiterungen. Sitzungsber. Heidelberg. Akad. Wiss. Math., naturwiss. Kl., No. 4 (1961), 23 pp.

396. S. Kass, "On a class of generalized alternative algebras," Doctoral dissertation, Ill. Inst. Technol. (1966), 40 pp.; Dissert. Abstr., B27(6):2032 (1966).

397. P. Katz and E. G. Straus, Infinite sums in algebraic structures. Pacif. J. Math., 15(1):181-190 (1965).

398. Y. Kawada, On Köthe's problem concerning algebras for which every indecomposable module is cyclic. III. Sci. Repts. Tokyo Kyoiku Daigaku A, 8(196-201): 1-250 (1965).

399. S. M. Kaye, Ring theoretic properties of matrix rings. Canad. Math. Bull., 10(3):365-374 (1967).

400. O. H. Kegel, On rings that are sums of two subrings. J. Algebra, 1(2):103-109 (1964).

401. A. Kertész, On the existence of a left unit element in a noetherian or in an artinian ring. Bull. Acad. polon. sci. Sec. sci. math., astron. et phys., 14(12): 671-672 (1966).

402. A. Kertész, Gyürük Jacobson-féle radikáljától. Magyar. tyd. akad. Mat. és fiz. tud. oszt. közl., 16(4):445-461 (1966).

403. T. P. Kezlan, A note on higher commutators of bounded nilpotence. Amer. Math. Monthly, 73(6):632-633 (1966).

404. T. P. Kezlan, Rings nil commutator ideals. Doct. diss. Univ. Kans. (1964), 26 pp.; Dissert. Abstr., 26(1):389 (1965).

405. T. P. Kezlan, Rings in which certain subsets satisfy polynomial identities. Trans. Amer. Math. Soc., 125(3):414-421 (1966).

406. A. A. Klein, Rings nonembeddable in fields with multiplicative semigroups imbeddable in groups. J. Algebra, 7(1):100-125 (1967).

407. I. Kleiner, A counterexample to a conjecture of D. F. Sanderson. Canad. Math. Bull., 9(4):517 (1966).

408. M. Knebusch, Der Begriff der Ordnung einer Jordanalgebra. Abhandl. Math. Semin. Univ. Hamburg, 28(3-4):168-184 (1965).

409. M. Knebusch, Eine Klasse von Ordnungen in Jordanalgebren vom Grade 3. Abhandl. Math. Semin. Univ. Hamburg, 28(3-4):185-207 (1965).

410. J. Knopfmacher, On the isomorphism problem for Lie algebras. Proc. Amer. Math. Soc., 18(5):898-901 (1967).

411. M. A. Knus, Sur une classe d'algèbres filtrées. Comment. math. helv., 42(2): 111-131 (1967).

412. D. E. Knuth, Finite semifields and projective planes. J. Algebra, 2(2):182-217 (1965).

413. S. Kobayashi and T. Nagano, On filtered Lie algebras and geometric structures. Ill. J. Math. and Mech., 14(4):697-707 (1965).

414. S. Kobayashi and T. Nagano, A Report on Filtered Lie Algebras, Proceedings U.S.-Japan Sem. Differential Geometry, Kyoto (1965); Tokoyo Nippon Hyoronsha Co., Ltd. (1966), pp. 63-70.

415. M. Koecher, On homogeneous algebras. Bull. Amer. Math. Soc., 72(3):347-357 (1966).

416. M. Koecher, Imbedding of Jordan algebras into Lie algebras. I. Amer. J. Math., 89(3):787-816 (1967).

417. K. Koh, On a semi-prime self-injective ring. Amer. Math. Monthly, 74(6):687-688 (1967).

418. K. Koh, On simple rings with uniform right ideals. Amer. Math., Monthly, 74(6): 685-687 (1967).

419. K. Koh, On some characteristic properties of self-injective rings. Proc. Amer. Math. Soc., 19(1):209-213 (1968).

420. K. Koh, On a semi-primary ring. Proc. Amer. Math. Soc., 19(1):205-208 (1968).

421. K. Koh, On very large one-sided ideals of a ring. Canad. Math. Bull., 9(2): 191-196 (1966).

422. K. Koh, On simple rings with maximal annihilator right ideals. Canad. Math. Bull., 8(5):667-668 (1965).

423. K. Koh, A note on a certain class of prime rings. Amer. Math. Monthly, 72(1):
      46-48 (1965).
424. K. Koh, On the class of rings which do not contain nonzero semi-singular ideals.
      Amer. Math. Monthly, 72(8):875-877 (1965).
425. K. Koh, On the set of zero divisors of a topological ring. Canad. Math. Bull.,
      10(4):595-596 (1967).
426. K. Koh, On properties of rings with a finite number of zero divisors. Math. Ann.,
      171(1):79-80 (1967).
427. K. Koh and A. Mewborn, The weak radical of a ring. Proc. Amer. Math. Soc.,
      18(3):554-559 (1967).
428. K. Koh and A. Mewborn, Prime rings with maximal annihilator and maximal
      complement right ideals. Proc. Amer. Math. Soc., 16(5):1073-1076 (1965).
429. C. W. Kohls, Properties inherited by ring extensions. Mich. Math. J., 12(4):
      399-404 (1965).
430. C. W. Kohls, On convex ideals. Proc. Amer. Math. Soc., 18(2):359-363 (1967).
431. B. Kolman, Semi-modular Lie algebras. J. Sci. Hiroshima Univ., Ser. A, Div. I.,
      29(2):149-163 (1961).
432. B. Kolman, Relatively complemented Lie algebras. J. Sci. Hiroshima Univ.,
      Ser. A, Div. I., 31(1):1-11 (1967).
433. L. Konguetsof, Certaines propositions sur les annoïdes. Constructions d'annoïdes.
      Bull. Soc. Math. Belg., 19(2):179-193 (1967).
434. F. Kosier, Certain algebras of degree one. Pacif. J. Math., 15(2):541-544 (1965).
435. B. Kostant and A. Novikoff, A homomorphism in exterior algebra. Canad. J.
      Math., 16(1):166-168 (1964).
436. H. F. Kreimer, Galois theory for noncommutative rings and normal bases. Trans.
      Amer. Math. Soc., 127(1):42-49 (1967).
437. H. F. Kreimer, A Galois theory for noncommutative rings. Trans. Amer. Math.
      Soc., 127(1):29-41 (1967).
438. H. F. Kreimer, The foundations for an extension of differential algebra. Trans.
      Amer. Math. Soc., 111(3):482-492 (1964).
439. E. Kreindler, Opérateurs additifs dans les modules unitares a base dénombrable.
      Bull. math. Soc. sci. math. RSR, 9(4):333-336 (1965).
440. J. L. Krivine, Anneaux préordonnés. J. analyse math., 12:307-326 (1964).
441. R. L. Kruse, "Rings with periodic additive group in which all subrings are ideal,"
      Doctoral dissertation, Calif. Inst. Technol. (1964) 56 pp.; Dissert. Abstr. 25(8):
      4725 (1965).
442. E. G. Kundert, Structure theory in s-d-rings. Atti Accad. naz. Lincei Rend. Cl.
      sci. fis., mat. e natur., 41(5):270-278 (1966).
443. E. Kunz, Gruppenringe und Differentiale. Math. Ann., 163(4):346-350 (1966).
444. H. Kupisch, Symmetrische Algebren mit endlich vielen unzerlegbaren Dar-
      stellungen. I. J. reine und angew. Math., 219(1-2):1-25 (1965).
445. H. Kupisch, Über Klasse von Ringen mit Minimalbedingung I. Arch. Math.,
      17(1):20-35 (1966).
446. Y. Kurata, On an additive ideal theory in a non-associative ring. Math. Z.,
      88(2):129-135 (1965).

447. J. P. Lafon, Spectre premier bilatère de l'anneaux des endomorphismes d'un module de type fini. C. r. Acad. sci., AB262(20):A1098-A1099 (1966).

448. J. P. Lafon, Représentation d'algèbres comme quotines d'anneaux d'endomorphismes. Bull. sci. math., 91(1-2):13-16 (1967).

449. J. Lambek, On the ring of quotients of a noetherian ring. Canad. Math. Bull., 8(3):279-290 (1965).

450. J. Lambek, Lectures on Rings and Modules, Blaisdell, Waltham (Mass.) (1966), Vol. 8, 183 pp.

451. W. M. Lambert, Jr., Effectiveness, elementary definability, and prime polynomial ideals. Doct. diss. Univ. Calif. Los Angeles, 26(6):3371-3372 (1965).

452. D. R. La Torre, On h-ideals and k-ideals in hemirings. Publs. Math., 12(1-4): 219-226 (1965).

453. D. R. La Torre, On the radical of a hemiring. Doct. diss. Univ.Tenn. (1964), 111 pp.; Dissert. Abstr., 25(5):3001 (1964).

454. D. A. Lawver, "Left ideal axioms for non-associative rings," Doctoral dissertation Univ. Wisconsin (1967) 71 pp.; Dissert. Abstr. B28(3):1002 (1967).

455. R. R. Laxton, Prime ideals and the ideal-radical of a distributively generated near-ring. Math. Z., 83(1):8-17 (1964).

456. W. G. Leavitt, Sets of radical classes. Publs. math., 14(1-4):321-324 (1967).

457. W. G. Leavitt and E. P. Armendariz, Nonhereditary semisimple classes. Proc. Amer. Math. Soc., 18(6):1114-1118 (1967).

458. L. C. A. van Leeuwen, Holomorphe von endlichen Ringen. Proc. Koninkl. nederl. akad. wet., A68(4):632-634 (1965); Indagationes Math., 27(4):632-645 (1965).

459. D. Legrand, Formes quadratiques et algebres quadratiques. C. r. Acad. sci., 265(23):A764-A767 (1967).

460. L. Lesieur, Sur les anneaux tels que $x^n = x$. Semin. Dubreil et Pisot. Fac. Sci. Paris, 19(2):13/01-13/08 (1967).

461. J. Lewin, Subrings of finite index in finitely generated rings. J. Algebra, 5(1): 84-88 (1967).

462. J. Lewin and T. Lewin, On ideals of free associative algebras generated by a single element. J. Algebra, 9(2):248-255 (1968).

463. Gen-dao Li, On Weyl groups of real semisimple Lie algebras and their application to the classification of maximal solvable subalgebras with respect to inner conjugation. Chinese Math., 8(1):74-89 (1966).

464. F. E. J. Linton, The obstruction to the localizability of a measure space. Bull. Amer. Math. Soc., 71(2):353-356 (1965).

465. Shao-hsueh Liu, On algebras in which each subalgebra is an ideal. Sci. Abstrs. China. Math. and Phys. Sci., Vol. 2, Nos. 4-7 (1964).

466. Hool-tong Loh, Notes on semirings. Math. Mag., 40(3):150-152 (1967).

467. O. Loos, Über eine Beziehung zwischen Malcev-Algebren und Lie-Tripelsystemen. Pacif. J. Math., 18(3):553-562 (1966).

468. H. P. Lorenzen, Mutationsinvariante Unteralgebren von Jordan-Algebren. Math. Ann., 171(1):54-60 (1967).

469. H. P. Lorenzen, Quadratische Darstellungen in Jordan Algebren. Abhandl. Math. Semin. Univ. Hamburg, 28(1-2):115-123 (1965).

470. N. Losey, Useful theorems on commutative non-associative algebras. Proc. Edinburgh Math. Soc., 15(3):203-208 (1967).

471. R. J. Loy, A note on the preceding paper by J. B. Miller. Acta scient. math., 28(1-2):233-236 (1967).

472. T. Luchian, "Observatii asupra inelor slabe," An. ştiinţ. Univ. Iaşi, Sec. Ia, 11b: 99-111 (1965).

473. T. Luchian, "Homomorfisme in inele slabe," An. ştiinţi.Univ. Iaşi, Sec. Ia, 13(1): 17-28 (1967).

474. T. Luchian, "Asupra algebrelor slabe de ordin 3, cu diviziune," An. ştiinţ. Univ. Iaşi, Sec. Ia, 10(1):29-42 (1964).

475. T. Luchian, "Asupra unor algebre slabe generalizate de ordin 2," An. ştiinţ. Univ. Iaşi, Sec. Ia, 12(1):19-36 (1966).

476. A. S.-T. Lue, Crossed homomorphisms of Lie algebras. Proc. Cambridge Philos. Soc., 62(4):577-581 (1966).

477. Jiang Luh, On the structure of J-rings. Amer. Math. Monthly., 74(2):164-166 (1967).

478. Jiang Luh, An elementary proof of a theorem of Herstein. Math. Mag., 38(2): 105-106 (1965).

479. Jiang Luh, On the commutativity of J-rings. Canad. J. Math., 19(6):1289-1292 (1967).

480. J. G.MacDonald, Jordan algebra with three generators. Proc. London Math. Soc., 10(3):395-408 (1960).

481. L. A. Machtinger, "Rings with not necessarily transitive order relation," Doctoral dissertation, Washington Univ. (1965), 90 pp.; Dissert. Abstr., 26(8):4693 (1966).

482. J. J. Malone, Jr., Near-rings with trivial multiplications. Amer. Math. Monthly, 74(9):1111-1112 (1967).

483. J.-M. Maranda, Injective structures. Trans. Amer. Math. Soc., 110(1):98-135 (1964).

484. E. Marshall, The Frattini subalgebra of a Lie algebra. J. London. Math. Soc., 42(3):416-422 (1967).

485. G. E. Martin, On Mammana division rings. Matematiche, 20(1):1-6 (1965).

486. W. S. Martindale, Jordan homomorphisms of the symmetric elements of a ring with involution. J. Algebra, 5(2):232-249 (1967).

487. H. Marubayashi, A remark on components of ideals in noncommutative rings. Proc. Japan Acad., 43(1):11-12 (1967).

488. K. Mathiak, Zur Theorie nicht endlichdimensionaler Jordanalgebren über einem Körper eine Charakteristik 2. J. reine und angew. Math., 224:185-210 (1966).

489. I. G. Maurer and L. Purdea, Unele pentru care submodulele sint subinele. Studia Univ. Babes-Bolyai. Ser. math. phys., 9(2):13-15 (1964).

490. J. P. May, The cohomology of restricted Lie algebras and of Hopf algebras. J. Algebra, 3(2):123-146 (1966).

491. K. O. May, The impossibility of a division algebra of vectors in three dimensional space. Amer. Math. Monthly, 73(3):289-291 (1966).

492. C. J. Maxson, On finite near-rings with identity. Amer. Math. Monthly, 74(10): 1228-1230 (1967).

493. P. J. McCarthy, Note on primary ideal decompositions. Canad. J. Math., 18(5): 950-952 (1966).

494. P. J. McCaithy, General graded rings. Monatsh. Math., 69(3):208-214 (1965).

495. J. McConnel, The intersection theorem for a class of noncommutative rings. Proc. London Math. Soc., 17(3):487-498 (1967).

496. N. H. McCoy, The Theory of Rings, Macmillan, New York (1964), 161 pp.; Brit. Nat. Bibliogr. No. 174, p. 12 (1964).

497. K. M. McCrimmon, A note on quasi-associative algebras. Proc. Amer. Math. Soc., 17(6):1455-1459 (1966).

498. K. M. McCrimmon, "Norms and non-commutative Jordan algebras," Doctoral dissertation, Yale Univ. (1965) 76 pp.; Dissert. Abstr., 26(8):4994 (1966).

499. K. M. McCrimmon, Norms and noncommutative Jordan algebras. Pacif. J. Math., 15(3):925-927 (1965).

500. K. M. McCrimmon, Structure and representations of noncommutative Jordan algebras. Trans. Amer. Math. Soc., 121(1):187-199 (1966).

501. K. M. McCrimmon, Macdonald's theorem with inverses. Pacif. M. Math., 21(2): 315-325 (1967).

502. K. M. McCrimmon, A proof of Schafer's conjecture for infinite-dimensional forms admitting composition. J. Algebra, 5(1):72-83 (1967).

503. K. M. McCrimmon, Finite power-associative division rings. Proc. Amer. Math. Soc., 17(5):1173-1177 (1966).

504. K. M. McCrimmon, A general theory of Jordan rings. Proc. Nat. Acad. Sci. U.S.A., 56(4):1072-1079 (1966).

505. K. M. McCrimmon and R. Schafer, On a class of noncommutative Jordan algebras. Proc. Nat. Acad. Sci. U.S.A., 56(1):1-4 (1966).

506. W. A. McWorter, Some simple properties of simple nil rings. Canad. Math. Bull., 9(2):197-200 (1966).

507. R. A. Melter, A note on geometry in a p-ring. Amer. Math. Monthly, 73(9): 983-986 (1966).

508. K. Meyberg, Ein Satz über Mutationen von Jordan-Algebren. Math. Z., 90(4): 260-267 (1965).

509. K. Meyberg, Über die Killing-Form in Jordan-Algebren. Math. Z., 89(1):52-73 (1965).

510. K. Meyberg, Über die Lie-Algebren der Derivationen und der linksregulären Darstellungen in zentral-einfachen Jordan-Algebren. Math. Z., 93(1):37-47 (1966).

511. G. Michler, Radikale und Sockel. Math. Ann., 157(1):1-48 (1966).

512. G. Michler, On maximal nilpotent subrings of right Noetherian rings. Glasgow Math. J., 8(2):89-101 (1967).

513. G. Michler, Halberbliche, fastlokale Ordnungen in eifachen Artinringen. Arch. Math., 18(5):456-464 (1967).

514. G. Michler, Der Satz von Artin-Wedderburn und der Satz von Goldie. Math. Ann., 163(4):299-304 (1966).

515. G. Michler, Klassische Quotientenringe von nicht notwentig endlichdimensionalen halbprimen Ringen. Math. Z., 91(4):314-335 (1966).

516. G. Michler, Kleine Ideale, Radikale und die Eins in Ringen. Publs. mat., 12(1-4): 231-252 (1965).

517. G. Michler, Charakterisierung einer Klasse von Noetherschen Ringen. Math. Z., 100(2):163-182 (1967).

518. J. B. Miller, Homomorphisms, higher derivations and derivations on associative algebras. Acta. scient. math., 28(1-2):221-231 (1967).

519. A. Mirbagheri, "Finiteness in radical algebras," Doctural dissertation, Ind. Univ. (1965), 35 pp.; Dissert. Abstr. 26(11):6746 (1966).

520. H. Y. Mochizuki, Finitistic global dimension for rings. Pacif. J. Math., 15(1): 249-258 (1965).

521. H. Y. Mochizuki, On the double commutator algebra of QF-3 algebras. Nagoya Math. J., vol. 25, pp. 221-230 (March, 1965).

522. R. L. Moore, "Derivation algebras of a class of Jordan algebras," Doctoral dissertation, Univ. of Virginia (1966) 60 pp.; Dissert. Abstr., B27(8):2786-2787 (1967).

523. K. Morita, Adjoint pairs of functors and Frobenius extensions. Sci. Perts. Tokyo Kyoiku Daigaku, A9(202-208):40-71 (1965).

524. K. Morita, On S-rings in the sense of F. Kasch. Nagoya Math. J., 27(2):687-695 (1966).

525. K. Morita, Duality for modules and its applications to the theory of rings with minimum condition. Sci. Repts Tokyo Kyoiku Daigaku, A6:83-142 (May 15 1958).

526. M. Moriya, On the automorphisms of a certain class of finite rings. Kodai Math. Semin. Repts., 18(4):357-367 (1966).

527. A. Morris, On a generalized Clifford algebra. Quart. J. Math., 18(69):7-12 (1967).

528. I. Murase, On extended cyclic algebras. Scient. Papers Coll. Gen. Educ. Univ. Tokyo, 17(1):1-16 (1967).

529. I. Murase, On the automorphisms of a quasi-matrix algebra. Scient. Papers Coll. Gen. Educ. Univ. Tokyo, 14(2):139-155 (1964).

530. I. Murase, On the derivations of a quasi-matrix algebra. Scient. Papers Coll. Gen. Educ. Univ. Tokyo, 14(2):157-164 (1964).

531. I. Murase, Generalized uniserial group rings. I. Scient. Papers Coll. Gen. Educ. Univ. Tokyo, 15(1):15-28 (1965).

532. I. Murase, Generalized uniserial group rings. II. Scient. Papers Coll. Gen. Educ. Univ. Tokyo, 15(2):111-128 (1965).

533. I. Murase, On the structure of generalized uniserial rings. III. Scient Papers Coll. Gen. Educ. Univ. Tokyo, 14(1):11-25 (1964).

534. T. Nagahara, On Galois conditions and Galois groups of simple rings. Trans. Amer. Math. Soc., 116(4):417-434 (1965).

535. T. Nagahara, Quasi-Galois extensions of simple rings. Math. J. Okayama Univ., 12(2):107-146 (1966).

536. T. Nagahara, Supplements to the previous paper "Quasi-Galois extensions of simple rings". Math. J. Okayama Univ., 12(2):159-196 (1966).

537. T. Nakano, A generalized valuation and its value groups. Comment. math. Univ. St. Pauli, 12(1):1-22 (1964).

538. T. Nakano, Rings and partly ordered systems. Math. Z., 99(5):335-376 (1967).

539. N. S. Natarajan, Rings with generalized distributive laws. J. Indian Math. Soc., 28(1):1-6 (1964).

540. M. Neumann and P. Dragomir, Asupra multiplilor intregi si rationali ai elemen-

tului unitate intr-un inel respectiv intr-un corp. An. Univ. Timişoara. Ser. ştiinţe mat.-fiz., 3:205-209 (1965).

541. A. Nijenhuis and R. W. Richardson, Jr., Cohomology and deformations in graded Lie algebras. Bull. Amer. Math. Soc., 72(1, Part I):1-29 (1966).

542. A. Nijenhuis and R. W. Richardson, Jr., Deformations of homomorphisms of Lie groups and Lie algebras. Bull. Amer. Math. Soc., 73(1):175-179 (1967).

543. A. Nijenhuis and R. W. Richardson, Jr., Deformations of Lie algebra structures. J. Math. and Mech., 17(1):89-105 (1967).

544. G. M. L. Noronha and C. A. Almeida, Sur le demi-anneau des nombres naturels. An. Fac. cienc. Univ. Porto, 48(1-2):35-39 (1965).

545. T. Ochiai, Classification of the finite nonlinear primitive Lie algebras. Trans. Amer. Math. Soc., 124(2):313-322 (1966).

546. R. H. Oehmke, On a class of Lie algebras. Proc. Amer. Math. Soc., 16(5):1107-1113 (1965).

547. R. H. Oehmke and E. Sandler, The collineation groups of division ring planes. II. Jordan division rings. Pacif. J. Math., 15(1):259-265 (1965).

548. A. Ofsti, On the structure of a certain class of locally compact rings. Math. scand., 18(2):134-142 (1966).

549. M. Okuzumi, On Galois conditions in division algebras. Kodai Math. Semin. Repts., 18(1):16-23 (1966).

550. M. Okuzumi, The Cartan-Brauer-Hua theorem for algebras. Kodai Math. Semin. Repts., 18(2):193-194 (1966).

551. T. Onodera, A characterization of strongly separable algebras. J. Fac. Sci. Hokkaido Univ., 19(2):71-73 (1966).

552. J. M. Osborn, Commutative algebras satisfying an identity of degree four. Proc. Amer. Math. Soc., 16(5):1114-1120 (1965).

553. J. M. Osborn, On commutative nonassociative algebras. J. Algebra, 2(1):48-79 (1965).

554. J. M. Osborn, Jordan algebras of capacity two. Proc. Nat. Acad. Sci. U.S.A., 57(2):582-588 (1967).

555. B. L. Osofsky, A non-trivial ring with non-rational hull. Canad. Math. Bull., 10(2):275-282 (1967).

556. B. L. Osofsky, A generalization of quasi-Frobenius rings. J. Algebra, 4(3):373-387 (1966).

557. D. L. Outcalt, Power-associative algebras in which every subalgebra is an ideal. Pacif. J. Math., 20(3):481-485 (1967).

558. D. L. Outcalt, Simple n-associative rings. Pacif J. Math., 17(2):301-309 (1966).

559. D. L. Outcalt and A. Yaqub, A generalization of Wedderburn's theorem. Proc. Amer. Math. Soc., 18(1):175-177 (1967).

560. R. Ouzieou, Sur un propriété des anneaux artiniens. C. r. Acad. sci., 260(20): 5155-5156 (1965).

561. S. Page and R. W. Richardson, Jr., Stable subalgebras of Lie algebras and associative algebras. Trans. Amer. Math. Soc., 127(2):302-312 (1967).

562. D. J. Palmer, On the structure of nilpotent algebras. Doct. diss. Univ. Md., 26(11):6747-6748 (1966).

563. G. Panella, Osservazione riguardante gli anelli che verificano una condizione di

massimo. Atti Accad. naz. Lincei Rend. Cl. sci. fis. mat. e natur., 42(6):755-756 (1967).

564. G. Panella, Un teorema di Golod-Šafarevič e alcune sue consequenze. Conf. semin. mat. Unov. Bari, No. 104 (1966), 17 pp.

565. J. Panvini, Su certi omomorfismi di tipo esponenziale. Rend. Semin. mat. Univ. Padova, 35(2):371-379 (1965).

566. B. Pareigis, Einige Bemerkungen über Frobenius-Ezweiterungen. Math. Ann., 153(1):1-13 (1964).

567. G. Pascual, "Estructura de D-anillo," Actas 5 Reun. anual. mat. esp. Valencia, 1964, Madrid (1967), pp. 225-227.

568. E. M. Patterson, The Wedderburn-Artin theorem for pseudo rings. Math. Z., 98(1):31-41 (1967).

569. E. M. Patterson, The Jacobson radical of a pseudo-ring. Math. Z., 89(4):348-364 (1965).

570. E. M. Patterson, Commutation problems involving rings of infinite matrices. Proc. Edinburgh Math. Soc., 14(1):55-60 (1964).

571. K. R. Pearson, Interval semirings on R_1 with ordinary multiplication. J. Austral. Math. Soc., 6(3):273-288 (1966).

572. R. E. Peinado, On finite rings. Math. Mag., 40(2):83-85 (1967).

573. R. E. Peinado, Una clasificación de anillos. Rev. Mat. hisp.-amer., 25(6):249-261 (1965).

574. R. E. Peinado, Ancestral rings. Proc. Edinburgh Math. Soc., 15(2):107-109 (1966).

575. A. Peister, Quadratische Formen in beliebigen Körpen. Invent. math., 1(2):116-132 (1966).

576. H. P. Petersson, Zur Theorie der Lie-Tripel-Algebren. Math. Z., 97(1):1-15 (1967).

577. H. P. Petersson, Über den Wedderburnschen Struktursatz für Lie-Tripel-Algebren. Math. Z., 98(2):104-118 (1967).

578. J.-C. Petit, Quasi-corps généralisant un type d'anneau quotient. C. r. Acad. sci., 265(22):A708-A711 (1967).

579. F. Pokropp, Dicksonsche Fastkörper. Abhandl Math. Semin. Univ. Hamburg, 30(3):188-219 (1967).

580. G. Pollak, Bemerkungen zur Holomorphentheorie der Ringe. Acta. scient. math., 25(3-4):181-185 (1964).

581. I. Popovici and C. Ghéorghe, Algèbres de Clifford généralisées. C. r. Acad. sci., AB262(12):A682-A685 (1966).

582. F. Poyatos, Decomposiciones irreducibles en suma directa interna de ciertas algebraicas. Rev. mat. hisp.-amer., 27(4):151-170 (1967).

583. C. Procesi, Non-commutative affine rings. Atti Accad. naz. Lincei, Mem. Cl. sci. fis., mat. e natur., Sez. I, 8(6):239-255 (1967).

584. C. Procesi, Non-commutative Jacobson-rings. An. Scuola norm. super. Pisa Sci. fis. e. mat., 21(2):281-290 (1967).

585. C. Procesi, Su un teorema di Faith-Utumi. Rend. mat. e applic., 24(3-4):346, 347 (1965).

586. C. Procesi, A non commutative Hilbert Nullstellensatz. Rend. math. e applic., 25(1-2):71-121 (1966).

587. J. Querre, Orders maximaux. Sémin. Dubreil et Pisot. Fac. Sci. Paris, 18(1):
     4/01-4/21, 1964-1965 (1967).

588. C. Raggi and F. Francisco, Localizacion en anillos no commutativos. An. Inst.
     mat. Univ. nac. Autónoma México, 6:1-6 (1966).

589. N. Ramabhadran, Coherent pairs of extensions of Lie algebras. (Thesis abstract.)
     Math. Student, 32(1-2):111 (1964).

590. R. von Randow, The involutory antiautomorphisms of the quaternion algebra.
     Amer. Math. Monthly, 74(6):699-700 (1967).

591. J. D. Reid, On subcommutative rings. Acta math. Acad. scient. hung., 16(1-2):
     23-26 (1965).

592. I. Reiner, Completion of primitive matrices. Amer. Math. Monthly, 73(4):380
     (1966).

593. C. Reischer and V. Tamas, "Asupra şirurilor recurente cu elemente dintr'o
     algebră liniară," An. ştiinţ. Univ. Iaşi, Sec. Ia, 11(2):249-258 (1965).

594. G. Renault, Étude des anneaux réduits et de leurs enveloppes injectives. C. r.
     Acad. sci., 264(2):A53-A55 (1967).

595. G. Renault, Anneaux self-injectifs. Sémin. Dubreil et Pisot. Fac. Sci. Paris,
     19(1):11/01-11/04 (1967).

596. G. Renault, Anneaux réduits non commutatifs. J. math. pures et appl., 46(2):
     203-214 (1967).

597. R. Rentschler, Eine Bemerkung, zu Ringen mit Minimalbedingung für Hauptideale.
     Arch. Math., 17(4):298-301 (1966).

598. R. Rentschler and P. Gabriel, Sur la dimension des anneaux et ensembles
     ordonnés. C. r. Acad. sci., 265(22):A712-A715 (1967).

599. R. W. Richardson, Jr., On the rigidity of semi-direct products of Lie algebras.
     Pacif. J. Math., 22(2):339-344 (1967).

600. R. W. Richardson, Jr., A rigidity theorem for subalgebras of Lie and associative
     algebras. Ill. J. Math., 11(1):92-110 (1967).

601. F. Richman, Generalized quotient rings. Proc. Amer. Math. Soc., 16(4):794-799
     (1965).

602. M. A. Rieffel, Burnside's theorem for representations of Hopf algebras. J. Algebra,
     6(1):123-130 (1967).

603. Ringe und Moduln, Tagungsbericht, 27 Feb.-5 märz, Math. Forschungsinst.
     Oberwolfach (1966), 14 pp.

604. D. W. Robinson, A note on additive commutators in a ring. Amer. Math.
     Monthly, 72(10):1106-1107 (1965).

605. J. C. Robson, Pri-rings and ipii-rings. Quart. J. Math., 18(70):125-145 (1967).

606. J. C. Robson, Do simple rings have unity elements? J. Algebra, 7(1):140-143
     (1967).

607. D. J. Rodabaugh, Antiflexible algebras which are not power-associative. Proc.
     Amer. Math. Soc., 17(1):237-239 (1966).

608. D. J. Rodabaugh, Some new results on simple algebras. Pacif. J. Math., 17(2):
     311-317 (1966).

609. D. J. Rodabaugh, A generalization of the flexible law. Trans. Amer. Math. Soc.,
     114(2):486-487 (1965).

610. B. R. Rodrigquez-Salinas and J. Garay de Pable, "Sobre los cuerpos no commutativas

de rango cuatro respecto de su centro," Actas 5 Reun. anual mat. esp. Valencia, 1964, Madrid (1967), pp. 137-152.

611. W. Romberg, Das Henselsche Lemma in potenzassoziativen Algebren. Math. Nachr., 29(5-6):375-380 (1965).

612. W. Romberg, Pseudobewertungen und Arithmetik in nichtassoziativen Algebren. I. Math. Nachr., 26(5):287-306 (1964).

613. G. Romier, Automorphismes et algèbre d'un schéma d'association. Caractérisation algébrique des correspondances partiellement équilibrees. C. r. Acad. sci., 258(22):5345-5348 (1968).

614. B. van Rootselaar, Zum ALE-Fasthalbringbegriff. Nieuw arch. wiskunde, 15(3): 247-249 (1967).

615. A. Rosenfeld, A note on two special types of rings. Scripta Math., 28(1):51-54 (1967).

616. L. E. Ross, Representations of graded Lie algebras. Trans. Amer. Math. Soc., 120(1):17-23 (1965).

617. L. E. Ross, "On representations and cohomology of graded Lie algebras," Doctoral dissertation, Univ. California, Berkeley (1964) 48 pp.; Dissert. Abstr., 25(10):5969 (1965).

618. J. Rouband, Sur un théorème de M. P. Schutzenberger. C. r. Acad. sci., 261(17): 3265-3267 (1965).

619. D. E. Rutherford, The Cayley-Hamilton theorem for semi-rings. Proc. Roy. Soc. Edinburgh, A66(4):211-215 (1965).

620. E. A. Rutter, Jr., A remark concerning quasi-Frobenius rings. Proc. Amer. Math. Soc., 16(6):1372-1373 (1965).

621. E. A. Rutter, Jr., "Characterizations of quasi-Frobenius rings," Doctoral dissertation, Iowa State Univ. Sci. and Technol. (1965), 60 pp.; Dissert. Abstr. 26(6): 3379-3380 (1965).

622. A. A. Sagle, On anti-commutative algebras and homogeneous spaces. J. Math. and Mech., 16(12):1381-1393 (1967).

623. A. A. Sagle, Simple algebras that generalize the Jordan algebra M_3^8. Canad. J. Math., 18(2):282-290 (1966).

624. A. A. Sagle, On simple extended Lie algebras over fields of characteristic zero. Pacif. J. Math., 15(2):621-648 (1965).

625. A. A. Sagle, Remarks on simple extended Lie algebras. Pacif. J. Math., 15(2): 613-620 (1965).

626. A. A. Sagle, On anti-commutative algebras and general Lie triple systems. Pacif. J. Math., 15(1):281-291 (1965).

627. A. A. Sagle, On simple algebras obtained from homogeneous general Lie triple systems. Pacif. J. Math., 15(4):1397-1400 (1965).

628. P. Salmon, Suele algebre simmetriche e di Rees di un ideale. Genova, Ed., scientif., 1964, 27 pp.; Bibliogr. naz. ital., 7(7):428 (1964).

629. San Román, J. Sancho, Contributión al estudio de la valoraciones no triviales subordinadas de una valoración dada. Publs. semin. math. Fac. cienc. Zaragoza, No. 3, pp. 133-135 (1962).

630. F. L. Sandomierski, Nonsingular rings. Proc. Amer. Math. Soc. 19(1):225-230 (1968).

631. F. I.. Sandomierski, Semisimple maximal quotient rings. Trans. Amer. Math. Soc. 128(1):112-120 (1967).

632. A. D. Sands, Primitive rings of infinite matrices. Proc. Edinburg. Math. Soc., 14(1):47-53 (1964).

633. E. Sasiada and P. M. Cohn, An example of a simple radical ring. J. Algebra, 5(3):373-377 (1967).

634. M. Satyanarayana, "Non-commutative local rings," Doctoral dissertation, Univ. Wisconsin, Milwaukee (1966), 67 pp.; Dissert. Abstr., B28(3):1019 (1967).

635. M. Satyanarayana, Semisimple rings. Amer. Math. Monthly, 74(9):1086 (1967).

636. R. D. Schafer, On the simplicity of the Lie algebras E_7 and E_8. Proc. Koninkl. nederl. akad. wet., A69(1):64-69 (1966).

637. R. D. Schafer, An Introduction to Non-associative Algebras, Academic Press, New York (1966), Vol. 10, 166 pp.

638. J. H. Scheuneman, "Two-step nilpotent Lie algebras," Doctoral dissertation, Purdue Univ. (1966) 60 pp.; Dissert. Abstr., B27:2449 (1967).

639. H. Schneider and J. Weissglass, Group rings, semigroup rings and their radicals. J. Algebra, 5(1):1-15 (1967).

640. J. R. Schue, Symmetry for the enveloping algebra of a restricted Lie algebra. Proc. Amer. Math. Soc., 16(5):1123-1124 (1965).

641. J. Selden, A note on compact semirings. Proc. Amer. Math. Soc., 15(16):882-886 (1964); Left zero simplicity in semirings. Proc. Amer. Math. Soc., 17(3): 649-698 (1966).

642. G. B. Seligman, Some results on Lie p-algebras. Bull. Amer. Math. Soc., 73(4): 528-530 (1967).

643. L. Silver, Noncommutative localizations and applications. J. Algebra, 7(1):44-76 (1967).

644. L. A. Skornjakov, Alternative rings. Rend. mat. e applic., 24(3-4):360-372 (1965).

645. M. Slater, Ideals in semiprime alternative rings. J. Algebra, 8(1):60-76 (1967).

646. M. Slater, Nucleus and center in alternative rings. J. Algebra, 7(3):372-388 (1967).

647. L. W. Small, On some questions in Noetherian rings. Bull. Amer. Math. Soc., 72(5):853-857 (1966).

648. L. W. Small, Orders in Artinian rings. J. Algebra, 4(1):13-41 (1966).

649. L. W. Small, Correction and addendum: "Orders in Artinian rings". J. Algebra, 4(3):505-507 (1966).

650. L. W. Small, Semihereditary rings. Bull. Amer. Math. Soc., 73(5):656-658 (1967).

651. L. W. Small, An example in Noetherian rings. Proc. Nat. Acad. Sci., U.S.A., 54(4):1035-1036 (1965).

652. D. A. Smith, Chevalley bases for Lie modules. Trans. Amer. Math. Soc., 115(3):283-289 (1965).

653. F. A. Smith, A structure for a class of lattice ordered semirings. Fundam. Math., 59(1):49-64 (1966).

654. F. A. Smith and A. Engelbert, "A structure theory for a class of lattice ordered semirings," Doctoral dissertation, Purdue Univ., (1965) 46 pp.; Dissert. Abstr., 26(3):1675 (1965).

655. D. Soda, Some groups of type D_4 defined by Jordan algebras. J. reine und angew. Math., 223:150-163 (1966).

656. L. Solomon and V. Daya-Nand, Sur le corps des quotients de l'algèbre envelloppante d'une algèbre de Lie. C. r. Acad. sci., 264(23):A986 (1967).

657. G. Sorani, Radical subrings of matrix rings. Amer. Math. Monthly, 73(9):989-991 (1966).

658. J. Stallings, Whitehead torsion of free products. Ann. Math., 82(2):354-363 (1965).

659. L. Stammler, Elementare Beweise zweier Sätze über Matrizenringe. Publs. math., 12(1-4):33-38 (1965).

660. A. Steger, Elementary factorization in π-regular rings. Canad. J. Math., 18(2): 307-313 (1966).

661. J. von den Steinen, Über Ausdehnungen von Derivationen. Math. Z., 92(3):234-255 (1966).

662. O. Steinfeld and R. Wiegandt, Über die Verallgemeinerungen und Analoga der Wedderburn-Artinschen und Noetherschen Struktursätze. Math. Nachr., 34(3-4): 143-156 (1965).

663. N. J. Sterling, Prime associator-dependent rings with idempotent. Trans. Amer. Math. Soc., 128(3):474-481 (1967).

664. N. Straumann, Branching rules and Clebsch-Gordan series of semi-simple Lie algebras. Helv. phys. acta, 38(5):481-498 (1965).

665. R. W. Stringall, Endomorphism rings of primary abelian groups. Pacif. J. Math., 20(3):535-557 (1967).

666. H. Subramanian, Ideal neighbourhoods in a ring. Pacif. J. Math., 24(1):173-176 (1968).

667. H. Subramanian, l-prime ideals in f-rings. Bull. Soc. math. France, 95(2): 193-203 (1967).

668. A. Sulinski, R. Anderson, and N. Divinsky, Lower radical properties for associative and alternative rings. J. Lond. Math. Soc., 41(3):417-424 (1966).

669. F. Szász, Einige Kriterien für die Existenz des Einselements in einem ring. Acta scient. math., 28(1-2):31-37 (1967).

670. F. Szász, Hinreichende Bedingung für die Existenz eines Rechtseinselementes in einem Ring. Publs. math., 14:151-152 (1967).

671. F. Szász, The sharpening of a result concerning primitive ideals of an associative ring. Proc. Amer. Math. Soc., 18(5):910-912 (1967).

672. F. Szász, Gyűrűk maximális jobbideáljairól. Magyár tud.akad. Mat. és fis. tud. oszt. közl., 17(4):473-476 (1967).

673. J. Szendrei, Megjegyzések a gyűrű centrumáról, Szegedi tanárképző főisk. tud. közl., 1966, 2 rész., Szegel (1966), pp. 173-175.

674. H. Tachikawa, On dominant dimensions of QF-3 algebras. Trans. Amer. Math. Soc., 112(2):249-266 (1964).

675. E. J. Taft, Invariant splitting in Jordan and alternative algebras. Pacif. J. Math., 15(4):1421-1427 (1965).

676. E. J. Taft, On certain d-groups of algebra automorphisms and antiautomorphisms. J. Algebra, 3(1):115-121 (1966).

677. E. J. Taft, Cohomology of algebraic groups and invariant splitting of algebras. Bull. Amer. Math. Soc., 73(1):106-108 (1967).

678. S. Takahashi, A characterization of group rings as a special class of Hopf algebras. Canad. Math. Bull., 8(4):465-475 (1965).

679. Y. Takeuchi, On Galois extensions over commutative rings. Osaka J. Math., 2(1):137-145 (1965).

680. Y. Takeuchi, Infinite outer Galois theory of non-commutative rings. Osaka J. Math., 3(2):195-200 (1966).

681. T. D. Talintyre, Quotient rings with minimum condition on right ideals. J. London Math. Soc., 41(1):141-144 (1966).

682. V. Tharmaratnam, Complete primitive distributively generated near-rings. Quart. J. Math., 18(72):293-313 (1967).

683. A. Thedy, Ringe mit x(yz) = (yx)z. Math. Z., 99(5):400-404 (1967).

684. G. Tjierrin, Anneaux metaprimitifs. Canad. J. Math., 17(1):199-205 (1965).

685. J. Tits, Algebres alternatives, algebres de Jordan et Algebres de Lie exceptionelles. I. Construction. Proc. Koninkl. nederl. acad. wet., A69(2):223-237 (1966).

686. S. Tôgô, Derivations of Lie algebras. Bull. Amer. Math. Soc., 72(4):690-692 (1966).

687. S. Tôgô, On some properties of t (n, Φ) and ƒt (n, Φ). J. Sci. Hiroshima Univ., Ser. A, Div. 1, 31(1):35-38 (1967).

688. S. Tôgô, Outer derivations of Lie algebras. Trans. Amer. Math. Soc., 1967, 128(2):264-276 (1968).

689. S. Tôgô, Dimensions of the derivation algebras of Lie algebras. J. Sci. Hiroshima Univ., Ser. A, Div. 1, 31(1):17-23 (1967).

690. S. Tôgô, Lie algebras which have few derivations. J. Sci. Hiroshima Univ., Ser. A, Div. 1, 29(1):29-41 (1965).

691. H. Tominaga, On Nagahara's theorem. J. Fac. Sci. Hokkaido Univ., Ser. 1, 18(3-4):153-157 (1965).

692. H. Tominaga, On q-Galois extensions of simple rings. Nagoya Math. J., 27(2):485-507 (1966).

693. H. Tominaga, On Galois extensions over a Field and abelian Galois groups. Math. J. Okayama Univ., 12(2):153-158 (1966).

694. B. R. Toskey, A system of canonical forms for rings on a direct sum of two infinite cyclic groups. Pacif. J. Math., 20(1):179-188 (1967).

695. C. E. Tsai, Flexible partially stable algebras. Trans. Amer. Math. Soc., 122(1):48-63 (1966).

696. C. E. Tsai, Von Neumann regularity in Jordan algebras. Proc. Amer. Math. Soc., 18(4):726-728 (1967).

697. A. Tversky, A general theory of polynomial conjoint measurement. J. Math. Psychol., 4(1):1-20 (1967).

698. Y. Utumi, On the continuity and self-injectivity of a complete regular ring. Canad. J. Math., 18(2):404-412 (1966).

699. Y. Utumi, On continuous rings and self-injective rings. Trans. Amer. Math. Soc., 118(6):158-173 (1965).

700. Y. Utumi, Self-injective rings. J. Algebra, 6(1):56-64 (1967).

701. D. Vaida, Idéaux isoliés gauches nilpotents et radical isolé. Bull. math. Soc.
      sci. math. et phys. RPR, 1962, 6(3-4):257-260 (1964).
702. W. V. Vasconcelos, On local and stable cancellation. An. Acad. brasil. ciênc.,
      37(3-4):389-393 (1965).
703. M. Vergne, Réductibilité de la variété des algèbres de Lie nilpotentes. C. r.
      Acad. sci., 263(1):A4-A6 (1966).
704. D.-N. Verma, "Structure of certain induced representations of complex semi-
      simple Lie algebras," Doctoral dissertation, Yale Univ. (1966), 107 pp.; Dissert.
      Abstr., B27(8):2799 (1967).
705. O. E. Villamayor, La theorie de Galois pour les anneaux associatifs. Publs. Fac.
      cienc. fisicomat. Univ. nac. La Plata, No. 213, pp. 173-184 (1956).
706. O. E. Villamayor, Sur une representation matricielle de l'anneau d'endomor-
      phismes d'un module quelconque. Publs. Fac. cienc. fisico mat. Univ. nac.
      La Plata, No. 213, pp. 185-190 (1956).
707. J. Vrabec, Delno urejene algebraične structure. Obz. mat. in fiz., 12(1):15-31
      (1965).
708. G. Vranceanu, Asupra unei forme canonice a unei algebre comutative cudonă
      elemente. Gaz. mat. (RPR), A70(6):201-203 (1965).
709. B. L. van der Waerden, On Clifford Algebras. Poc. Koninkl. nederl. akad. wet.,
      A69(2):78-83 (1962); Indagationes math., 28(2):78-83 (1966).
710. W. von Waldenfels, Zur charakterizierung Liescher Elemente in freien Algebren.
      Arch. Math., 17(1):44-48 (1966).
711. C. T. C. Wall, Graded Brauer groups. J. reine und angew. Math., 213(3-4):187-
      199 (1964).
712. D. W. Wall, Basic algebras of algebras with unique minimal faithful representa-
      tions. Duke Math. J., 32(1):53-63 (1965).
713. D. W. Wall, Characterisations of generalized uniserial algebras. III. Proc. Edin-
      burgh Math. Soc., 15(1):37-42 (1966).
714. D. A. R. Wallace, On the commutativity of the radical of a group algebra. Proc.
      Glasgow Math. Assoc., 7(1):1-8 (1965).
715. E. W. Wallace, A note on Levi's theorem. Amer. Math. Monthly, 74(10):1238
      (1967).
716. N. R. Wallach, "A classification of real simple Lie algebras," Doctoral disserta-
      tion, Washington Univ. (1966), 41 pp.; Dissert. Abstr., B27(6):2049 (1966).
717. L. J. Wallen, On the magnitude of x^n-1 in a normed algebra. Proc. Amer.
      Math. Soc., 18(5):956 (1967).
718. A. P. J. van der Walt, Rings with dense quasi-centre. Math. Z., 97(1):38-44 (1967).
719. A. P. J. van der Walt, On the Levitzki nil radical. Arch. Math., 16(1):22-24 (1965).
720. A. P. J. van der Walt, Prime ideals and nil radicals in near-rings. Arch. Math.,
      15(6):408-414 (1964).
721. S. Warner, Compactly generated algebras over discrete fields. Bull. Amer. Math.
      Soc., 73(2):227-230 (1967).
722. S. Warner, Locally compact simple rings having minimal left ideals. Trans.
      Amer. Math. Soc., 125(3):395-403 (1966).
723. Y. Watanabe, Simple algebras over a complete local ring. Osaka J. Math., 3(1):
      13-20 (1966).

724. E. C. Weinberg, On the scarcity of lattice-ordered matrix ring. Pacif. J. Math., 19(3):561-571 (1966).

725. H. J. Weinert, On the theory of pre-p-rings. Amer. Math. Monthly, 74(4):378-384 (1967).

726. H. J. Weinert, Ein Struktursatz für idempotente Halbkörper. Acta math. Acad. scient. hung., 15(3-4):289-295 (1964).

727. H. J. Weinert, Über Halbringe und Halbkörper. III. Acta math. Acad. scient hung., 15(1-2):177-194 (1964).

728. H. J. Weinert, Zur Theorie der Algebren und monomialen Ringe. Acta scient. math., 26(1-2):171-186 (1965).

729. H. J. Weinert, Eine Bemerkung zur Difinition halbgeordneter Mengen. Wiss. Z. Pädagog. Hochschule Potsdam. Math.-naturwiss. Reihe, 8(1-2):87-88 (1964).

730. J. Weissglass, "Group rings, semigroup rings and their radicals," Doctoral dissertation, Univ. Wisconsin (1967), 62 pp.; Dissert. Abstr., B28(3):1024 (1967).

731. R. H. Wenger, "Semigroups having quasi-Frobenius algebras," Doctoral dissertation, Michigan State Univ. (1965), 70 pp.; Dissert. Abstr., 26(8):4703 (1966).

732. R. Wiegandt, Über transfinit nilpotente Ringe. Acta math. Acad. scient. hung., 17(1-2):101-114 (1966).

733. R. Wiegandt, Über halbeinfache linear Kompakte Ringe. Studia scient math. hung., (ex Akad. Mat. kutato int. közl.), 1(1-2):31-38 (1966).

734. R. Wiegandt, Vizsgálatok a lineárisan kompakt gyűrűk elméletében. II. Magyar tud. akad. Math. és. fiz. tud. oszt. közl., 16(3):333-363 (1966).

735. R. Wiegandt, Über lokal linear kompakte Ringe. Acta scient. math., 28(3-4): 255-260 (1967).

736. R. Wiegandt, Über linear kompakte reguläre Ringe. Bull. Acad. polon. sci. Sér. sci. math. astron. et phys., 13(7):445-446 (1965).

737. P. Wilker, Doppeloops und Ternärkörper. Math. Ann., 159(3):172-196 (1965).

738. R. E. Williams, A note on near-rings over vector spaces. Amer. Math. Monthly, 74(2):173-175 (1967).

739. R. E. Williams, Simple near-rings and their associate rings. Doct. diss., Univ. Missouri (1965); 83 pp.; Dissert. Abstrs., 26(9):5470 (1966).

740. S. Williamson, Crossed products and ramification. Nagoya Math. J., 28:85-111 (1966).

741. Ling-Erl Eileen Ting Wu, Bicontinuous isomorphisms between two closed left ideals of a compact dual ring. Canad. J. Math., 18(6):1148-1151 (1966).

742. Ling-Erl Eileen Ting Wu, H. Y. Mochizuki, and J. P. Jans, A characterization of QF-3 rings. Nagoya Math. J., 27(1):7-13 (1966).

743. Yong-hua Xu, L^2-radical and L^2-direct sum. Chinese Math., 8(4):538-549 (1966).

744. S. Yamamuro, Ideals and homomorphisms in some near-algebras. Proc. Japan. Acad., 42(5):427-432 (1966).

745. K. Yamazaki, On a duality between gratations and automorphisms of algebras. Scient. Papers Coll. Gen. Educ. Univ. Tokyo, 14(1):37-50 (1964).

746. A. Zaks, Residue rings of semi-primary hereditary rings. Nagoya Math. J., 30:279-283 (1967).

747. A. Zaks, A note on semi-primary hereditary rings. Pacif. J. Math., 23(3):627-628 (1967).

748. H. Zassenhaus, Orders as endomorphisms rings of modules of the same rank. J. London Math. Soc., 42(1):180-182 (1967).

749. H. Zassenhaus and W. Eichhorn, Herleitung von Acht- und Sechzehn-Quadrate-Identität mit Hilfe von Eigenschaften der verallgemeinerten Quaternionen und Cayley-Dicksonschen Zahlen. Arch. Math., 17(6):492-496 (1966).

750. J. L. Zemmer, Near-fields, planar and non-planar, Math. Student., 32(3-4): 145-150 (1964).

Modules

A. V. Mikhalev and L. A. Skornyakov

In the present article we consider the papers on the theory of modules reviewed in the Mathematics section of Referativnyi Zhurnal during 1966-1968. Unless we specify otherwise all the modules to be considered in the survey are taken to be left and and unitary. Therefore, as a rule we shall set forth the left-sided versions of the results regardless of which version was chosen by the author or by the reviewer. When we speak of work which has already been mentioned in previous issues of "Itogi Nauki" (Progress in Mathematics), we shall make reference to these issues.

The injective hull of a module A is systematically denoted by $\hat{A}$, the base ring by Λ, and its Jacobson radical by $J(\Lambda)$.

Now some words on the books. During the period being considered there appeared the small book by Lambek [232] in which ring and module methods were successfully combined. The axiomatic definitions of the concepts of purity, torsion, and completeness (divisibility) are given and investigated in the monograph by Mishina and Skornyakov [27]. The fundamentals of homological algebra, including the structure of the extension group, are presented in the lectures by Skornyakov [44]. The monographs by Mitchell [267] and by Freyd [149] on the theory of categories are of interest from the point of view of modules. Finally, we note the appearance of the Russian translations of the books by Lang [25], MacLane [26], and Curtis and Reiner [23].

§ 1. Radicals

It is very well known that a radical r can be given by indicating the corresponding radical or semisimple class. To specify torsion ([42], p. 180) we can indicate the pair of classes $(\mathfrak{T}, \mathfrak{P})$ with the following properties: 1) $\mathfrak{T} \cap \mathfrak{P} = 0$; 2) $\mathfrak{T}$ is closed relative to the factor module and $\mathfrak{P}$ is closed relative to the submodule; 3) for every module A there exists an exact sequence $0 \to T \to A \to P \to 0$ such that $T \in \mathfrak{T}$ and $P \in \mathfrak{P}$. This approach was developed by Dickson [122] and was applied by Jans [198] and Walker [406]. Jans [198] noted that the r-radical class defined by the radical filter $\mathfrak{E}$ (i.e., by the set with the properties 1)-4) mentioned on p. 180 of [42]) turns out to be semisimple (for some other radical r') if and only if the system $\mathfrak{E}$ contains the smallest left ideal I_0. This left ideal I_0 turns out to be two-sided. The relations $r(A) = 0$ and $r'(A) = A$ are equivalent if and only if r is a splittable radical, i.e, $r(A)$ serves as a retract of module A. Also considered was the class $\mathfrak{T}_0 = \{X, \text{Hom}_\Lambda(X, \hat{\Lambda}) = 0\}$, which turns out to be radical and semisimple in the case of a perfect ring Λ ([42], p. 194). Gorbachuk [10] proposed several criteria for the nonsplittability of torsion.

A submodule A of a module B is called small if $A + S = B$ implies $S = B$. Leonard [241] proved that a periodic module A over a commutative integral domain is small in $\hat{A}$ if and only if its primary components are bounded. The tendency to clarify the role of small modules in the theory of radicals is a natural one. Miyashita [270] proved that the sum $r(A)$ of small submodules of module A coincides with the intersection of its maximal submodules. Under certain additional conditions the submodule $r(A)$ itself turns out to be small. Pareigis [307] deals with the same series of questions. Beidleman [75] has considered a radical, defined as the intersection of maximal submodules, in the category of modules over nearrings.

Dickson [125] has proposed a generalization of the primary decomposition of a periodic Abelian group, which has attracted attention. This idea is applied to the case of a radical with a radical class consisting of all modules such that all their nonzero factor modules contain a simple submodule [121, 124]. Koifman [21] has characterized the rings the projective modules over which form a semisimple class. The Goldie radical Z_2 ([42], p. 182) has been studied in detail by Alin and Dickson [54, 55]. They established that it is a torsion. The functor $\text{Gold}(A) = A/Z_2(A)/(A/Z_2(A))$ and its derivatives $\{\text{Gold}^k(A)\}$ have proved to be useful in studying it. Here, $\text{Gold}^k(A) = 0$ if $k \geq 2$. It was proven that the exact sequence

$0 \to A \to B \to C \to 0$ generates the exact sequence

$$0 \to Z_2(A) \to Z_2(B) \to Z_2(C) \to$$
$$\to \text{Gold}(A) \to \text{Gold}(B) \to \text{Gold}(C) \to 0,$$

and that the splittability of the Goldie torsion is equivalent to the identical vanishing of the functor Gold. We remark that the derivative functors of the radicals of modules were analyzed by Dickson [125] and by Govorov [9]. Harada [173] established that the Goldie radical of a quasi-injective module turns out to be a retract in it. Sandomierski [354] proved that Goldie-semisimple finitely-generated modules over a selfinjective ring are projective and injective. It can be shown that the Goldie torsion is defined by the radical filter $\mathfrak{C} = \{I, \forall_\rho \in \Lambda \setminus I, \exists_\sigma \in \Lambda \setminus (I : \rho); (I : \sigma\rho)$ is dense in $\Lambda\}$ which was considered by Dlab in [128]. Various applications of radical filters have been given by Helzer [188], Sanderson [352], Stănașilă [377], and Chew [102]. In particular, Helzer proved that in a hereditary Noetherian ring the radical filter forms left ideals which are dense in Λ. Stenström [379] has proposed a category-theoretic generazation of the description of a torsion with the aid of a radical filter.

The Λ-module A is called a torsion-free module in the sense of Bass if it is imbedded in the direct product of some set of copies of the ring Λ. A number of properties of such modules were established by Bass [69] and Chase [101]. A cyclic module Λ/I is torsion-free in the sense of Bass if and only if $l(r(I)) = I$ [211]. Colby and Rutter [116] pointed out conditions necessary and sufficient that a class of modules, torsion-free in the sense of Bass, over a semi-primary ring be semisimple for some torsion. Therein also it was clarified as to when this torsion coincides with the Goldie torsion.

A submodule A of a module B is called initially-closed if $Z(B/A) = B/A$ and if from $A \subseteq C \subseteq B$ and $\lambda C \subseteq A$ it follows that $\lambda B \subseteq A$. The functor which picks out the intersection of all the initially-closed submodules of every module turns out to be the radical studied by Feller and Swokowski [141]. Matlis [261], by considering modules over a commutative Noetherian ring Λ, defined r(A) as the collection of all the elements of the Λ-module A such that the prime ideals containing their annihilators is maximal. Dickson [123] showed that the splittability of this radical is equivalent to the ring Λ being Artinian, and in the case $r(A) = A$ for every Λ-module A.

An original approach to the concept of a torsion is outlined in Pleasant's dissertation [312].

The papers of Banaschewski [67] and Butler [90] are on the classical torsion of modules over a commutative integral domain. In particular, the former proves the existence of the torsion-free coverings introduced by Enochs ([42], p. 187). Butler's investigation is devoted to the generalization of the theory of torsion-free Abelian groups of rank one (also see [265]).

A unified exposition of the theory of radicals in modules is given in [27].

§ 2.   Projection, Injection, etc.

Levy [243] considered the conditions which ensure the possibility of "lifting" the direct decomposition of the factor module of a projective module up to the direct decomposition of the projective module itself. Strooker [382] studied the feasibility of obtaining the projective Λ/I-module, where I is a two-sided quasiregular ideal, from some projective Λ-module. A criterion for the projection of a module over a Dedekind domain was proposed by Kulkarni [227]. Finitely-generated projective modules of finite rank over a Dedekind ring, equipped with a nonsingular quadratic form, were considered by Flamant [144], Kasch and Mares [210], as well as by Miyashita [270] who proved that all factor modules of a projective module A have projective coverings if and only if the set $\{V, U + V = A\}$ contains a minimal element for every submodule $U \subseteq A$. Miyashita also pointed out other properties equivalent to this. Koh [224] described a ring all irreducible modules over which possess a projective covering. Miyashita [270] and Wu and Jans [413] call a module M quasi-projective if every epimorphism $\varphi \colon M \to N$, where N is a factor module of module M, is induced by an epimorphism of module M. A number of properties of quasiprojection and of dual quasiprojection have been established. It has turned out that the retract of a quasiprojective module is quasiprojective, while each complement submodule of it is an absolute retract (see p. 66). Wu and Jans [413] considered a quasiprojective covering (cf. [42], p. 188) and proved that a finitely-generated undecomposable quasi-

projective module over a semiperfect ring Λ (a ring Λ is called
semiperfect if all finitely-generated modules possess a projective
covering) has the form $\Lambda e/(J(\Lambda) \cap \Lambda e)$, where e is the primitive
idempotent in Λ. If here Λ is Artinian, then the infiniteness of the
set of nonisomorphic undecomposable quasiprojective Λ-modules
is equivalent to the lattice of two-sided ideals of ring Λ being
infinite.

If I is a finitely-generated ideal of a commutative ring Λ,
then the Λ-module A is called formally projective (weakly I-adic
free) if all the factor modules $A/I^k A$ (k = 1,2, ...) are projective
(free). Suzuki [385] established that both these properties are
equivalent for local rings.

Jensen [202] noted that under certain conditions the exterior
product* of a flat module over a commutative integral domain
equals zero.

Passing on to injection, we note the paper by Fort [146] of-
fering a new criterion for this property. The characterization of
injective modules, connected with the specifying of some topologization, occurs in Narita [286]. Osofsky [298] proved that every
simple injective module over a ring of linear transformations of
a vector space over a field Φ is isomorphic to a left ideal if the
cardinality of Φ does not exceed $2^{\aleph_0}$. Dickson [123] pointed out the
properties of a commutative Noetherian ring Λ, under which the
equality $A = \mathfrak{M} \cdot A$ is valid for every injective Λ-module A and for
any maximal ideal $\mathfrak{M}$ of ring Λ. From Bumby's results [88] it follows the isomorphism of the injective (and even the quasi-injective)
hulls of modules each of which is isomorphic to a submodule of
the other. Fleischer [145] proposed a new construction for obtaining the injective hull. Certain properties of these hulls were
established by Renault [330]. Tiwari [392] showed that if the quo-
tient ring of ring Λ relative to a maximal ideal $\mathfrak{M}$ is a ring of prin-
cial ideals, then the injective hull for $\Lambda/\mathfrak{M}$ is isomorphic with any
of its nonzero factor modules. Faith [134] studied injective submodules the direct product of any number of copies of which is injective. The injective hulls of cyclic modules over Dedekind rings

---

*Translator's Note: This is Jensen's term for the original "puissance extérieure"
used by N. Bourbaki, Algèbre Multilineare, 2nd ed., Hermann et Cie. Paris (1958); the
Russian text uses "exterior power."

were examined by Banaschewski [68]. Osofsky [297] constructed a cyclic module all of whose cyclic submodules are injective but do not form a lattice. Snapper [373] and Chaptal [99] touched upon the preservation of injection and quasi-injection under a replacement of rings. Harada [173] described the structure of quasi-injective modules over Dedeking domains (also see [99, 313]). Ishikawa [193], in the situation ($_\Lambda$A, $_\Lambda$B$_\Gamma$, $_\Gamma$C), investigated the natural homomorphism τ: Hom_Λ(A, B) $\otimes_\Gamma$C $\to$ Hom_Λ(A, B $\otimes_\Gamma$C) which, in particular, led to the establishment of certain connections between faithful injection and faithful flatness ([42], pp. 183-184) of the modules B and Hom_Γ(B, E), where E is a faithfully injective ring Γ-module. See section 6 for the endomorphism rings of injective and quasi-injective modules.

Ravel [324] calls a module semi-injective if it is a factor module of any module in which it is contained. Semiprojective modules are defined in dual manner. A semi-injective projective module is injective. In the case of an integral domain semi-injective semiprojective modules are injective and projective. Certain conditions which ensure that semi-injective modules are injective can be extracted from Miyata [271]. Other generalizations of the concepts of injection and projection are considered on p.

We note further the paper by Ribenboim [34] wherein projective and injective objects of ordered modules are studied.

The question of the possibility of converting the injective hull $\hat{\Lambda}$ of a ring Λ into a ring has been considered previously ([42], p. 194). Some analogs of the ring $\hat{\Lambda}$, connected with the generalizations of the concepts of injection, have been analyzed by Helzer [188] and Sanderson [352]. Renault [330, 334] showed that the ring $\hat{\Lambda}$ does not contain nilpotent elements if and only if there are no such elements in the ring Λ and for x, y $\in \Lambda$ the relation xy = 0 follows from Λx $\cap$ Λy = 0. Ishikawa [193] derived a number of conditions equivalent to the injective hull of a commutative Noetherian ring coinciding with its faithful quotient ring. Osofsky [301] constructed a Jacobson-semisimple primary ring for which $\hat{\Lambda}$ is not a rational extension of the Λ-module of Λ (see [39], pp. 63-64). The connections between injection and quotient rings were drawn in Faith's lectures [137]. Here we should also mention Tiwary [393]. See pp. 75-77 for the generalization of projection and injection in the framework of relative homological algebra. This aspect has been considered also in Sandomierski's dissertation [353].

If the ring Λ has a left generalized quotient ring $\Lambda(S)$ relative to a multiplicatively closed system S (see [39], p. 62) and φ is the natural homomorphism of ring Λ into $\Lambda(S)$, the $\Lambda(S)$ is a right $\varphi(\Lambda)$-flat module. A study of subrings Γ such that $\varphi(\Lambda) \subseteq \Gamma \subseteq \Lambda(S)$ and Γ is a $\varphi(\Lambda)$-flat module, has been carried out in the commutative case by Akiba [52, 53] and by Richman [341] and in the noncommutative case by Elizarov [18] who proved that all right ideals of the ring Γ are extended. Skornyakov [43] obtained the necessary and sufficient condition under which $\Lambda(S)$ is a Λ-flat module and characterized strongly regular rings as rings in which the ring $\Lambda(S)$, where S = $\Lambda \setminus M$, exists for any maximal left ideal and is a division algebra and a flat right Λ-module. Larsen [234] considered a class of commutative rings Λ in which all the overrings Γ in a total quotient ring are integrally closed and obtained a number of conditions equivalent to this one. He also studied the flat extensions of ring Λ. We mention here the work of Lazard [236] and Olivier [295]. Quotient modules were treated by Elizarov [17] and Budach [87].

By a local Goldie ring [160, 161] we mean a ring Γ with a unique maximal ideal M, for which Γ/M is a simple Artinian ring and $\bigcap\limits_{n=0}^{\infty} M^n = 0$. If, furthermore, M = J(Γ), then Γ is called a J-local ring. Goldie proposed a construction which permits us to construct a J-local extension $\overline{\Gamma}$ of a Noetherian ring Λ by knowing the primary ideal P of ring Λ and by using the set S = $\{\lambda \in \Lambda,$ $\lambda x \in P, x \in \Lambda \Rightarrow x \in P\}$. Now let $\Gamma = \cap \Gamma_\alpha$, where Γ_α is a J-local ring a maximal ideal M_α and, moreover, $\Lambda \subseteq \Gamma_\alpha \subseteq \overline{\Gamma}$, $P \subseteq M_\alpha \subseteq \overline{M}$, M is a maximal ideal in Γ. Goldie proved that Γ is a J-local ring and its maximal ideal M = $\cap M_\alpha$.

Let an ideal P possess the Artin-Riesz property (i.e., for any right ideal I of ring Λ we can find an n such that I $\cap$ $P^n \subseteq$ IP). McConnel [248] studied whether in this case Goldie's construction leads to a classical quotient ring. In particular, he proved that this holds if Λ is a universal enveloping finite-dimensional solvable algebra over an algebraically closed field of zero characteristic.

Silver [362] defined a left localization of a ring as the ring homomorphism $\alpha: \Lambda \rightarrow \Gamma$ which induces on Γ the structure of a flat right Λ-module and studied its properties. Localizations in noncommutative rings have been considered also by Raggi [322].

Ferrand [142] obtained certain results on modules which are projective limits of their own localizations.

The concept of dual rationality (compare [39], pp. 63-64) of a module A over A/B, where A $\supseteq$ B, and, in particular, the concept of a maximal corational extension A of module A/B were studied in [117].

§ 3. Homological Classification of Rings

As before ([40], pp. 80-81; [42], p. 190) we start by considering the question of the freedom of projective modules. The first part of Serre's report [359] is devoted to the proof of Seshadri's theorem: if Λ is a Dedekind ring, then every finitely-generated projective module over the polynomial ring $\Lambda[x]$ has the form $\Lambda[x] \otimes_{\Lambda} P$, where P is a projective Λ-module. This theorem has been proven also by Bass [70] and, for certain special cases, by Seshadri [360, 361]. In the second part of the report the connection is pointed out between projective modules (or, equivalently, vector bundles over affine varieties) and total intersections. In particular, if every projective module over a ring of polynomials in three variables over a field Φ is free, then every nonsingular curve of genus 0 or 1 in Φ^3 is a total intersection. Endo [131] found conditions necessary and sufficient for Seshadri's theorem to be valid for a one-dimensional Noetherian domain. Therein were obtained conditions on a semilocal one-dimensional domain Λ, under which every projective module over the ring $\Lambda[x, y]$ is free. Murthy [284] pointed out the properties of a Noetherian domain Λ which ensure the decomposition of every finitely-generated projective $\Lambda[x]$-module into a direct sum of a free module and of some ideal of the ring $\Lambda[x]$. Horrocks [190] and Murthy [285] proved that every finitely-generated projective $\Lambda[x]$-module, where Λ is a two-dimensional regular ring, is free. Swan [387], and then Uchida [395], analyzed the conditions on a total quotient ring Q of a commutative ring Λ, under which the tensor product $Q \otimes_{\Lambda} P$ is a free module over the group ring $Q\pi$ of an nth-order finite group π, for every finitely-generated projective Λ_π-module P. Here it was assumed that the prime divisors of the number n are noninvertible in Λ. Rings over which all projective modules are free were examined also in Cassel's dissertation.

Subsequently we shall say that the sequence $\{a_1, ..., a_k\}$ of elements of ring Λ is a sequence of minors if it coincides with the set of $(k-1)$th-order minors of some $[(k-1)\times k]$-matrix over Λ. A commutative ring is called an OP-ring if every finite sequence of elements of Λ is a sequence of minors. Lissner [245, 246] proved that all Dedekind domains, as well as the ring $\Lambda[x]$, where Λ is the domain of principal ideals, are OP-rings. He remarked, however, that the latter assertion had been proven in Tauber's dissertation for the case when Λ is a Dedekind ring. We note that all stably free Λ-modules (i.e., projective modules which can be complemented upto a free finitely-generated module by a free module) are free if and only if every finite system of generators of ring Λ, as Λ-modules, is a sequence of minors. This property is possessed by rings of power series of three or more variables over a field, which are not OP-rings. Closely related here is the paper by Mount [277] proving the equivalence of the following properties of the polynomial ring Λ over some field: 1) finitely-generated projective Λ-modules of rank $\geq$ n are free; 2) a sequence of length $\geq$ n + 1 of elements of Λ is a sequence of minors if the homological dimension of the ideal generated by it does not exceed 1. We mention further the papers by Burch [89] and Steger [378]. The latter considered commutative rings over which every idempotent matrix is similar to a diagonal one, which in the absence of nontrivial idempotents is equivalent to the freedom of finitely-generated projective modules.

Passing on to other aspects, we mention the characterization, noted by Ravel [324], of hereditary rings as rings for the modules over which projection and semiprojection (see p. 62) are equivalent. Under the assumption of being Noetherian, the same thing is valid for semi-injection. Harada [172] proved that a semiprimary hereditary ring is isomorphic to the ring of generalized triangular matrices over classically semisimple rings Λ_i. Here, by a generalized triangular matrix we mean a matrix in which zeros occur on the diagonal, the elements of ring Λ_i occur at the location (i, i), and elements of the Λ_i-Λ_j-bimodule occur at the location (i, j) with i > j. In another paper [176] he has proven that the classes of left and right PP-rings ([40], p. 82) coincide in the class of primary rings. This same result was obtained by Small [368] for finite-dimensional rings ([40], p. 86). Clark [105] also occupied himself with semiprimary PP-rings. From the somewhat more general results of Jensen [203] it ensues that the ideals of the center of a semihereditary local ring form a chain. He also noted that for local rings right semiheredity follows from left semiheredity. The same thing was proved for finite-dimensional rings by Sandomierski [354]. Certain properties of local semihereditary rings were established by Cohn [109]. Therein was constructed a non-hereditary domain of principal right ideals with a single-valued expansion. Numerous characterizations of Prüfer rings have been collected by Jensen [201], Butts and Smith [92], Richman [341], and Larsen [234]. Certain tests for a domain to be Dedekind were given by Matlis [262], Butts and Wade [93], and Arnold and Gilmer [57]. Hereditary rings appear in Kil'p [19], Koifman [21], Zaks [418, 420], Kaye [215], Heller [186], and Quentel [315]. See p. 87 for hereditary orders. In Michler [266] we encounter rings all of whose uniform left ideals (a module is called uniform if any two nonzero submodules of it have a nonzero intersection) are projective. Satyanarayana [356] made use of the projection of the maximal ideals of a commutative ring. Jensen [202] pointed out the properties of an integral domain, equivalent to the vanishing of the functor Tor_2^Λ. Leavitt [237, 238] continued the investigations connected with the classification of rings suggested by him ([40], p. 82). Peinado [310, 311] worked on this same theme. A certain refinement of the results involved here was given by Cohn [112]. Elizarov [16] considered completely measurable rings (i.e., rings over which every finitely-generated free module is measurable). Cohn has carried out further investigations on completely measurable rings. As a matter of fact, let us call a completely measurable ring Λ a (locally) Cohn ring if all its (finitely-generated) left ideals are free. All such rings have no zero divisors [41]. A number of properties of Cohn and locally Cohn rings were established in [108]. In particular, it turned out that the property of a ring to be locally Cohn is equivalent to it being its own right analog. Conversely, this is not so for Cohn rings [108, 109]. A ring Λ is isomorphic to a complete ring of matrices over a (locally) Cohn ring if and only if it is hereditary (semihereditary) and P-trivial (i.e., the class group of projective Λ-modules is cyclic). If, in addition, Λ is Noetherian, then it is isomorphic to a complete matrix ring over the domain of principal left ideals [110]. It has been proven [111] that a semihereditary local ring is locally Cohn and, in particular, is a domain (cf. [203]). From the results of [113] it follows that the free product of locally Cohn rings over a division algebra is once again locally Cohn. A free associative algebra over a field and the group algebra of a free algebra over a field are examples of Cohn rings [108].

Satyanarayana [357] remarked that for the classical semisimplicity of a ring it suffices that its maximal left ideals be injective. A complete exposition of Renault's results [328, 333] on complement and isotypic modules ([42], pp. 182, 202) has appeared. Therein it has been proven that the classical semisimplicity of a ring Λ is equivalent to every complement submodule of a finitely-generated Λ-module being an absolute retract (the submodule $A \subseteq B$ is called an absolute retract if $B = A + X$ for any X complementary relative to A). A ring Λ being Artinian is equivalent to the possibility of representing every injective Λ-module as a direct sum of injective hulls of simple modules, which, in its own turn, is equivalent to the imbeddability of any Λ-module in a direct sum of finitely-generated ones. In the commutative case "imbeddability" can be replaced by "decomposability." These results were obtained by Faith and Walker [140] and also by Vámos [400]. Noetherian rings are characterized as rings over which every injective module decomposes into a direct sum of modules of some fixed cardinality [138]. Colby [114, 115] indicated several sufficient conditions for an Artinian ring Λ to possess the following property: for every positive integer n we can find d > n such that there exists an infinite set of nonisomorphic finitely-generated Λ-modules with a composition series of length d.

Hamsher [167] has given a new characterization of commutative perfect rings (a ring is called perfect if all modules over it are perfect ([42], p. 186)). In particular, it turns out that every nonzero module over such a ring possesses a maximal submodule. Later Renault [331] and Hamsher [168] independently proved that for commutative rings the latter property is equivalent to the Jacobson radical being T-nilpotent ([40], p. 81), while the factor ring with respect to it is regular. Osofsky [297] established the equivalence of the following properties of a ring Λ: 1) Λ is regular; 2) the finitely-generated submodules of the projective Λ-modules are retracts; 3) in the category of Λ-modules there exists a projective generator all of whose finitely-generated submodules are retracts. We note further the paper by Rentschler [336] giving a ring-theoretic proof of certain properties equivalent to the perfectness of the ring.

A ring Λ is called coherent if all its finitely-generated left ideals are finitely connected (cf. [42], p. 192). Harris [179] proved the coherency of the factor ring of a coherent ring Λ being a coherent Λ-module. Therein it was shown that not every coherent ring is a direct limit of Noetherian rings. Commutative coherent rings have been treated by Radu [318] and Quentel [314, 315].

See §4 for Köthe rings. Selfinjective rings will be considered there.

Matlis [264] proved that all finitely-generated torsion-free modules over a commutative integral domain Λ are reflexive ([42], p. 196) if and only if all the ideal are reflexive and inj. dim.$_\Lambda \Lambda = 1$. The second condition can be dropped in the case of Noetherian rings. A complete one-dimensional local ring Λ is characterized by the fact that it is Noetherian and that all the torsion-free reduced Λ-modules of finite rank have no Bass-torsion (see p. 59 or [42], p. 196). Other properties equivalent to the one mentioned have been noted. In another paper [263] he characterized commutative integral domains over which every torsion-free module of finite rank decomposes into a direct sum of modules of rank 1. Fort [147, 148] showed that every module is an essential extension of the direct sum $K \oplus S$, where K is the maximal one among the submodules not containing homogeneous modules (see p. 66) and S is the direct sum of the homogeneous modules. Fort, as well as Chamard [97], established a number of properties of modules for which $K = 0$. Năstăsescu and Popescu [288, 289] proved that the term K is absent in all Λ-modules if and only if every left ideal of ring Λ is either Λ-irreducible or representable as the intersection of two Λ-irreducible ones.

Szász [389] determined when a ring, in all modules over which the trivial submodule is a retract, has a unity.

Skornyakov has published [45] a survey article devoted to the homological classification of rings.

§ 4. Quasi-Frobenius Rings and Their Generalizations

The theory of quasi-Frobenius rings has been set forth in the books by Curtis and Reiner [23] and Jans [197]. Faith and Walker [136, 138] proved the equivalence of the following properties of ring Λ: 1) Λ is quasi-Frobenius; 2) all injective Λ-modules are projective; 3) all projective Λ-modules are injective. Still earlier Tol'skaya [46] had shown that in the last requirement projection can be replaced both by freedom as well as by flatness. During the proof Faith studied rings with the maximality condition for annihilator ideals. The equivalence of a number of well-known definitions of quasi-Frobenius rings was noted in [347]. Jans [199] found the conditions necessary and sufficient for a classical quotient ring to be quasi-Frobenius. Roy [346] remarked that if for a filtered algebra Λ over a commutative ring Φ the associated algebra is isomorphic with the algebra of truncated polynomials $\Phi[x_1, ..., x_n]/(x_1^{\alpha_1}, ..., x_n^{\alpha_n})$, then Λ is a Frobenius algebra (herein is contained the known result that a universal enveloping finite-dimensional restricted Lie algebra over a field of characteristic p is Frobenius). Next, Bongale [78-80] proved that a filtered ring (algebra) is a quasi-Frobenius ring (a Frobenius algebra) if and only if the associated graded ring (algebra) is quasi-Frobenius (Frobenius). A special class of quasi-Frobenius rings was studied by Feller [139]. Wenger [409] and Ponizovskii [32] described a semigroup whose semigroup algebras over a field are quasi-Frobenius. If Φ is a regular local ring, m is its maximal ideal, Λ is a Φ-algebra of finite type and torsionless over Φ, then, as Vasconcelos [403] has shown, the Φ-algebra Λ being Frobenius is equivalent to each of the following conditions: a) Λ is Φ-free and $\Lambda/m\Lambda$ is a Frobenius Φ/m-algebra; b) Λ is a Gorenshtein ring.

A ring Λ is called a Köthe ring if any Λ-module decomposes into a direct sum of cyclic submodules, and is called a hypercyclic ring if the injective hulls of the cyclic Λ-modules are cyclic. By studying Köthe rings Faith [135] arrived at a description of rings all proper factor rings of which are universal (i.e., are decomposed into a direct sum of primary rings, while each of their nondecomposable left ideals possess a unique composition series). The latter is equivalent to all the proper factor rings being quasi-Fro-

benius. The class of rings defined by this property was examined
by Levy [242]. Injective modules over a quasi-Frobenius ring de-
compose into a direct sum of cyclic ones. The converse is also
valid for commutative rings [135]. Jans [200] indicated the properties of a finite-dimensional algebra over a field, equivalent to
the decomposition of injective modules into a direct sum of cyclic
ones. Faith [135] noted certain connections between uniserial and
quasi-Frobenius rings and the rings of principal ideals. In the
same paper it was proven that uniseriality is equivalent to hypercyclicity for Artinian and Noetherian rings. Caldwell [94, 95]
proved the same thin for perfect rings. He gave a hypercyclicity
criterion for certain classes of commutative rings. Osofsky [304]
reduced the study of hypercyclic rings with an Artinian factor ring
by a Jacobson radical to the study of local hypercyclic rings. There
is some information also on the structure of the latter.

A ring Λ is called a QF-3 ring if it contains a faithful projective injective left ideal, and is called a QF-3$^+$ ring if the injective hull $\hat{\Lambda}$ is projective. Jans [196] showed that a semiprimary
QF-3 ring Λ (a ring Λ is semiprimary if it contains a nilpotent
ideal N such that the factor ring Λ/N is Artinian) possesses a
projective injective left ideal, being the retract of any faithful Λ-
module. This left ideal is called a minimal faithful module. Therein it was proven that the concepts of QF-3 and of QF-3$^+$ rings coincide in the class of Artinian rings. The class of semiprimary
rings possessing this property was pointed out by Colby and Rutter
[116]. A number of theorems on the structure of semiprimary
QF-3 rings occur in Harada [174, 175]. In particular, he clarified
the structure of the base ring and constructed a QF-3 ring which
is not a right QF-3 ring. The structure of QF-3 rings has been
studied also in Fuller's dissertation [151]. Wu, Mochizuki, and
Jans [414] proved that Artinian QF-3$^+$ rings are characterized by
the conditions: a) the class of modules without Bass-torsion (see
p. 76) is closed relative to extensions; b) the class $\{A, \text{Hom}_\Lambda$
$(A, \Lambda) = 0\}$ is closed relative to submodules.

For a finite-dimensional algebra Λ over a field Nakayama
proposed to consider the dominant dimension dom.dim.Λ: dom.dim.
$\Lambda \geq n$ if there exists the exact sequence $0 \to \Lambda \to X_1 \to \ldots \to X_n$,
where the X_i are projective and injective Λ-Λ-bimodules ([42], p.
196). Tachikawa [390] analyzed an analogous dimension, taking the

X_i to be Λ-modules. Muller [282] showed that the right and the left Tachikawa dimensions coincide with the dominant dimension. This same paper also bear on the Nakayama problem: does the condition dom. dim. $\Lambda = \infty$ imply that the algebra Λ is quasi-Frobenius? It also contains a description of algebras Λ for which dom. dim. $\Lambda \geq 2$, as the endomorphism rings of completely faithful modules (i.e., modules for which every nondecomposable projective or injective module serves as a retract) over some algebra. If X is a minimal faithful module of a QF-3 algebra Λ and $E = \mathrm{Hom}_\Lambda(X, X)$, then $\Lambda' = \mathrm{Hom}_E(X, X)$ turns out to be a QF-3 algebra and $\Lambda \subseteq \Lambda'$. The questions concerning the coincidence of Λ and Λ', as well as the classification of algebras Λ in accordance with the properties of algebra Λ', were taken up by Tachikawa [390], Mochizuki [272], and Muller [281, 282].

A number of other papers have been devoted to other generalizations of quasi-Frobenius rings. Morita [275] studied Artinian S-rings, i.e., rings with nonzero left annihilators of any proper right ideal. These rings are characterized, in particular, by the fact thay any faithful projective module is the generator of the category of all modules. We remark that the class of quasi-Frobenius rings coincides with the intersection of the classes of S and QF-3 rings. Osofsky [299] examined a ring Λ which is an injective cogenerator of the category of all Λ-modules and proved that such a ring decomposes into a direct sum of nondecomposable left ideals, while the factor ring $\Lambda/J(\Lambda)$ proves to be Artinian. Rings over which all faithful modules are generators were considered by Utumi [398], Azumaya [59], Sugano [383], and Kato [212, 213]. The stated property of ring Λ is equivalent to any of the following: 1) Λ is selfinjective, $\Lambda/J(\Lambda)$ is Artinian, every nonzero left ideal of ring Λ contains a minimal one; 2) $\Lambda/J(\Lambda)$ is Artinian and every faithful Λ-module is a cogenerator; 3) Λ is selfinjective and the right annihilator of any maximal left ideal is nonzero; 4) Λ is an injective cogenerator. Endo [132] was occupied with the same questions for finitely-generated modules. Kato [211] showed that a ring Λ is a right- and left-selfinjective ring satisfying the annihilator condition for left and right ideals if and only if all right and left cyclic Λ-modules are reflexive (see [42], p. 196). Therein he also accertains the connection between annihilator conditions and the reflexivity of cyclic modules. Chandler and Koh [98] investigate the possibility of prolonging any homomorphism f: $C \to \Lambda$, where C is a cyclic submodule of any free Λ-module F, upto the homomorphism f: $F \to \Lambda$, and also the connection of this condition with the annihilator conditions and with the properties of selfinjective rings. These investigations are tied in with the topological characterization of selfinjective rings proposed by Wu [412]. Gentile [155] and Kaye [215] have proven that selfinjection is preserved under transition to a matrix ring.

It is well known that a semisimple selfinjective ring is continuous, but not vice versa ([39], p. 66). Utumi ([396]; also see [329]) pointed out a number of conditions ensuring the selfinjection of a continuous ring. The generalization of the notions of injection and continuity to the case when in the respective definitions there occur only ideals of a specified cardinality, and the connection between these notions, was considered in another of his papers [397]. Chaptal [99] noted that a (Noetherian) selfinjective ring is classically semisimple if and only if the (finite) direct sum of the quasi-injective modules over it is quasi-injective. Gemintern [7] proved that a ring Λ is a complete direct sum of full matrix rings over selfinjective strictly-regular rings if and only if Λ is a semisimple ring and every ideal of ring Λ contains an ideal the nilpotency index of whose elements is bounded. Hence once again there was obtained a characterization of complete direct sums of matrix rings over division algebras (see [42], p. 193). Satyanarayana [358] showed that in the class of selfinjective rings the localness of a ring Λ is equivalent to the absence of nontrivial idempotents. If at the same time Λ is Noetherian, then Λ is a local ring with a nilpotent maximal ideal if and only if there are no projective proper principal left ideals in Λ. Năstăsescu [287] has described the structure of selfinjective rings subject to certain additional conditions.

Gabriel and Oberst [152] noted that the endomorphism ring Λ of a Grothendieck category $\mathfrak{G}$ with invertible mono- and epimorphisms is regular and selfinjective. The category $\mathfrak{G}$ itself is equivalent to the category of retracts of the complete direct sums of some set or other of copies of ring Λ. This latter is a Grothendieck category with the above-mentioned properties for every regular selfinjective ring.

Muller [278, 279], 279], by generalizing the concept of a Frobenius extension introduced by Kasch and studied by Nakayama and Tsuzuku ([42], p. 190) and by Pareigis [306], called a ring Λ a quasi-Frobenius extension of a subring Γ if Λ is a finitely-generated projective Γ-module while the bimodule ${}_\Gamma\Lambda_\Lambda$ decomposes into a a direct sum of a certain number of copies of the bimodule Hom_Γ $(\Lambda_\Gamma, \Gamma_\Gamma)$. In this notation the ring $\text{Hom}_\Gamma({}_\Gamma\Lambda, {}_\Gamma\Lambda)$ turns out to be a quasi-Frobenius extension of the ring Λ. The necessary and sufficient conditions for the validity of the converse theorem were found by Muller [278, 279] and, in a more convenient form, by Morita [276]. Onodera [296] found the condition on the ring Γ and on its subring Λ equivalent to Γ being a projective Frobenius extension of ring Λ. Its use leads to a simplification of the proofs and, in a number of cases, to a generalization of well-known results on Frobenius extensions. Subsequent development, using conjugate functions and the theory of Frobenius and quasi-Frobenius extensions was obtained by Morita in [274, 276]. The work by Harase [178] as well as the research by Pareigis [308] on forgetful functors relates to these same matters.

§ 5. Some Aspects of Homological Algebra

In this paragraph we consider those aspects of homological algebra most closely related to the general theory of modules. We shall denote the category of all Λ-modules by ${}_\Lambda\mathfrak{M}$.

Let Λ and Γ be rings, A and C be Λ-modules. Solian [374, 375] studies the functor T: ${}_\Lambda\mathfrak{M} \rightarrow {}_\Gamma\mathfrak{M}$ such that $C = \Sigma\text{Im} f_\alpha$, where $f_\alpha \in \text{Hom}(A, C)$ implies $\cap \text{Ker}(f_\alpha) = 0$. Sakhaev [37] noted that if for an additive covariant functor T the exact sequence $T(K) \rightarrow T(F) \rightarrow T(A) \rightarrow 0$ follows from the exact sequence of Λ-modules $0 \rightarrow K \rightarrow F \rightarrow A \rightarrow 0$, where F is a free module, then T is right exact. The dual assertion also is proved. The conditions under which the inverse limit of exact sequence remains exact and the theory of arbitrary functors of the inverse limit are contained in Brandenburg's dissertion [82]. Muller [280] calls the sequence T^k of categories of the functor $T^{-1} = \text{Ext}^1_\Lambda(A, -)$ a complete cohomology sequence for the Λ-module A if it vanishes on all projective and injective modules. The connection of these concepts with the theory of Frobenius and quasi-Frobenius algebras and extensions was

pointed out. Balcerzyk [64] modified somewhat the definitions of
the functors ⊗ and Hom and examined their relation to the functors
introduced standardly. Rudyk [35] studied extensions in the category of all modules over all rings with semilinear mappings as
morphisms. Radu [321] considered certain subcategories of the
category $_\Lambda\mathfrak{M}$, connected with the ring homomorphism $\varphi: \Lambda \to \Gamma$.
For example, $\{ M, \Gamma \otimes {}_\Lambda M = 0 \}$. Lam [231] characterized the categories of all finitely-generated modules over Artinian and Noe-
therian rings. Roos [345] characterized the categories equivalent
to the factor-category of a category of the form $_\Lambda\mathfrak{M}$ by a localizing
subcategory of the form $_{\Lambda_1}\mathfrak{M}$. Kleiner [220] investigated free and
injective modules over a Lie ring.

Let us now dwell on the papers related to Morita-equivalence
and to duality.

We shall say that the rings Λ and Γ are Morita-equivalent
if there exist covariant functors S: $_\Lambda\mathfrak{M} \to {}_\Gamma\mathfrak{M}$ and T: $_\Gamma\mathfrak{M} \to {}_\Lambda\mathfrak{M}$
between the categories of all modules over these rings such that
the functors ST and TS are equivalent to the identity functor. A
property of a ring is called a Morita-property if it is preserved
under transition to a Morita-equivalent ring. The properties of
being a primary ring and of being an order in an Artinian ring
turn out to be Morita-properties [367, 369, 370]. The property of
being the ring of principal left ideals is not a Morita-property.
Cohn [110] found the properties of rings which are Morita-equivalent to Cohn rings, to locally-Cohn rings (see p. 68), and to domains of principal left ideals. Kaye [215] used the fact that selfinjection, (semi)heredity, (semi)perfection, regularity, primitivity,
and semisimplicity are Morita-properties to establish that they
are preserved under transition to matrix rings.

Let $\mathfrak{S}_i$ be a complete subcategory of the category $_\Lambda{}_i\mathfrak{M}$. with
generators V_i, being a Grothendieck category, and, moreover, let
the imbedding $0 \to A \to V_i$ induce the isomorphism $\mathrm{Hom}_{\mathfrak{S}_i}(V_i, V_i) \to$
$\mathrm{Hom}_{\mathfrak{S}_i}(A, V_i)$. if and only if $A = V_i$ (i = 1,2). Walker and Walker
[407] have proven that then the isomorphism of the rings $E(V_1)$
and $E(V_2)$ is equivalent to the existence of an equivalence T: $\mathfrak{S}_1 \to$
$\mathfrak{S}_2$ such that $T(V_1) \cong V_2$.

Let us once again consider the functors S: $_\Lambda\mathfrak{M} \to {}_\Gamma\mathfrak{M}$ and
T: $_\Gamma\mathfrak{M} \to {}_\Lambda\mathfrak{M}$. By generalizing Morita's results [273] Willard

[410, 411] proved that the existence in $_\Lambda\mathfrak{M}$ of a generator P such that $\Lambda = \mathrm{Hom}_\Lambda$ (P, P) $\cong \Gamma$ and such that the functors S and T are naturally equivalent, respectively, to the functors Hom_Λ (P, $-$) and Hom_Γ (P*, $-$) where P* = Hom_Λ (P, Λ), is equivalent to the fact that S and T are adjoint, TS (Λ) $\cong \Lambda$, and ST is naturally equivalent to the identity functor of category $_\Lambda\mathfrak{M}$. If P is a projective Λ-module, then Ker (P $\otimes$ P* $\rightarrow \Lambda$) = 0, while $\mathfrak{X}$ = lm (P* $\otimes$ $_\Lambda$P $\rightarrow \Delta$) is dense in Λ. Walker and Walker [407] showed (without the assumption that P is projective) that the following assertions are valid: a) $\mathfrak{X}$ is a finitely-generated two-sided ideal; b) $\mathfrak{X}$P = P; c) $\mathfrak{X}^2 = \mathfrak{X}$; d) $l(\mathfrak{X}) = 0$; d) $\mathfrak{X}$ is essentially imbedded in P. In this connection we mention also the dissertation by Beckwith [74]. Adjoint functors of modules have been treated in [178, 274] also. Vámos [399] obtained metatheorems on the preservation of the properties of modules under transition from one category to another.

Let Λ be a complete commutative local ring with maximal ideal $\mathfrak{m}$, let $\Lambda^* = \widehat{\Lambda/\mathfrak{m}}$, let M be a linearly topologized (i.e., possessing a basis of neighborhoods of zero, consisting of submodules) Λ-module, and let M* = Hom_Λ (M, Λ^*) be a Λ-module of continuous homomorphisms, topologized in suitable fashion. We say that the module M is linearly discrete if every $\mathfrak{m}$-primary factor-module of module M is discrete, is linearly compact if every centered system of cosets by a closed submodule has a nonzero intersection, and, finally, is locally compact if its factor-module by some linearly compact submodule is linearly discrete. Macdonald [249] proved a duality theorem for locally compact modules, and, moreover, linearly compact and linearly discrete modules turn out to be dual. Conditions were ascertained under which the coupling between M and M* is nondegenerate. The category of linearly compact modules proves to be Abelian and possesses a sufficient supply of projective objects. Harte [182] studied the aspects of duality for topological modules over a topological ring. Eisenreich [130] considered duality for the submodules of a free module over a regular local ring. Duality relative to a specially chosen module has been treated in [169].

In what follows we shall adhere to the following notation: [r.]gl.dim.Λ is the (right) global dimension of ring Λ; [r.]w.gl.dim.Λ is the weak [right] global dimension of ring Λ; h.dim.$_\Lambda$M is the projective dimension of the Λ-module M; inj.dim.$_\Lambda$M is the injective dimension of the Λ-module M. Roy [346] proved that the global dimension of a graded ring Λ coincides with the maximum of the homological dimensions of the graded Λ-modules. The left and right global dimensions of the ring of triangular second-order matrices were analyzed by Small [363]. It turned out that they may not coincide. Jensen [204] proved that w.gl.dim.Λ[x] = w.gl.dim.Λ + 1 for every ring Λ. Osofsky [302], and in a weak form Jensen [204, 205], established that gl.dim.Λ $\leq$ w.gl.dim.Λ + n + 1 if all the left ideals of ring Λ are generated by $\aleph_n$ elements. It was noted in [205] that if this condition is fulfilled also for the right ideals, then |gl.dim.Λ − r.gl.dim.Λ | $\leq$ n + 1. An example by Small [365], constructing a right-hereditary ring of global dimension three, shows that the restrictions imposed are essential. Another example by Small [366] proves the existence of a right- and left-Noetherian ring of global dimension two with a non-Artinian two-sided quotient ring. He has also proven [371] that the equality gl.dim.Λ = r.w.dim$_\Lambda$ (Λ/N) + gl.dim (Λ/N) is valid for every nilpotent ideal N. If Λ is Noetherian, then this equality holds for every ideal N$\subseteq$J (Λ). As a corollary there is obtained a generalization of the Auslander-Buchsbaum theorem: if Λ is a Noetherian ring and if gl.dim.Λ = n, then the global dimension of the power series ring Λ[[x]] equals n + 1. This theorem occurs also in Strooker [382]. If Λ is a Noetherian, Λ/J (Λ) is right-Noetherian, and gl.dim.Λ = 1, then r.gl. dim.Λ $\leq$ 2. This estimate cannot be improved. The global dimension of group rings $\Lambda\pi$ of Abelian groups π was studied by Balcerzyk [61, 62]. In particular, if Λ is a commutative ring, π is a torsion-free Abelian group of rank r, and

$$\dim. \Lambda\pi \underset{\text{def.}}{=} \begin{cases} r, & \text{if } \pi \text{ is finitely generated} \\ r+1, & \text{otherwise,} \end{cases}$$

then gl.dim.$\Lambda\pi$ = gl.dim.Λ + dim$\Lambda\pi$. If $\Lambda = \lim\limits_{\longrightarrow} \Lambda_\alpha$ and the Λ-module $A = \lim\limits_{\longrightarrow} A_\alpha$, and, moreover, if the cardinality of the set of indices α equals $\aleph_n$ then as Osofsky [302] has proved, h. dim.$_\Lambda$ $(A) \leqslant n+1+\sup$ h. dim$_{\Lambda_\alpha} M_\alpha$ and gl. dm. $\Lambda \leq n + 1 + \sup.$ gl. dimΛ_α, and, moreover, the estimates stated may not be improved. A particular case of these results was obtained by Balcerzyk [63]. As an application he proved the inequality gl.dim $\Lambda\pi$ $\leq$ gl.dim.Λ + r + s + 1, where π is an Abelian group of rank r of cardinality $\aleph_s$ the orders of whose elements are invertible in a commutative ring Λ. A generalization of Balcerzyk's results, as well as of other results on the global dimension of a ring, is given by Mitchell [268, 269] who examined the connection between the global dimension of the category $\mathfrak{A}^\Pi$ of covariant functors from Π into $\mathfrak{A}$ and the global dimension of the category $\mathfrak{A}$, and, moreover, gl.dim.$\mathfrak{A} = \max\{n,$ $\mathrm{Ext}^n (A, -) \neq 0$ for some $A \in \mathfrak{A}\}$. As Π we can take a free Abelian monoid, a free Abelian group, a finitely-generated Abelian group, or a finite partially-ordered set.

If the quotient field Q of a commutative integral domain Λ is a countably-generated Λ-module, then h.dim.$_\Lambda Q \leq 1$. Kaplansky [208] proved the converse for the case when Λ is local, and Small [364], for certain other cases. In the same place it is shown that h.dim$_\Lambda Q \leq n + 1$, If Q is an $\aleph_n$-generated Λ-module. Osofsky [300] constructed valuation rings $\Lambda^{(k)}$ ($1 \leq k \leq \infty$) such that gl.dim.$\Lambda^{(k)} =$ k but sup$\{$inj.dim.$_\Lambda I\}$ = sup$\{$inj.dim.$_\Lambda \Lambda / I\}$ = 1 (in both cases I ranges over all left ideals of ring Λ). She also proved the coincidence of these three numbers for Noetherian and perfect rings. By using the techniques of Kaplansky [208] and her own [300], Osofsky [305] proved that h.dim.$_\Lambda Q = \min\{k + 1, m\}$ if Λ is a regular local ring of dimension m and, moreover, either the cardinalities of Λ and $\Lambda/J(\Lambda)$ coincide or Λ is a complete ring and the Λ-module Q is generated by precisely $\aleph_k$ elements. This leads to the following assertion: if the cardinality of a field Φ equals $2^{\aleph_n}$, Q_m is the quotient field of the ring of polynomials $\Phi[x_1, ..., x_m]$, m > n + 3, then the equalities h.dim$_{\Phi[x_1, ..., x_n]} Q_m = n + 2$ and $2^{\aleph_n} = \aleph_{n+1}$ ensue one from the other. If Φ is the real number field, then the equality h.dim.$_{\Phi[x, y, z]} Q_3 = 2$ is equivalent to the continuum hypothesis. We can find a module over $\Phi[x, y, z]$ which is free if and only if the continuum hypothesis is valid.

Brumer [84, 85] generalized the very well known results on the cohomological dimension of complete semilocal rings to the case of topological rings possessing a base of neighborhoods of zero consisting of ideals the factor rings by which are Artinian. If such a ring Λ is commutative, then the global dimension of the ring of noncommutative power series over it equals gl.dim.$\Lambda + 1$. This result was obtained by Cohn [107] for the case when Λ is a field. The homological dimension of group algebras of profinite groups also was studied. The global dimension of polynomial rings with special relations was examined in [376].

Zaks [418] proved the equivalence of the following properties of a semiprimary ring Λ (see p. 70): 1) all factor rings of ring Λ have a finite global dimension; 2) gl.dim.Λ/J^2 (Λ) $< \infty$; 3) Λ is a factor ring of a semiprimary hereditary ring. He also obtained [419] an estimate for the global dimension of an Artinian ring, using the projective dimension of a left ideal selected in a special way. Clark [104] described the structure of a class of finite-dimensional algebras of global dimension ≤ 1. The homological dimension of Ore extensions is discussed in [162]. The homological dimensions of Lie algebras are considered in [103].

A general scheme for the construction of a relative homological algebra in a pointed category, worked out by Hochschild [189], Heller [185], Buchsbaum [86], and Butler and Horrocks [91] (also see [5], p. 205), has been proposed by Eilenberg and Moore [129]. If E a some class of exact sequences $l: A \overset{i}{\to} B \overset{\pi}{\to} C$ of a pointed category $\mathfrak{A}$, then by $\mathfrak{P}(E)$ we denote the class of all objects P such that the sequence $\mathrm{Hom}_{\mathfrak{A}}(P, l)$ is exact for all $l \in E$. Objects from $\mathfrak{P}(E)$ are said to be E-projective. E-injective objects are defined analogously. Conversely, the above-mentioned condition allows us also to determine from a given class $\mathfrak{P}$ of objects a class of exact sequences said to be $\mathfrak{P}$-exact. Closed classes of objects and sequences, which are in a one-to-one correspondence, are determined in a natural way. A fundamental concept is that of a projective class of sequences, i.e., of a closed class E such that for any morphism φ: A $\to$ B we can find a sequence P $\to$ A $\to$ B in the class E, where P $\in \mathfrak{P}(E)$. Being given a projective class permits us to construct resolvents in the usual manner and to determine the derived functors. In particular, we can speak of a relative homological dimension of rings and modules [355]. The conjugacy theorems make it possible to construct from a projective class in one category the projective class in another. For example, if S: $\mathfrak{B} \to \mathfrak{A}$ and T: $\mathfrak{A} \to \mathfrak{B}$ are a pair of conjugate functors, while E is a projective class in $\mathfrak{B}$, then $T^{-1}E$ is a projective class in $\mathfrak{A}$. This construction has been applied to the obtaining of projective classes in categories of modules as well as in certain non-Abelian categories. The study of the category of complexes over an Abelian category shows that the so-called dual resolvents* are the usual resolvents relative to some projective class. Stenström [379] indicated the conditions which ensure the existence of a sufficient number of E-projective objects and constructed a relative theory of essential extensions. The latter generalizes the theory constructed by Sanderson [352] for the case when $\mathfrak{S}$-pure sequences (see p. 59) are taken as E.

Helzer [187, 188] investigated E-injection for the case when a certain set of sequences of the form $0 \rightarrow I \rightarrow \Lambda \rightarrow \Lambda/I \rightarrow 0$ was marked as E. Later he denotes by E' a subset of E such that the corresponding left ideals I are finitely generated. It was proven that E'-injection is equivalent to injection if and only if the ring Λ is Noetherian and hereditary. Certain results of Levy ([42], p. 188) and Hattori ([40], p. 83) were generalized. A criterion for the coincidence of E- and E'-injections was indicated. The connection with the theories of radicals and quotient rings was established.

We consider further the class E of sequences b such that i is a monomorphism and π is an epimorphism. We denote the corresponding classes of mono- and epimorphisms by E_m and E_l. The class E is called proper if: 1) E contains all the splitting sequences; 2) if $\alpha, \beta, \in E_m$ and $\alpha\beta$ exists, then $\alpha\beta \in E$; 2') if $\alpha, \beta \in E_l$ and $\alpha\beta$ exists then $\alpha\beta \in E_l$; 3) if $\alpha\beta \in E_m$ and β is a monomorphism, then $\alpha \in E_m$; 3') if $\alpha\beta \in E_l$; and β is a monomorphism, then $\alpha \in E_m$; 3') if $\alpha\beta \in E_l$ and α is an epimorphism, then $\beta \in E_l$. These definitions were proposed by Butler and Horrocks [91]. They also pointed out a method for constructing a proper class, depending on the chosen system of pairs $\{(F_\alpha, U_\alpha)\}$, where F_α is a free module and U_α is a submodule marked in it. Namely, the class E_m consists of all monomorphisms i: $A \rightarrow B$ such that for every number α and for any commutative diagram

$$\begin{array}{ccc} & \chi & \\ U_a & \!\!\to\!\! & F_a \\ \varphi\downarrow & & \downarrow \\ & i & \\ A & \to & B \end{array}\ ,$$

where χ is the natural imbedding under a suitable homomorphism $\psi: F_\alpha \rightarrow A$, the equality $\varphi = \chi\psi$ is valid. Kielpiński [216] established that for such a system E there exist sufficiently many E-

---

*See: H. Cartan and S. Eilenberg, Homological Algebra, Princeton Univ. Press, Princeton, N. J. (1956), Chapter XVII.

injective objects if all the F_α are countable and all the U_α are
finitely generated. A number of questions touching on universal
purity (A is universally pure in B if the natural imbedding of A
into B belongs to the class E_m where the F_α range over all free
modules and the U_α over all their submodules) were examined by
Cohn [106], Kuz'minov [22], Maddox [251, 252], and Miyata [271].
The most important special case of the definition of Butler and
Horrocks is the case when all the F_α are equal to Λ. Then E =
$\{U_\alpha\}$ is the set of left ideals, and sequences from E are said to be
E-pure.

Richman and Walker [342] proved that $\mathfrak{P}$-exact sequences
form a proper class, where $\mathfrak{P}$ is the class of all periodic groups.
Stenström [380] noted that the class consisting of all sequences
$0 \to A \xrightarrow{l} B \to C \to 0$, such that A is a complementary submodule in B.
Walker [406] pointed out another way of obtaining proper classes.
Namely, a class $\mathfrak{X}$ of modules is picked out, closed relative to the
factor modules, and in the class E_m are collected all monomorphisms of A and B such that A serves as a direct summand for
every submodule S such that $A \subseteq S \subseteq B$ and $S/A \in \mathfrak{X}$. A proper
class can be obtained also in dual fashion.

All these aspects of the general theory of purity are discussed in the monograph by Mishina and Skornyakov [27].

A module A over a commutative integral domain Λ with a
quotient field Q is called coperiodic if $\mathrm{Hom}_\Lambda (Q, A) = 0$ (such a
module is called reduced) and $\mathrm{Ext}(Q, A) = 0$. This definition was
proposed by Matlis [262]. Radu [319] generalized it by replacing
the field Q by a quotient ring relative to some system S, and, by
establishing a number of properties of such generalized coperiodic
modules, he applied the results obtained to construct the completion of a topological module A with a basis of neighborhoods of
zero consisting of all submodules of the form sA, where $s \in S$.
Matlis has developed his own theory considerably further. In particular, he investigated the category $\mathfrak{R}$ of coperidic Λ-modules.
This category has a sufficient supply of projective objects and $\mathfrak{R}$ =
gl.dim.Λ. The category of coperiodic torsion-free modules is
isomorphic with the category of torsion-free modules, being the
homomorphic images of injective modules. It was shown also that
every reduced module can be imbedded in a coperiodic one. The
connection with topological aspects also was not disregarded.

§ 6. Endomorphism Rings

In this section we shall denote by $E_\Lambda(A)$ the endomorphism ring of the Λ-module A. A number of authors have continued the research on the isomorphisms and anti-isomorphisms of endomorphism rings (see [42], pp. 198-199). Glushkin [8] showed that every nonclycic free module is determined to within isomorphism by its own semigroup of endomorphisms into any class of free modules in which the cyclic direct summands (and only these) are not decomposable into a direct sum of their own submodules (here enter the free modules over the domain of principal ideals, indecomposable into a direct sum of left ideals, and free torsion-free modules over the domain of principal ideals). He also studied the structure of the endomorphism semigroup of a free module. Mikhalev [28, 29] found the sufficient conditions on Λ_i-modules A_i and on endomorphism semigroups $D_i \subseteq E(A_i)$ under which from the presence of an isomorphism Φ of semigroups D_1 and D_2 there ensues the existence of a semilinear isomorphism of the modules, inducing Φ. In particular, the results indicated hold if Φ is the isomorphism of rings $E(A_1)$ and $E(A_2)$, the ring Λ is not isomorphic to the matrix ring Γ_h, $n > 1$, for any ring Γ whatsoever, while all the projective modules over the ring Λ_2 free (for example, when Λ_1 and Λ_2 are Cohn or locally Cohn rings (see p. 66).* Analogous results have been established also for the case when the A_i are partially-ordered modules and the D_i are semigroups of positive endomorphisms. Fayans [49, 50] has obtained the necessary and sufficient conditions for the definability of a finite-dimensional ordered linear space of the semigroup of all positive endomorphisms of rank ≤ 1. Gentile [156] proved that from the isomorphism of matrix rings over selfinjective rings there follows the isomorphism of the rings themselves. Gewirtzman [158] continued ([42], p. 199) the examination of anti-isomorphisms of the endomorphism rings of free and local-

---

*In his dissertation (W. Stephenson, "Characterization of rings and modules by means of lattices") and papers ("Lattice isomorphisms between modules (I)," "Endomorphism rings," "Unique addition rings") Stephenson investigated isomorphisms Φ of a lattice $L(_{\Lambda_i} A_i)$ of submodules of modules $_{\Lambda_i} A_i$, $i = 1, 2$, in particular, in a rather general case he showed that the presence of implies the isomorphism of $E(A_1)$ and $E(A_2)$. He has found the conditions under which Φ is induced by a semilinear transformation or, in a more general case, by a Morita-equivalent one. This allows us to generalize a number of the results presented above. When studying rings whose multiplicative structure defines an additive one, Stephenson showed, in particular, that an isomorphism of the semigroups of all endomorphisms of free modules is a ring isomorphism.

ly free modules over the domain of principal left ideals, as well
as of torsion-free modules over complete discrete valuation rings.

A number of results touch on the representation of rings as
the endomorphism rings of some module. Koh [222] noted that a
simple ring with unity containing a maximal annihilator left ideal
is isomorphic to the endomorphism ring of a torsion-free module
over an integral domain. Zassenhaus [421] showed that a ring Λ
with a reduced additive torsion-free group of rank n is isomorphic
to the endomorphism ring of a torsion-free Abelian group of rank
$\leq n$. Thierrin [391] proved the equivalence of the following properties of ring Λ: 1) Λ is isomorphic to a locally dense endomorphism ring Δ of a left unitary Γ-module M (Δ is locally dense if
there exists a Γ-module $N \subset M$ such that $N\Delta \subseteq N$ and if for any
$x_1, \ldots, x_n \in M$ with the property that the sum $\Sigma \Gamma x_i$ is direct and
for any $y_1 \ldots, y_n \in N$ there exists a $\in \Delta$ for which $x_i a = y_i$), where
Γ is a complete primary ring; 2) there exists a faithful Λ-module
A which is subdirectly indecomposable and $\Lambda C \neq 0$, where C is the
intersection of nonzero submodules of module A; 3) there exists an
injective faithful Λ-module A which is subdirectly indecomposable
and $\Lambda C \neq 0$. The representation of a classical quotient ring of the
ring Λ in the form $E(\widetilde{\Lambda})$, where $\widetilde{\Lambda}$ is a minimal $\mathfrak{C}$-injective extension of the module $_\Lambda\Lambda$ (see p. 78) and $\mathfrak{C}$ is the set of left ideals
of ring Λ, containing divisors of zero, was obtained by Sanderson
[352]. If G is an Abelin p-group, then in the ring E (G) the subset
$H(G) = \{\alpha \in EG, px = 0, (\text{height } x) < \infty \Rightarrow (\text{height } \alpha(x)) > (\text{height } x)\}$
is a two-sided ideal containing J (E (G)). Stringall [381] derived
conditions on the ring Λ under which it is isomorphic with the ring
E (G)/H (G).

Now let A be a Λ-module, $\Gamma = E_\Lambda(A)$ and $\Delta = E_\Gamma(A)$. Then
the natural ring homomorphism $\lambda: \Lambda \to \Delta$ holds. Rieffel [343]
and Faith [136] proved that the module A is a generator of all Λ-
modules if and only if λ is an isomorphism and A is a finitely-
generated projective right Γ-module. In particular, λ is an isomorphism if Λ is a simple ring and A is its nonzero left ideal. If
Γ also is simple, then the categories of Λ- and Γ-modules are
isomorphic. Closely related here is the paper by Hart [181] showing that a simple ring Λ with a uniform left ideal A (see p. 66) is
isomorphic with the ring $eE_n e$, where e is an idempotent matrix
and the ring E = E (A) itself is a domain satisfying the Ore condition. It is ascertained when E (A) is a simple ring.

Azumaya [59] analyzed a Λ-module A such that A is isomorphic with the direct sum of its factor modules. It turned out that Λ coincides with $E_{E(A)}$ ($A_{E(A)}$). Conditions were found under which every faithful Λ-module possesses this property. This same paper investigates when the covariant functors $\text{Hom}_\Lambda(P, -)$ and $P \otimes_{E(P)}(-)$, where P is some Λ-module, effect the isomorphism of the categories of Λ- and E (P)-modules.

Feller and Swokowski [140] call a Λ-module A prime if $\{\lambda \in \Lambda, \lambda N = 0\} = 0$ for any nonzero submodule $N \subseteq A$ and if Z (A) = 0 (see [42], p. 182). If A is a finitely-generated torsion-free module over an Ore domain Λ, then E (A) is a prime ring and A is a prime right E (A)-module. The uniformity of a prime Λ-module is equivalent to $E(\hat{A})$ being a division ring. Hart [180] proved the (semi)primeness of the ring E (A) for the case when Λ is a (semi) · prime ring with a two-sided quotient ring and A is a module torsion-free in the sense that $\lambda a = 0$, where $\lambda \in \Lambda$, $0 \neq a \in A$, implies $\xi \lambda = 0$ or $\lambda \xi = 0$ for some $0 \neq \xi \in \Lambda$. Conditions were indicated under which the ring $E_\Lambda(A)$ possesses a quotient ring. In particular, they are fulfilled if Λ is a semiprime ring and A is its nonzero left ideal. Zelmanowitz [422] studied a ring $E_\Lambda(A)$, where A is a module torsion-free in the sense of Bass (see p. 60), Z (A) = 0 (see [42], p. 182), and Λ is semiprime. His results are: a) E (A) is semiprime and $Z_{E(A)}(E(A)) = 0$; b) if Λ is prime, then E (A) is prime also; c) $\widehat{E(A)}$ is a selfinjective regular ring, isomorphic to $E(\hat{A})$; d) the uniformity of a Λ-module A is equivalent to $\widehat{E}(A)$ being a division ring; e) if A is finite dimensional ([40], p. 86) and Λ possesses a classical semiprime quotient ring Q, then the quotient ring Q, then the quotient ring for E (A) is isomorphic to $E_Q(Q \otimes_\Lambda A)$.

Osofsky [298] investigated the left ideals of the ring $E_\Phi(V)$, where V is an infinite-dimensional vector space over a field Φ. In particular, she established that the ring E (V) is not always self-injective. Lafon [228, 229] has proposed a characterization of one class of two-sided ideals of a ring E (A), where A is a faithful Noetherian module over a commutative ring. Renault [329] describes a radical $J(E_\Lambda(A))$, where A is an injective Λ-module.

Tsukerman [51] obtained a characterization of certain classes of rings by the properties of the endomorphism rings of free modules over them. For example, the regularity of the ring $E_\Lambda(F)$

for any free Λ-module F is equivalent to the classical semisimplicity of the ring Λ. The fact that $E_\Lambda(F)$ is a Rickart ring and a Baer
ring ([40], p. 66) is equivalent to the following system of conditions:
Λ is a hereditary ring, $J(\Lambda)$ is T-nilpotent ([40], p. 81), $\Lambda/J(\Lambda)$ is
classically semisimple, and every right ideal of ring Λ is isomor-
phic to F/K, where F and K are finitely-generated right Λ-modules.

Gailleau [153, 154] (also see Fort [148]) connected the ring
$S(A) = E(A)/N$ with a module A, where $N = \{ \varphi;\ \varphi \in E(A),\ \text{Ker } \varphi$
is essential in A$\}$. If $A = \Sigma A_i$, where the A_i are indecomposable
injective modules, then $S(A) = \Pi S(M_k)$, where M_k is the sum of all
interisomorphic A_i and every $S(M_k)$ is isomorphic with the endomorphism ring of a vector space over the division ring $S(A_i)$,
where $A \subseteq M_k$. If A is a quasi-injective module, then $N = J(E(A))$.
In this case Renault [332, 335] and Osofsky [303] have proved that
$S(A)$ is right-selfinjective. Furthermore, Osofsky showed that
every orthogonal system of idempotents from $S(A)$ can be lifted
in $E(A)$. With every system $\{e_i\}$ of orthogonal idempotents from
$E(A)$ there is associated a homomorphism $\varphi: M \to \Pi Me_i$, under
which $\varphi(m) = (me_i)$. If all such φ are endomorphisms, then $S(A)$ is
left-selfinjective. If $Z(A) = 0$, then the latter property is equivalent to module A being injective. A remark in [170, 171] is useful in the study of hereditary orders: the endomorphism ring of a
projective module over an hereditary order is hereditary. Let Λ
be commutative local ring and let A be a Λ-module of finite projective dimension, whose localizations by simple ideals P other
than maximal are free Λ_p-modules. Horrocks [191, 192] noted
that $E_\Lambda(A)$ is a completely primary ring.

Let $\Lambda^a = \Lambda/[\Lambda, \Lambda]$, where $[\Lambda, \Lambda]$ is a subgroup of the additive
group of the group of the ring Λ, generated by all commutators
$xy-yx$. For every endomorphism f of a finitely-generated pro-
jective Λ-module P we can define the trace $\text{Tr}_\Lambda(f) \in \Lambda^a$, and then
introduce the rank of module P as $r_\Lambda(P) = \text{Tr}_\Lambda(1p)$. Hattori [183]
shows that in a number of cases the isomorphism $P_1 \cong P_2$ is equivalent to the equality $r_\Lambda(P_1) = r_\Lambda(P_2)$.

Levich [24] considers a linear space V over a commutative
ring Φ and, taking a certain transformation, converts V into a
$\Phi[x]$-module in the usual manner. For this module he proves theorems analogous to theorems of primary Abelian groups (for example, Ulm's theorem). Along the way he gives a definition of a

simple Jordan module, coinciding in the commutative case with that of the subspace corresponding to a Jordan cell.

Let A, A_1, A_2 be Λ-modules, $A = A_1 \dotplus A_2$, $\alpha_{ij} \in \mathrm{Hom}_\Lambda(A_i, A_j)$, $i = 1, 2$, α_{11} an isomorphism. In this case Karták [209] observed that the (α_{ij}) yield a monomorphism (epimorphism, isomorphism) of module A if and only if $\varphi = \alpha_{22} - \alpha_{21}\alpha_{11}^{-1}\alpha_{12}$ is a monomorphism (epimorphism, isomorphism).

Yohe [417] ascertained the commutative rings over which certain very well-known properties of matrices over fields are valid. For example, over a Noetherian ring Λ every nilpotent matrix is similar to an upper triangular matrix if and only if Λ decomposes into a direct sum of commutative domains of principal ideals.

Matrix representations of endomorphism rings were considered by Villamayor [405]. The representation of algebras as the quotients of endomorphism rings was studied by Lafon [230]. Suzuki [386] showed that if an injective module A is such that $Ae \cong A$, where $e^2 = e \in E(A)$, implies $e = 1$, then any isomorphism of two submodules of the module can be continued up to an automorphism of this module. Gentile [155] proved that every submodule of a free Λ-module of rank n is closed relative to the taking of a double annihilator by a bilinear form $f\,|(a_i),\,(b_i)| = \sum\limits_{i=1}^{n} a_i b_i$ if and only if every submodule of the Λ_m-module of $(m \times n)$-matrices is closed relative to the taking of a double annihilator in $(n \times 1)$-matrices.

§ 7. Other Aspects

Ballet [65, 66] studied Artinian modules over commutative rings. Modules isomorphic to their own proper submodules were considered by Varela [401]. Charnow [100] studied the extension of the ground field Φ for $\Phi[x_1, \ldots, x_n]$-modules. Gentile [157] and Kreindler [225] were concerned with the representation of a module A over a ring Λ in the form $A \cong \Lambda/I_1 \oplus \ldots \oplus \Lambda/I_n$, where I_k are ideals in Λ and $\Lambda \supseteq I_1 \supseteq \ldots \supseteq I_n \supseteq 0$. Gentile proved the uniqueness of such a representation in the case of commutativity, without using exterior algebras. Kreindler proved its existence by assuming that Λ is a domain of principal ideals and of left ideals. Analogous results for orders of a finite-dimensional algebra were obtained by Knebusch [221]. Dauns [118] examined Loewy series.

Ravel [325] noted that submodules of a given module form a lattice
not only relative to an ordinary imbedding but also to an essential
one. This lattice also is a subject for research. Dlab [126-128]
constructed a general theory of linear dependence and applied it to
modules. Malliavin–Brameret [81, 256] ([42], p. 201) studied the
width of a module A, defined as the smallest integer n such that
every finitely-generated submodule of module A is generated by
$\leq$ n elements. Jouanlou [206] considered the depth $\mathrm{prof}_\Lambda(A)=$
$\min\{q,\ \mathrm{Ext}^q_\Lambda\ (\Lambda/\mathfrak{m},\ A)\neq 0\}$ of a finitely-generated module over a local
Noetherian ring Λ with a maximal ideal $\mathfrak{m}$. Rentschler and Grabriel
[337] defined a Krull dimension over any ring. In the case of Noetherian rings this coincides with the known dimension. For a
module A over a commutative Noetherian ring Λ in which $P_1, \ldots, P_s$
are minimal simple ideals, the introduction of the concept of a
length $l_\Lambda(A) = \Sigma l(A_{P_i})$, where $l(A_{P_i})$ is the length of module A_{P_i}
over an Artinian ring Λ_{P_i}, allowed Guerindon [165, 166] to define
a characteristic function of a graded module $A = \sum\limits_{n=0}^{\infty} A_n$ over a
graded Noetherian ring $\Lambda = \sum\limits_{n=0}^{\infty} \Lambda_n$ as $\sum\limits_{n=0}^{\infty} l_{\Lambda_0}(A_n) \cdot z^n$. The speciali-
zations of polynomial modules were examined by Northcott and
Reufel [292]. In another paper [293] they generalized the concept
of the length of a module. Let Φ be a commutative ring the spectrum of whose maximal ideals is Noetherian, let Λ be a finitely-
generated Φ-algebra, and let A be a finitely-generated Λ-module.
For simple ideals P of ring Φ we denote by $\mu(P, A)$ the minimal
number of generators of the Λ_p-modules A_p. In [388] Swan, with
the aid of $\mu(P, A)$ and of the dimensions of the simple ideals P,
being the intersection of the maximal ideals, estimated the number
of generators of the Λ-module A. Klein [219] looked at the so-
called Schur functions connected with the decomposition of a module
over a commutative domain of principal ideals into a direct sum
of cyclic ones.

Legrand [239] studied free modules of rank 2 over commutative rings equipped with a quadratic form.

Let Λ be a ring of characteristic 0 without divisors of zero,
$\tau: \Lambda \to \Lambda$ be a nonzero ring endomorphism, $\delta: \Lambda \to \Lambda$ be a τ-
differentiation (i.e., $\delta(\alpha\beta) = \tau(\alpha)\delta(\beta) + \delta(\alpha)\beta$, $\alpha, \beta \in \Lambda$), and A
be a Λ-module. Kreindler [226] studies the operators d: A $\to$ A
for which $d(\alpha a) = \tau(\alpha) d(a) + \delta(\alpha)$, $a, \alpha \in \Lambda$, $a \in A$. Nobusawa
[291] studies systems (V, f) consisting of the bimodule $_\Lambda V_\Lambda$ and

the bilinear mapping f: $V \times V \rightarrow \Lambda$. Vasconoolos [102, 404] found
the conditions under which A $\cong$ B ensues from the isomorphism of
Λ-modules $A \oplus C \cong B \oplus C$. If Λ is a finite-dimensional algebra
over a commutative ring Φ, then he also ascertained when $A \cong B$
follows from the isomorphism $A_p \cong B_p$ for every simple ideal P
of ring Φ.

Samuel [348-351] proved that a symmetric algebra S(A) of a
finitely-generated module A over a commutative Noetherian ring Λ
being factorial is equivalent to the ring Λ being factorial and to all
symmetric powers S^n(A) being reflexive ([42], p. 196). Reflexive
modules over graded rings are studied and it is ascertained to what
extent the class of reflexive modules is wider than the class of
projective ones. If $\Lambda = \sum_{n=0}^{\infty} \Lambda_n$ is a factorial graded ring, then the
ring Λ_0 turns out to be factorial while the A_0-modules Λ_n turn out
to be reflexive. Lazard [235] found the conditions on flat Λ-modules
A and B over a commutative ring Λ and a monomorphism φ: $A \rightarrow B$,
under which the induced mapping S(φ) of the corresponding powers
S(A) $\rightarrow$ S(B) of modules A and B is an imbedding.

At the beginning of a course of lectures on local algebra
Serre [38] proposed a theory of a primary decomposition of modules
over commutative rings and of Hilbert-Samuel polynomials (also
see [316]). Radu [320] showed that for the existence of a primary
decomposition for all Λ-modules it is sufficient that the finitely-
generated modules have a primary decomposition. To such rings
carry over the results on the completion of modules in an $\mathfrak{m}$-adic
topology, very well known in the Noetherian case. A unique presentation of the additive theory of ideals in rings, modules and
groupoids has been given by Andrunakievich and Ryabukhin [2]
(also see [1], § 1).

Modules over regular local rings were studied by Malliavin-
Brameret [253-258] and Lichtenbaum [244]. Modules over one-
dimensional Noetherian rings were considered by Andrews [56] and
Northover [294]. Closely related here are the investigations of
Ratliff [323]. Results concerning modules over Cohen-Macaulay
and Gorenstein rings may be found in the papers by Bass [71, 72],
Auslander [58], Iversen [194], Vasconcelos [403], Samuel [348-351],
and Nazarova and Roiter [30].

Now let Δ be a Noetherian integral domain, Γ its quotient
field, Σ a semisimple finite-dimensional Γ-algebra. We call a Δ-

algebra Λ an order if Λ is finitely generated over Δ and $\Gamma\Lambda = \Sigma$. Hereditary orders have been studied by Harada [170, 171, 177], and Drozd, Kirichenko, and Roiter [12, 13]. A systematic study of modules over orders (the multiplicative theory of modules of representations, questions connected with the finiteness of the number of nonisomorphic indecomposable finitely-generated projective modules or of modules belonging to one genus, etc.) was made by Faddeev [47, 48], Borevich and Faddeev [3, 4], Jacobiński [195], Drozd and Roiter [14], Gudivok [11], Roiter [34], Drozd and Turchin [15], Kirichenko [20], Reiner [327], and Brooks [83]. Various invariants of finitely-generated modules over orders were considered by Fröhlich [150]. Here we can mention also the paper [290] by Nobusawa.

Hattori [184] calls an algebra Λ over a commutative ring Φ semi-simple if every finitely-generated Λ-module is (Λ, Φ)-projective in the sense of Hochschild. If, furthermore, it is indecomposable and if there exist a finitely-generated Λ-module P, the elements $x_i \in P$, and homomorphism $\xi_i \in \mathrm{Hom}_\Lambda(P, \Lambda)$ such that $\sum_{i=1}^{n} \xi_i(x_i)=1$ and P does not contain nontrivial Λ-submodules which are Φ-direct summands in P, then it is called simple. In particular, Hattori found the conditions under which semisimple and separable algebras are simple.

Watanabe [408] examined a number of weaker conditions for the simplicity of an algebra. He noted that the property that an algebra be simple is a Morita-property (see p. 76). He looked at the case when Φ is a complete local ring separately. Endo and Watanabe [133] proved that the separability of an algebra Λ is equivalent to the separability of Λ_m over Φ_m for any maximal ideal m, and this in its own turn is equivalent to the separability of $\Lambda/m\Lambda$ over Φ/m by all m. Sugano [384] analyzed the extensions of a ring Γ, for which $\Lambda \oplus {}_\Gamma\Lambda$ is a finitely-generated projective Λ-bimodule, and also the extensions $\Lambda \supset \Gamma$ where every Λ-module is (Λ, Γ)-projective. Ferrand has considered the epimorphisms of rings and of separable algebras [143]. The papers by Kanzaki [207], Kishimoto [217], and DeMeyer [119] are devoted to the Galois theory of separable algebras. Murase has obtained a refinement of one separability criterion.

Wuytack and Depunt [415, 416] have considered the construction for modules, analogous to a sheaf of semigroups. Baumgartner

[73] ascertained when the isomorphism of modules which are rings implies a ring isomorphism. Smith [372] proved the existence of the so-called Chevalley base in a Lie module.

Nearmodules over nearings were considered by Plotkin [31] and Beidleman [75-77], and modules over pseudorings by Patterson [309]. Categories of semimodules over a certain semigroup were studied by Give'on [159].

Vol'vachev's note [6] is devoted to the elementary theory of modules. Reufel [338] examined the algorithmic aspects of the theory of modules over polynomial rings.

The very general discussion by Maranda in [259] is applied to the investigation of topological modules whose topologizations are defined by a filter of the ideals of the ground ring. Mascart [260] studied a topology on a module, connected with a certain bilinear form. Baker [60] proved a theorem on the solvability of a system of linear equations over a topological module. Further, if A is a topological Λ-module and $\mathfrak{P}$ is some property of its subsets, then the module A is called $\mathfrak{P}$-compact if each of its $\mathfrak{P}$-filters (i.e., a filter admitting of a base consisting of sets possessing property $\mathfrak{P}$) has an adherent point. Gouyon [163, 164] investigated this situation by proving, for example, the completeness of a $\mathfrak{P}$-compact module with a $\mathfrak{P}$-base of neighborhoods of zero. The results obtained were applied to the case when Λ is the complex number field and $\mathfrak{P}$ is convexity. Radu [317] considered the category of filtered modules, defined by a covering filtering $\{F_i\}$ on the ring Λ. If a Λ-module A is complete in a topology with the system $\{F_i A\}$ as a base of neighborhoods of zero and if $\cap F_i A = 0$, then it is called separable. Properties of the functor taking A into $\varprojlim (A/F_i A)$. were studied. The results obtained were applied to the case when $\{F_i\}$ consists of the powers of a certain ideal (cf. [240]).

Lugowski [247] discusses ordered semimodules.

BIBLIOGRAPHY

1.   V. A. Andrunakievich, V. I. Arnautov, and Yu. M. Ryabukhin, "Rings," in: R. V. Gamkrelidze (Ed.), Progress in Mathematics, Vol. 5: Algebra, Plenum Press, New York (1969), pp. 127-177.

2. V. A. Andrunakievich and Yu. M. Ryabukhin, "On the additive theory of ideals in rings, modules, and groupoids," Dokl. Akad. Nauk SSSR, 168(3):495-498 (1966).

3. Z. I. Borevich and D. K. Faddeev, "A remark on orders with a cyclic index," Dokl. Akad. Nauk SSSR, 164(4):727-728 (1965).

4. Z. I. Borevich and D. K. Faddeev, "Representations of orders with a cyclic index," Tr. Mat. Inst. Akad. Nauk SSSR, 80:51-65 (1965).

5. B. B. Venkov, "Homological algebra," in: Algebra, 1964 [in Russian], Science Summaries, VINITI, Akad. Nauk SSSR, Moscow (1966), pp. 203-235.

6. R. T. Vol'vachev, "Elementary theory of modules," Vestsi Akad. Nauk BSSR, Ser. Fiz.-Mat. Nauk, No. 4, 7-12 (1967).

7. V. I. Gemintern, "Semiprimary rings with an atomic structure of left ideals," Sibirsk. Mat. Zh., 8(4):947-951 (1967).

8. L. M. Gluskin, "Endomorphisms of modules," in: Algebra and Mathematical Logic [in Russian], Kiev Univ., Kiev (1966), pp. 3-20.

9. V. E. Govorov, "Derived radicals of modules," Mat. Zametki, 3(6):633-642 (1968).

10. E. L. Gorbachuk, "Splittability of torsion and pretorsion in the category of right λ-modules," Mat. Zametki, 2(6):681-688 (1967).

11. P. M. Gudivok, "Representations of finite groups over number rings," Izv. Akad. Nauk SSSR, Ser. Mat., 31(4):799-834 (1967).

12. Yu. A. Drozd and V. V. Kirichenko, "Hereditary orders," Ukrainsk. Mat. Zh., 20(2):246-248 (1968).

13. Yu. A. Drozd, V. V. Kirichenko, and A. V. Roiter, "Hereditary and Bass orders," Izv. Akad. Nauk SSSR, Ser. Mat., 31:1415-1436 (1967).

14. Yu. A. Drozd and A. V. Roiter, "Commutative rings with a finite number of indecomposable integral representations," Izv. Akad. Nauk SSSR, Ser. Mat., 31(4):783-798 (1967).

15. Yu. A. Drozd and V. M. Turchin, "The number of modules of representations in genus for second-order integral matrix rings," Mat. Zametki, 2(2):133-138 (1967).

16. V. P. Elizarov, "Nonsingularly dimensional rings," Sibirsk. Mat. Zh., 6(5):1181-1184 (1965).

17. V. P. Elizarov, "Quotient modules," Sibirstk. Mat. Zh., 7(1):221-226 (1966).

18. V. P. Elizarov, "Flat extensions of rings," Dokl. Akad. Nauk SSSR, 175(4):759-761 (1967).

19. M. Kol'p, "Growth rings," Sibirsk. Mat. Zh., 8(5):1193-1196 (1967).

20. V. V. Kirichenko, "Orders all of whose representations are completely decomposable," Mat. Zametki, 2(2):139-144 (1967).

21. L. A. Coifman, "A radical of a module," Sibirsk. Mat. Zh., 7(5):1204-1207 (1966).

22. V. I. Kuz'minov, "Derived functors of a projective limit," Sibirsk. Mat. Zh., 8(2):333-345 (1967).

23. C. W. Curtis and I. Reiner, Theory of Representations of Finite Groups and of Associative Algebras [Russian translation], Nauka, Moscow (1969), 668 pp.

24. E. M. Levich, "Modules over the ring of polynomials in one endomorphism,"
     Latvian Math. Annual, Vol. 2 [in Russian], Riga (1966), pp. 127-174.

25. S. A. Lang, Algebra [Russian translation], "Mir," Moscow (1968), 564 pp.

26. S. MacLane, Homology [Russian translation], "Mir," Moscow (1966), 543 pp.

27. A. P. Mishina and L. A. Skornyakov, Abelian Groups and Modules [in Russian],
     "Nauka," Moscow (1969), 151 pp.

28. A. V. Mikhalev, "Isomorphisms of endomorphism semigroups of modules,"
     Algebra i Logika. Seminar, 5(5):59-67 (1966).

29. A. V. Mikhalev, "Isomorphisms of endomorphism semigroups of modules. II,"
     Algebra i Lokika. Seminar, 6(2):35-47 (1967).

30. L. A. Nazarova and A. V. Roiter, "Refinement of a theorem of Bass," Dokl.
     Akad. Nauk SSSR, 176(2):266-268 (1967).

31. B. I. Plotkin, "Certain aspects of the general theory of group representations,"
     Izv. Akad. Nauk SSSR, Ser. Mat., 27(4):855-882 (1963).

32. I. S. Ponizovskii, "The Frobeniusness of the semigroup algebra of a finite com-
     mutative semigroup," Izv. Akad. Nauk SSSR, Ser. Mat., 32(4):820-836 (1968).

33. A. V. Roiter, "Analog of a theorem by Bass for the modules of representations
     of noncommutative orders," Dokl. Akad. Nauk SSSR, 168(6):1261-1264 (1966).

34. A. V. Roiter, "Integral representations belonging to one genus," Izv. Akad.
     Nauk SSSR, Ser. Mat., 30(6):1315-1324 (1966).

35. B. M. Rudyk, "Extensions of modules," Dokl. Akad. Nauk SSSR, 179(3):545-547
     (1968).

36. Yu. M. Ryabukhin, "Radicals in categories," in: Mathematical Researches,
     Vol. 2, No. 3 [in Russian], Kishinev (1967), pp. 107-165.

37. I. I. Sakhaev, "Exactness of functors," Izv. Vyssh. Uchebn. Zavedenii. Matematika,
     No. 4, 104-108 (1966).

38. J.-P. Serre, "Local algebra and the theory of multiplicities," Period. Sb. Perev.
     Statei, 7(5):3-93 (1963).

39. L. A. Skornyakov, "Rings," in: Algebra. Topology. 1962 [in Russian], Science
     Summaries, VINITI, Akad. Nauk SSSR, Moscow (1963), pp. 59-79.

40. L. A. Skornyakov, "Modules," in: Algebra. Topology. 1962 [in Russian], Science
     Summaries, VINITI, Akad. Nauk SSSR, Moscow (1963), pp. 80-89.

41. L. A. Skornyakov, "Cohn rings," Algebra i Logika. Seminar, 4(3):5-30 (1965).

42. L. A. Skornyakov, "Modules," in: R. V. Gamkrelidze (Ed.), Progress in Mathe-
     matics. Vol. 5: Algebra, Plenum Press, New York - London (1969), pp. 179-214.

43. L. A. Skornyakov, "Elizarov quotient ring and the localization principle," Mat.
     Zametki, 1(3):263-268 (1967).

44. L. A. Skornyakov, "Lectures on homological algebra," Mat. Vesn., 5(1):71-113
     (1968).

45. L. A. Skornyakov, "Homological classification of rings," Mat. Vesn., 4(4):415-
     434 (1967).

46. T. S. Tol'skaya, "Injection and freedom," Sibirsk. Mat. Zh., 6(5):1202-1207
     (1965).

47. D. K. Faddeev, "Introduction to the multiplicative theory of modules of integral
     representations," Tr. Mat. Inst. Akad. Nauk SSSR, 80:145-182 (1965).

48. D. K. Faddeev, "On the theory of cubic Z-rings," Tr. Mat. Inst. Akad. Nauk SSSR, 80:183-187 (1965).

49. V. G. Fayans, "Endomorphisms of ordered linear spaces," Herzen Lectures, XIX, Math. [in Russian], Leningrad. Gos. Ped. Inst., Leningrad (1966), pp. 10-12.

50. V. G. Fayans, "Endomorphism semigroups of ordered linear spaces," Uch. Zap. Leningrad. Gos. Ped. Inst. im. A. I. Gertsen [Herzen], 302:124-141 (1967).

51. G. M. Tsukerman, "Endomorphism rings of free modules," Sibirsk. Mat. Zh., 7(5):1161-1167 (1966).

52. T. Akiba, Remarks on generalized rings of quotients. Proc. Japan Acad., 40(10): 801-806 (1964).

53. T. Akiba, Remarks on generalized rings of quotients. II. J. Math. Kyoto Univ., 5(1):39-44 (1965).

54. J. S. Alin, "Structure of torsion modules," Doctoral dissertation, Univ. of Nebraska, (1967), 59 pp.; Dissert. Abstr., B28(7):2925-2926 (1968).

55. J. S. Alin and S. E. Dickson, Goldie's torsion theory and its derived functor. Pacif. J. Math., 24(2):195-203 (1968).

56. A. P. Andrews, Modules over a one-dimensional Noetherian ring. Quart. J. Math., 18(69):65-80 (1967).

57. J. T. Arnold and R. Gilmer, Idempotent ideals and unions of nets of Prüfer domains. J. Sci. Hiroshima Univ., Ser. A, Div. 1, 31(2):131-145 (1967).

58. M. Auslander, Remarks on a theorem of Bourbaki, Nagoya Math. J., 27(1):361-369 (1966).

59. G. Azumaya, Completely faithful modules and self-injective rings. Nagoya Math. J., 27(2):697-708 (1966).

60. A. C. Baker, Systems of linear equations over a topological module. Quart. J. Math., 15(60):327-336 (1964).

61. S. Balcerzyk, The global dimension of the group rings of Abelian groups. Fundam. math., 55(3):293-301 (1964).

62. S. Balcerzyk, The global dimension of the group rings of Abelian groups. II. Fundam. math., 58(1):67-73 (1966).

63. S. Balcerzyk, On projective dimension of direct limit of modules. Bull. Acad. Polon. sci. Sér. math., astron. et phys., 14(5):241-244 (1966).

64. S. Balcerzyk, On functors $\otimes_\Lambda$ and Hom $_\Lambda$. Bull. Acad. polon. sci. Sér. sci. math., astron. et phys., 15(4):235-238 (1967).

65. B. Ballet, Idéaux maximaux associés a un module artinien. C. r. Acad. sci., 263(21):A758-A761 (1966).

66. B. Ballet, Structure des modules artiniens (cas commutatif). C. r. Acad. sci., 266(1):A1-A4 (1968).

67. B. Banaschewski, On coverings of modules. Math. Nachr., 31(1-2):57-71 (1966).

68. B. Banaschewski, On the injective hulls of cyclic modules over Dedekind domains. Canad. Math. Bull., 9(2):183-186 (1966).

69. H. Bass, Finistic dimensions and a homological generalisation of semi-primary rings. Trans. Amer. Math. Soc. 95(3):466-488 (1960).

70. H. Bass, Torsion free and projective modules. Trans. Amer. Math. Soc., 102(2): 319-327 (1962).

71. H. Bass, On the ubiquity of Gorenstein rings. Math. Z., 82(1):8-28 (1963).

72. H. Bass, Anneaux de Gorenstein. Sémin. Dubreil et Pisot. Fac. Sci. Paris, 16(1):
    5/01-5/08 (1967).

73. K. Baumgartner, Bemerkungen zum Isomorphieproblem der Ringe. Monatsh.

74. H. B. Beckwith, "On trace for modules with applications to algebras," Doctoral.
    dissertation, Univ. California, San Diego (1967), 68 pp.; Dissert. Abstr., F.28(7):
    2926 (1968).

75. J. C. Beidleman, A radical for near-ring modules. Mich. Math. J., 12(3):3⁻7-
    383 (1965).

76. J. D. Beidleman, "On near-rings and near-ring modules," Doctoral. dissertation.,
    Pennsylvania State Univ. (1964), 178 pp.; Dissert. Abstr., 25(8):4616-4617
    (1965).

77. J. C. Beidleman, Strictly prime distributively generated near-rings. Math. Z.,
    100(2):97-105 (1967).

78. P. R. Bongale, Filtered Frobenius algebras. Math. Z., 97(4):320-325 (1967).

79. P. R. Bongale, Filtered Frobenius algebras. II. J. Algebra, 9(1):79-93 (1968).

80. P. R. Bongale, Filtered Quasi-Frobenius rings. Math. Z., 106(3):191-196 (1968).

81. M.-P. Brameret, "Largeurs d'anneaux et de modules," Doctoral dissertation, Univ.
    Paris (1965), 106 pp.

82. J. Brandenburg, "Über die Rechtsderivierten des inversen Limes von Modul- und
    Gruppenfamilien," Doctoral dissertation, Friedrich Wilhelms-Univ., Bonn; Bonn
    math. Schr., Vol. 11, No. 27 (1967), 65 pp.

83. J. O. Brooks, "Classification of representation modules over quadratic orders,"
    Doctoral dissertation, Univ. Michigan (1964), 64 pp.; Dissert. Abstr., 25(2,
    Part 1):7287-7288 (1965).

84. A. Brumer, Pseudocompact algebras, profinite groups and class formations. J.
    Algebra, 4(3):442-470 (1966).

85. A. Brumer, Pseudocompact algebras, profinite groups and class formation. Bull.
    Amer. Math. Soc., 72(2):321-324 (1966).

86. D. A. Buchsbaum, A note on homology in categories. Ann. Math., 69(1):66-74
    (1959).

87. L. Budach, Zum Begriff des Modulquotienten. Monatsber. Dtsch. Akad. Wiss.
    Berlin, 6(2):81-85 (1964).

88. R. T. Bumby, Modules which are isomorphic to submodules of each other. Arch.
    Math., 16(3):184-185 (1965).

89. L. Burch, A note on ideals of homological dimension one in local domains. Proc.
    Cambridge Philos. Soc., 63(3):661-662 (1967).

90. M. C. R. Butler, Torsion-free modules and diagrams of vector spaces. Proc.
    London Math. Soc., 18(4):635-652 (1968).

91. M. C. R. Butler and G. Horrocks, Classes of extensions and resolutions. Phil.
    Trans. Royal. Soc. London, 254:155-202 (1961).

92. H. S. Butts and W. Smith, Prüfer rings. Math. Z., 95(3):196-211 (1967).

93. H. S. Butts and L. I. Wade, Two criteria for Dedekind domains. Amer. Math.
    Monthly, 73(1):14-21 (1966).

94. W. H. Caldwell, "Rings for which cyclic modules have cyclic injective hulls,"
    Doctoral Dissertion, Rutgers State Univ. (1966), 60 pp.; Dissert. Abstrs.,
    B27(6):2015 (1966).

95. W. H. Caldwell, Hypercyclic rings. Pacif. J. Math., 24(1):29-44 (1968).

96. D. W. Cassel, "Rings over which projective modules are free," Doctoral dissertation, Syracuse Univ. (1967), 81 pp.; Dissert. Abstr., B28(4):1607-1608 (1967).

97. J.-Y. Chamard, Modules riches en co-irréductibles et sousmodules compléments. C. r. Acad. sci., 264(23):A987-A990 (1967).

98. R. E. Chandler and K. Koh, Applications of a function topology on rings with unit. Ill. J. Math. 11(4):580-585 (1967).

99. N. Chaptal, Sur les modules quasi-injectifs. C. R. acad. sci., 264(4):A173-A175 (1966).

100. A. Charnow, Extension of the ground fields in a $k[X_1, ..., X_n]$ module. J. London Math. Soc., 41(3):425-432 (1966).

101. S. U. Chase, Direct products of modules. Trans. Amer. Math. Soc., 97(3):457-473 (1960).

102. K.-L. Chew, "Extensions of rings and modules," Doctoral dissertation, Univ. Brit. Columbia (1965), 129 pp.; Dissert. Abstr., 26(8):4681 (1966).

103. B.-S. Chwe, Relative homological algebra and homological dimension of Lie algebras. Trans. Amer. Math. Soc., 117(5):477-493 (1965).

104. W. E. Clark, Algebras of global dimension one with a finite ideal lattice. Pacif. J. Math., 23(3):463-471 (1967).

105. W. E. Clark, A note on semiprimary PP-rings. Osaka J. Math., 4(1):177-178 (1967).

106. P. M. Cohn, On the free product of associative rings. Math. Z., 71(4):380-398 (1959).

107. P. M. Cohn, Factorization in non-commutative power series rings. Proc. Cambridge Philos. Soc., 58(3):452-464 (1962).

108. P. M. Cohn, Free ideal rings. J. Algebra, 1(1):47-49 (1964).

109. P. M. Cohn, Sur une classe d'anneaux héréditaires. Sémin. Dubreil et Pisot. Fac. Sci. Paris, 19(1):7/01-7/05 (1967).

110. P. M. Cohn, A remark on matrix rings over free ideal rings. Proc. Cambridge Philos. Soc., 62(1):1-4 (1966).

111. P. M. Cohn, Hereditary local rings. Nagoya Math. J., 27(1):223-230 (1966).

112. P. M. Cohn, Some remarks on the invariant basis property. Topology, 5(3):215-228 (1966).

113. P. M. Cohn, On the free product of associative rings. III. J. Algebra, 8(3):376-383 (1968).

114. R. R. Colby, "On indecomposable modules over rings with minimum condition," Doctoral dissertation, Univ. of Washington (1965), 55 pp.; Dissert. Abstr., 26(2):1059 (1965).

115. R. R. Colby, On indecomposable modules over rings with minimum condition. Pacif. J. Math., 19(1):23-33 (1966).

116. R. R. Colby and E. A. Rutter, Semi-primary QF-3 rings. Nagoya Math. J., 32:253-258 (June, 1968).

117. R. C. Courter, The maximal co-rational extension by a module. Canad. J. Math., 18(5):953-962 (1966).

118. J. Dauns, Chains of modules with completely reducible quotients. Pacif. J. Math., 17(2):235-242 (1966).

119. F. R. DeMeyer, The trace map and separable algebras. Osaka J. Math., 3(1):7-11 (1966).

120. R. Desq, Sous-modules isotypiques et sous-modules tertiaires d'un module. Sémin. Dubreil et Pisot. Fac. Sci. Paris, 17(2):25/01-25/13 (1967).

121. S. E. Dickson, Decomposition of modules. I. Classical rings. Math. Z., 90(1): 9-13 (1965).

122. S. E. Dickson, A torsion theory for Abelian categories. Trans. Amer. Math. Soc., 121(1):223-235 (1966).

123. S. E. Dickson, Noetherian splitting rings are Artinian. J. London Math. Soc., 42(4):732-736 (1967).

124. S. E. Dickson, Decomposition of modules. II. Rings without chain conditions. Math. Z., 104(5):349-357 (1968).

125. S. E. Dickson, Direct Decomposition of Radicals, Proc. Conf. Categorical Algebra, La Jolla, Calif., 1965, Springer, New York (1966), pp. 366-374.

126. V. Dlab, On the dependence relation over modules (Prelim. commun.). Comment. math. Univ. Carolinae, 6(1):115-117 (1965).

127. V. Dlab, Dependence over modules. Czech. Math. J., 16(1):137-157 (1966).

128. V. Dlab, The concept of rank and some related questions in the theory of modules. Comment. Math. Univ. Carolinae, 8(1):39-47 (1967).

129. S. Eilenberg and J. C. Moore, Foundations of relative homological algebra. Mem. Amer. Math. Soc., No. 55 (1965), 39 pp.

130. G. Eisenreich, Eine Dualitätsbeziehung zwischen S-Moduln. Math. Nachr., 34(5-6):351-359 (1967).

131. S. Endo, Projective modules over polynomial rings. J. Math. Soc. Japan, 15(3): 339-352 (1963).

132. S. Endo, Completely faithful modules and quasi-Frobenius algebras. J. Math. Soc. Japan, 19(4):437-456 (1967).

133. S. Endo and Y. Watanabe, On separable algebras over a commutative ring. Osaka J. Math., 4(2):233-242 (1967).

134. C. Faith, Rings with ascending condition on annihilators. Nagoya Math. J., 27(1):179-191 (1966).

135. C. Faith, On Köthe rings. Math. Ann., 164(3):207-212 (1966).

136. C. Faith, A general Wedderburn theorem. Bull. Amer. Math. Soc., 73(1):65-67 (1967).

137. C. Faith, Lectures on Injective Modules and Quotient Rings (1967).

138. C. Faith and E. A. Walker, Direct-sum representations of injective modules. J. Algebra, 5(2):203-221 (1967).

139. E. H. Feller, A type of quasi-Frobenius ring. Canad. Math. Bull., 10(1):19-27 (1967).

140. E. H. Feller and E. W. Swokowski, Prime modules. Canad. J. Math., 17(6):1041-1052 (1965).

141. E. H. Feller and E. W. Swokowski, Semi-prime modules. Canad. J. Math., 18(4): 823-831 (1966).

142. D. Ferrand, Sur les modules qui sont limite projective de leurs localisés. C. r. Acad. sci., AB262(11):A609-A611 (1966).

143. D. Ferrand, Epimorphismes d'anneaux et algebres séparables. C. r. Acad. sci., 265(15):A411-A414 (1967).

144. M. Flamant, Indice d'une forme quadratique définie sur un module projectif de
     rang fini sur un anneau de Dedekind. C. r. Acad. sci., 261(18):3513-3515 (1965).
145. I. Fleischer, A new construction of the injective hull. Canad. Math. Bull., 11(1):
     19-21 (1968).
146. J. Fort, Contribution a l'étude des éléments tertiaires et isotypiques dans les
     modules et les L-algebres. Bull. Soc. math. France, Vol. 92, no. 1, suppl.,
     99 pp. (1964).
147. J. Fort, Sommes directes de sous-modules co-irreductibles d'un module. C. r.
     Acad. sci., A262(22):1239-1242 (1966).
148. J. Forth, Sommes directes de sous-modules co-irréductibles d'un module. Math.
     Z., 103(5):363-388 (1968).
149. P. Freyd, Abelian Categories. An Introduction to the Theory of Functors, Harper,
     New York (1964), 164 pp.
150. A. Fröhlich, Invariants for modules over commutative separable orders. Quart. J.
     Math., 16(63):193-232 (1965).
151. K. R. Fuller, "On the structure of QF-2 and QF-3 rings," Doctoral dissertation
     Univ. Oregon (1967), 85 pp.; Dissert. Abstr., B28(6):2518 (1967).
152. P. Gabriel and U. Oberst, Spectralkategorien und reguläre Ringe in von-Neumann-
     schen Sinn. Math. Z., 92:389-395 (1966).
153. A. Gailleau, Anneau associé a un module injectif riche en co-irréductibles. C. r.
     Acad. sci., 264(24):A1040-A1042 (1967).
154. A. Gailleau, Anneau associé a un module injectif riche en co-irréductibles.
     Sémin. Dubreil et Pisot. Fac. sci. Paris, 20(2):19/01-19/10 (1968).
155. E. R. Gentile, Homomorphismes et fermeture algebrique des modules a coef-
     ficients dans un anneau associatif. Publs. Fac. cienc. fisicomat. Univ. nac. La
     Plata, No. 213, pp. 191-200 (1966).
156. E. R. Gentile, A uniqueness theorem on rings of matrices. J. Algebra, 6(1):131-
     134 (1967).
157. E. R. Gentile, A note on the uniqueness of the invariant factors. Rev. Union mat.
     argent. y Aroc. fis. argent., 23(1):31-34 (1966).
158. L. Gewirtzman, Anti-isomorphisms of the endomorphism rings of torsion-free
     modules. Math. Z., 98(5):391-400 (1967).
159. Y. Give'on, Categories of semimodules: the categorical structural properties of
     transition systems. Math. Syst. Theor., 1967, 1(1):67-78 (1967).
160. A. W. Goldie, Localization in non-commutative Noetherian rings. J. Algebra,
     5(1):89-105 (1967).
161. A. W. Goldie, A note on non-commutative localization. J. Algebra, 8(1):41-44
     (1968).
162. N. S. Gopalakrishnan and R. Sridharan, Homological dimension of Ore-extensions.
     Pacif. J. Math., 19(1):67-75 (1966).
163. L. Gouyon, Sur le choix de certaines "P-compacités". C. r. Acad. sci., 263(10):
     A345-A347 (1966).
164. L. Gouyon, Sur le choix de quelques topologies sur un espace vectoriel ou un
     module. Ann. Fac. sci. Univ. Toulouse sci. math. et sci. phys., 29:17-34 (1965).
165. J. Guerindon, Sur la fonction caractéristique d'un module gradué. Bol. Soc. mat.
     São Paulo, 16(1-2):115-127 (1965).

166. J. Guerindon, Quelques applications des modules gradues. Sémin. Dubreil et Pisot. Fac. Sci. Paris., 17(1):4/01-4/11 (1967).

167. R. M. Hamsher, Commutative, noetherian rings over which every module has a maximal submodule. Proc. Amer. Math. Soc., 17(6):1471-1472 (1966).

168. R. M. Hansher, Commutative rings over which every module has a maximal submodule. Proc. Amer. Math. Soc., 18(6):1133-1137 (1967).

169. A. Hanna, Dualität in Moduln. J. reine und angew. Math., 224:113-117 (1966); correction, ibid., 228:220 (1967).

170. M. Harada, Multiplicative ideal theory in hereditary orders. J. Math. Osaka City Univ., 14(2):83-106 (1963).

171. M. Harada, Hereditary orders which are dual. J. Math. Osaka City Univ., 14(2): 107-115 (1963).

172. M. Harada, On semi-primary PP-rings. Osaka J. Math., 2(1):153-161 (1965).

173. M. Harada, Note on quasi-injective modules. Osaka J. Math., 2(2):351-356 (1965).

174. M. Harada, QF-3 and semi-primary PP-rings. I. Osaka J. Math., 2(2):357-368 (1965).

175. M. Harada, QF-3 and semi-primary PP-rings. II. Osaka J. Math., 3(1):21-27 (1966).

176. M. Harada, Hereditary semi-primary rings and triangular matrix rings. Nagoya Math. J., 27(2):463-484 (1966).

177. M. Harada, Note on orders over which an hereditary order is projective. Osaka J. Math., 4(1):151-156 (1967).

178. T. Harase, Adjoint pairs of functors on abelian categories. J. Fac. Sci., Univ. Tokyo, Sec. 1, 13(2):175-182 (1966).

179. M. E. Harris, Some results on coherent rings. Proc. Amer. Math. Soc., 17(2): 474-479 (1966).

180. R. Hart, Endomorphisms of modules over semi-prime rings. J. Algebra, 4(1):46-51 (1966).

181. R. Hart, Simple rings with uniform right ideals. J. London Math. Soc., 42(4): 614-617 (1967).

182. R. E. Harte, A theorem of isomorphism. Proc. London Math. Soc., 16(4):753-765 (1966).

183. A. Hattori, Rank element of a projective module. Nagoya Math. J., 25:113-125 (March, 1965).

184. A. Hattori, Simple algebras over a commutative ring. Nagoya Math. J., 27(2): 611-616 (1966).

185. A. Heller, Homological algebra in Abelian categories. Ann. Math., 68(3):484-525 (1958).

186. A. Heller, Homological functors. Math. Z., 87(2):283-298 (1965).

187. G. A. Helzer, "On divisibility and injectivity," Doctoral dissertation, Northwestern Univ. (1964), 70 pp.; Dissert. Abstr., 25(11):6656-6657 (1965).

188. G. A. Helzer, On divisibility and injectivity. Canad. J. Math., 18(5):901-919 (1966).

189. G. Hochschild, Relative homological algebra. Trans. Amer. Math. Soc., 82(1): 246-269 (1956).

190. G. Horrocks, Projective modules over an extension of a local ring. Proc. London Math. Soc., 14(56):714-718 (1964).

191. G. Horrocks, Vector bundles on the punctured spectrum of a local ring. Proc. London Math. Soc., 14(56):689-713 (1964).

192. G. Horrocks, An application of projective homotopy theory to reduction of base. Proc. London Math. Soc., 16(1):40-52 (1966).

193. T. Ishikawa, On injective modules and flat modules. J. Math. Soc. Japan, 17(3): 291-296 (1965).

194. B. Iversen, On flat extension of Noetherian rings. Proc. Amer. Math. Soc., 16(6): 1401-1406 (1965).

195. H. Jacobiński, Sur les ordres commutatifs avec un nombre fini de réseaux indécomposables. Acta math., 118(1-2):1-31 (1967).

196. J. P. Jans, Projective injective modules. Pacif. J. Math., 9(4):1103-1108 (1959).

197. J. P. Jans, Rings and Homology, Holt, Rinehart and Winston, New York (1964), 88 pp.

198. J. P. Jans, Some aspects of torsion. Pacif. J. Math., 15(4):1249-1259 (1965).

199. J. P. Jans, On orders in quasi-Frobenius rings. J. Algebra, 7(1):35-43 (1967).

200. J. P. Jans, A note on injectives. Math. Ann., 175(3):239-242 (1968).

201. C. U. Jensen, On characterizations of Prüfer rings. Math. Scand., 13(1):90-98 (1963).

202. C. U. Jensen, A remark on flat and projective modules. Canad. J. Math., 18(5): 943-949 (1966).

203. C. U. Jensen, A remark on semi-hereditary local rings. J. London Math. Soc., 41(3):479-482 (1966).

204. C. U. Jensen, On homological dimensions of rings with countably generated ideals. Math. Scand., 18(2):97-105 (1966).

205. C. U. Jensen, Homological dimensions of $\aleph_0$-coherent rings. Math. scand., 20(1): 55-60 (1967).

206. J.-P. Jouanlou, Sur la profondeur des modules de type fini. C. r. Acad. sci., 258(8):2245-2247 (1964).

207. T. Kanzaki, On commutator rings and Galois theory of separable algebras. Osaka J. Math., 1(1):103-115 (1964).

208. I. Kaplansky, The homological dimension of a quotient field. Nagoya math. J., 27(1):139-142 (1966).

209. K. Karták, On endomorphisms of the direct sum of two modules. Casop. pěstov. mat., 93(1):117-120 (1968).

210. F. Kasch and E. A. Mares, Eine Kennzeichnung semi-perfecter Moduln. Nagoya Math. J., 27(2):525-529 (1966).

211. T. Kato, Duality of cyclic modules. Tôhoku Math. J., 19(3):349-356 (1967).

212. T. Kato, Self-injective rings. Tôhoku Math. J., 19(4):485-495 (1967).

213. T. Kato, Some generalizations of QF-rings. Proc. Japan Acad., 44(3):114-119 (1968).

214. T, Kato, Torsionless modules. Tôhoku Math. J., 20(2):234-243 (1968).

215. S. M. Kaye, Ring theoretic properties of matrix rings. Canad. math. Bull., 10(3): 365-374 (1967).

216. R. Kielpiński, On I'-pure injective modules. Bull. Acad. polon. sci. Sér sci. math. astron. et phys., 15(3):127-131 (1967).

217. K. Kishimoto, On zeros of polynomials and Galois extensions of simple rings. J. Fac. Sci. Shinshu Unov., 2(2):117-122 (1967).

218. K. Kishimoto, T. Onodera, and H. Tominaga, On the normal basis theorems and the extension dimension. J. Fac. Sci. Hokkaido Univ., Ser. 1, 18(1-2):8188 (1964).

219. T. Klein, The multiplication of Schur-functions and extensions of p-modules. J. London Math. Soc., 43(2):280-284 (1968).

220. I. Kleiner, Free and injective Lie modules. Canad. Math. Bull., 9(1):29-42 (1966).

221. M. Knebusch, Flementarteilertheorie über Maximalordnungen. J. reine und angew. Math., 226:175-183 (1967).

222. K. Koh, On simple rings with maximal annihilator right ideals. Canad. Math. Bull., 8(5):667-668 (1965).

223. K. Koh, On some characteristic properties of self-injective rings. Proc. Amer. Math. Soc., 19(1):209-213 (1968).

224. K. Koh, On a semiprimary ring. Proc. Amer. Math. Soc., 19(1):205-208 (1968).

225. E. Kreindler, Remarques sur le théorème des facteurs invariants. Bull. Math. Soc. sci. mat. RSR, 11(2):181-186 (1967).

226. E. Kreindler, Opérateurs pseudo-linéaires dans les modules unitaires. Rev. roumaine math. pures et appl., 13(2):201-234 (1968).

227. R. S. Kulkarni, On a theorem of Jensen. Amer. Math. Monthly, 74(8):960-961 (1967).

228. J.-P. Lafon, Spectre primier bilatère de l'anneau des endomorphismes d'un module de type fini. C. r. Acad. sci., 262(20):A1098-A1099 (1966).

229. J.-P. Lafon, Spectre premier bilatère de l'anneau des endomorphismes d'un module de type fini. Bull Soc. Math. France, 94(4):269-275 (1967).

230. J.-P. Lafon, Représentation d'algèbres comme quotients d'anneaux d'endomorphismes. Bull sci. math., 91(1-2):13-16 (1967).

231. T.-Y. Lam, The category of Noetherian modules. Proc. Nat. Acad. Sci. USA, 55(5):1038-1040 (1966).

232. J. Lambek, Lectures on rings and modules, Blaisdell, Waltham (Mass.), (1966), 183 pp.

233. M. L. Laplaza, "Sobre la ordenacion de modulos," Actas 5 reun. anual mat. esp. Valencia, 1964; Madrid (1967), pp. 221-224.

234. M. D. Larsen, Equivalent conditions for a ring to be a P-ring and a note on flat overrings. Duke Math. J., 34(2):273-280 (1967).

235. D. Lazard, Algèbre symètrique et platitude. C. r. Acad. sci., 264(5):A228-A230 (1967).

236. D. Lazard, Epimorphismes plats d'anneaux. C. r. Acad. sci., 266(6):A314-A316 (1968).

237. W. G. Leavitt, The module type of homomorphic images. Duke Math. J., 32(2):305-311 (1965).

238. W. G. Leavitt, Type radicals. Glasgow Math. J., 9(1):22-29 (1968).

239. D. Legrand, Formes quadratiques et algèbres quadratiques. C. r. Acad. Sci., 265(23):A764-A767 (1967).

240. C. Lemaire, Sur le complété d'un anneau-module topologique. Bull. cl. sci. Acad. roy. Belg., 52(3):390-394 (1966).

241. W. W. Leonard, Small modules. Proc. Amer. Math. Soc., 17(2):527-531 (1966).

242. L. S. Levy, Commutative rings whose homomorphic images are selfinjective. Pacif. J. Math., 18(1):149-153 (1966).

243. L. S. Levy, Decomposing pairs of modules. Trans. Amer. Math. Soc., 122(1):64-80 (1966).

244. S. Lichtenbaum, On the vanishing of Tor in regular local rings. III. J. Math., 10(2):220-226 (1966).

245. D. Lissner, Outer product rings. Trans. Amer. Math. Soc., 116(4):526-535 (1965).

246. D. Lissner, OP-rings and Seshadri's theorem. J. Algebra, 5(3):362-366 (1967).

247. H. Lugowski, Die Charakterisierung gewisser geordneter Halbmoduln mit Hilfe der Erweiterungstheorie. Publs. math., 13(1-4):237-248 (1966).

248. J. C. McConnel, Localisation in enveloping rings. J. London Math. Soc., 43(3): 421-428 (1968).

249. I. G. Macdonald, Duality over complete local rings. Topology, 1:213-235 (July, 1962).

250. S. MacLane, Homology, Springer, Berlin (1963), 422 pp.

251. B. H. Maddox, "Absolutely pure modules," Doctoral dissertation, Univ. of South Carolina (1965), 31 pp.; Dissert. Abstr., 25(11):6658-6659 (1965).

252. B. H. Maddox, Absolutely pure modules. Proc. Amer. Math. Soc., 18(1):155-158 (1967).

253. M.-P. Malliavin, Modules sur les anneaux locaux réguliers. C. r. Acad. sci., 262(13):A736-A739 (1966).

254. M.-P. Malliavin, Modules sur les anneaux locaux réguliers non ramifiés. C. r. Acad. sci., 262(14):A811-A812 (1966).

255. M.-P. Malliavin-Brameret, Modules sur les anneaux locaux réguliers. Bull. Soc. math. France, 94(4):261-268 (1966).

256. M.-P. Malliavin-Brameret, Largeurs d'anneaux et de modules. Bull Soc. math. France, 1966, 94(8):3-76 (1967).

257. M.-P. Malliavin-Brameret, Modules de longueur finie sur un anneau régulier. C. r. Acad. sci., 264(3):A91-A94 (1967).

258. M.-P. Malliavin-Brameret, Sous-anneaux non ramifiés d'anneaux réguliers. C. r. Acad. Sci., 264(9):A385-A387 (1967).

259. J.-M. Maranda, Completions of modules and rings. Trans. Roy. Soc. Canada, Sec. 1-3, Vol. 3, pp. 271-291 (1965).

260. H. Mascart, Sur l'utilisation de la polarité dans la théorie des modules topologiques. C. r. Acad. sci., 262(1):A16-A19 (1966).

261. E. Matlis, Modules with descending chain condition. Trans. Amer. Math. Soc., 97:495-508 (1960).

262. E. Matlis, Cotorsion modules. Mem. Amer. Math. Soc., No. 49 (1964), 66 pp.

263. E. Matlis, Decomposable modules. Trans. Amer. Math. Soc., 125(1):147-179 (1966).; correction, ibid., 134:315-324 (1968).

264. E. Matlis, Reflexive domains. J. Algebra, 8(1):1-33 (1968).

265. C. K. Megibben, Modules over an incomplete discrete valuation ring. Proc. Amer. Math. Soc., 19:450-452 (1968).

266. G. Michler, Charakterisierung einer Klasse von Noetherschen Ringen. Math. Z., 100(2):163-182 (1967).

267. B. Mitchell, Theory of Categories, Academic Press, New York (1965), 273 pp.

268. B. Mitchell, On the dimension of objects and categories. I. Monoids. J. Algebra, 9(3):314-340 (1968).

269. B. Mitchell, On the dimension of objects and categories. II. Finite ordered sets. J. Algebras, 9:341-368 (1968).

270. Y. Miyashita, Quasi-projective modules, perfect modules and a theorem for modular lattices. J. Fac. Sci. Hokkaido Univ., Ser. 1, 19(2):86-100 (1966).

271. T. Miyata, Note on direct summands of modules. J. Math. Kyoto Univ., 7:65-69 (1967).

272. H. Y. Mochizuki, On the double commutator algebra of QF-3 algebras. Nagoya Math. J., 25:221-230 (March, 1965).

273. K. Morita, Duality for modules and its applications to the theory of rings with minimum condition. Sci. Repts. Tokyo Kyoiku Daigaku, A6:83-142 (May 15, 1958).

274. K. Morita, Adjoint pairs functors and Frobenius extensions. Sci. Repts. Tokyo Kyolku Daigaku, A9(202-208):40-71 (1965).

275. K. Morita, On S-rings in the sense of F. Kasch. Nagoya Math. J., 27(2):687-695 (1966).

276. K. Morita, The endomorphism ring theorem for Frobenius extensions. Math. Z., 102(5):385-404 (1967).

277. K. R. Mount, On polynomial rings. Mich. Math. J., 13(2):161-163 (1966).

278. B. Muller, Quasi-Frobenius Erweiterungen. I. Math. Z., 85(4):345-368 (1964).

279. B. Muller, Quasi-Frobenius Erweiterungen. II. Math. Z., 88(5):380-409 (1965).

280. B. Muller, Über vollständige (Ko-) Homologie. Math. Ann., 165(3):223-235 (1966).

281. B. Muller, On algebras of dominant dimension one. Nagoya Math. J., 31:173-183 (1968).

282. B. Muller, The classification of algebras by dominant dimension. Canad. J. Math., 20(2):398-409 (1968).

283. I. Murase, On the derivations of a quasi-matrix algebra. Scient. Papers Coll. Gen. Educ. Univ. Tokyo, 14(2):157-164 (1964).

284. M. P. Murthy, Projective modules over a class of polynomial rings. Math. Z., 88(2):184-189 (1965).

285. M. P. Murthy, Projective A(X)-modules. J. London Math. Soc., 41(3):453-456 (1966).

286. M. Narita, Weak topologies and injective modules. Proc. Japan. Acad., 43(1): 6-10 (1967).

287. C. Năstăsescu, Sur une classe d'anneaux. C. r. Acad. sci., 266(19):A966-A969 (1968).

288. C. Năstăsescu and N. Popescu, Sur la structure des objets de certaines categories abéliennes. C. r. Acad. sci., AB262(24):A1295-A1297 (1966).

289. C. Năstăsescu and N. Popescu, Quelques observations sur les topos abéliens. Rev. roumaine math. pures et appl., 12(4):551-563 (1967).

290. N. Nobusawa, On lattices in a module over a matrix algebra. Canad. Math. Bull., 9(1):57-61 (1966).

291. N. Nobusawa, Structure theorems of a module over a ring with a bilinear mapping. Canad. Math. Bull., 10(5):649-652 (1967).

292. D. G. Northcott and M. Reufel, Contributions to the specialization theory of polynomial modules. Proc. Roy Soc., A281(1386):291-309 (1964).

293. D. G. Northcott and M. Reufel, A generalization of the concept of length. Quart. J. Math., 16(64):297-321 (1965).

294. A. P. Northover, Modules over a one-dimensional local ring. Quart. J. Math., 17(67):195-210 (1966).

295. J.-P. Oliver, Anneaux absolument plats universles et épimorphismes d'anneaux. C. r. Acad. sci., 266(6):A317-A318 (1968).

296. T. Onodera, Some studies on projective Frobenius extensions. J. Fac. Sci. Hokkaido Univ., Ser. 1, 18(1-2):89-107 (1967).

297. B. L. Osofsky, A counter example to a lemma of Skornjakov. Pacif. J. Math., 15(3):985-987 (1965).

298. B. L. Osofsky, Cyclic injective modules of full linear rings. Proc. Amer. Math. Soc., 17(1):247-253 (1966).

299. B. L. Osofsky, A generalization of quasi-Frobenius rings. J. Algebra, 4:373-387 (1966); correction, ibid.. 8:41-44 (1968).

300. B. L. Osofsky, Global dimension of valuation rings. Trans. Amer. Math. Soc., 127(1):136-149 (1967).

301. B. L. Osofsky, A non-trivial ring with non-rational injective hull. Canad. Math. Bull., 10(2):275-282 (1967).

302. B. L. Osofsky, Upper bounds on homological dimensions. Nogoya Math. J., 32: 315-322 June, (1968).

303. B. L. Osofsky, Endomorphisms rings of quasi-injective modules. Canad. J. Math., 20(4):895-903 (1968).

304. B. L. Osofsky, Noncommutative rings whose cyclic modules have cyclic injective hulls. Pacif. J. Math., 25:331-340 (1968).

305. B. L. Osofsky, Homological dimension and the continuum hypothesis. Trans. Amer. Math. Soc., 132:217-230 (1968).

306. B. Pareigis, Einige Bemerkungen über Frobenius Erweiterungen. Math. Ann., 153(1):1-13 (1964).

307. B. Pareigis, "Radikale und kleine Moduln," Sitzungsber. Bayer. Akad. Wiss. München, Math.-naturwiss. Kl. (1965), München (1966), pp. 185-199.

308. B. Pareigis, Vergessende Funktoren und Ringhomomorphismen. Math. Z., 93(4): 265-275 (1966).

309. E. M. Patterson, The Jacobson radical of a pseudo-ring. Math. Z., 89(4):348-364 (1965).

310. R. E. Peinado, Una clasificación de anillos. Rev. mat. hisp.-amer., 25(6):249-261 (1965).

311. R. E. Peinado and W. G. Leawitt, The maxit and minit of a ring. Proc. Glasgow Math. Assoc., 7:128-135 (1966).

312. J. C. Pleasant, "Certain relations between objects and morphisms in arbitrary categories and module categories," Doctoral dissertation, Univ. of South Carolina (1965), 51 pp.; Dissert. Abstr., 26(8):4697 (1966).

313. N. Popescu and A. Radu, La structure des modules injectifs sur un anneau à ideal principal. Bull. Math. Soc. sci. math. et phys. RPR, 8(1-2):67-73 (1964).

314. Y. Quentel, Sur une caractérisation des anneaux de valuation de hauteur. C. r. Acad. Sci., 265(21):A659-A661 (1967).

315. Y. Quentel, Sur une caractérisation des anneaux semihérédiaires. commutatifs. C. r. Acad. Sci., No. 5, pp. A266-A267 (1968).

316. A. Radu, Funcţia caracteristică a lui Hilbert. Studii şi cercetări mat. Acad. RSR, 19(4):537-547 (1967).

317. N. Radu, Asupra functorului de completare, Studii şi cercetări mat. Acad. RSR, 17(7):1077-1087 (1965).

318. N. Radu, Sur les anneaux coherénts laskériens. Rev. roumaine de math. pures et appl., Acad. RSR, 11(7):865-867 (1966).

319. N. Radu, Module de S-cotorsiune. An. Univ. Bucureşti. Ser. ştiinţ. natur. Mat. mecan., 15(2):25-32 (1966).

320. N. Radu, Sur la décomposition primaire des modules. Bull math. Soc. sci. math. RSR, 10(1-2):143-149 (1966).

321. N. Radu, Asupra unor categorii de module. Studii şi cercetări mat. Acad. RSR, 19(3):393-402 (1967).

322. C. F. F. Raggi, Localizaction en anillos no commutativos. An. Inst. mat. Univ. nac. autónoma México, 6:1-6 (1966).

323. L. J. Ratliff, On finite modules over a Noetherian domain. III. J. Math., 10(1): 56-60 (1966).

324. J. Ravel, Objects spéciaux dans certaines catégories. C. r. Acad. sci., 264(5): 226-227 (1967).

325. J. Ravel, Extensions essentielles dans un module, C. r. Acad. sci., 264(15): A657-A659 (1967).

326. M. Raynaud, Solution d'un problème universel relatif aux modules projectifs de type fini. C. r. Acad. sci., 260(17):4391-4394 (1965).

327. I. Reiner, Module extensions and blocks. J. Algebra, 5(2):157-163 (1967).

328. G. Renault, Etude des sous-modules compléments dans un A-module. Sémin. Dubreil et Pisot. Fac. Sci. Paris, 16(2):16/01-16/12 (1967).

329. G. Renault, Anneaux self-injectifs. Sémin. Dubreil et Pisot. Fac. Sci. Paris, 19(1):11/01-11/04 (1967).

330. G. Renault, Etude des anneaux réduits et de leurs enveloppes injectives. C. r. Acad. sci., 264(2):A53-A55 (1967).

331. G. Renault, Sur les anneaux A tels que tout A-module a gauche non nul contient un soud-module maximal. C. r. Acad. sci., 264(14):A622-A624 (1967).

332. G. Renault, Anneaux associé à un module injectif. C. r. Acad. sci., 264(26): A1163-A1164 (1967).

333. G. Renault, "Etude des sous-modules compléments dans un module," Bull. Soc. Math. France, Vol. 95, Mem. No. 9, 79 pp. (1967).

334. G. Renault, Anneaux réduits non commutatifs. J. Math. pures et appl., 46(2): 203-214 (1967).

335. G. Renault, Anneau associé a un module injectif. Bull. sci. math., 92(1-2):53-58 (1968).

336. R. Rentschler, Eine Bemerkung zu Ringen mit Minimalbedingung für Hauptideale. Arch. Math., 17(4):298-301 (1966).

337. R. Rentschler and P. Gabriel, Sur la dimension des anneaux et ensembles ordonnés. C. r. Acad. sci., 265(22):A712-A715 (1967).

338. M. Reufel, Konstruktionsverfahren bei Modulnüber Polynomringen. Math. Z., 90(3):231-250 (1965).

339. P. Ribenboim, On the completion of a valuation ring. Math. Ann., 155(5):392-396 (1964).

340. P. Ribenboim, On ordered modules. J. reine und angew. Math., 225:120-146 (1967); correction, ibid., 238:132-134 (1969).

341. F. Richman, Generalized quotient rings. Proc. Amer. Math. Soc., 16(4):794-799 (1965).

342. F. Richman and E. A. Walker, Cotorsion free, an example of relative injectivity. Math. Z., 102(2):115-117 (1967).

343. M. A. Rieffel, A general Wedderburn theorem. Proc. Nat. Acad. Sci. USA, 54(6): 1513 (1965).

344. "Ringe und Moduln," Tagungsbericht 27 Feb.-5 März, 1966, Math. Forschungsinst., Oberwolfach (1966), 14 pp.

345. J.-E. Roos, Caractérisation des catégories qui sont quotients de catégories de modules par des sous-catégories bilocalisants. C. r. Acad. sci., 261(23):4954-4957 (1965).

346. A. Roy, A note on filtered rings. Arch. Math., 16(6):421-427 (1965).

347. E. A. Rutter, Jr., A remark concerning quasi-Frobenius rings. Proc. Amer. Math. Soc., 16(6):1372-1373 (1965).

348. P. Samuel, Anneaux gradués factoriels et modules réflexifs. Bull. Soc. math. France, 92(2):237-249 (1964).

349. P. Samuel, Modules réflexifs et anneaux factoriels. Colloq. internat. Centre nat. rech. scient., 143:219-223 (1966).

350. P. Samuel, "Modules réflexifs et anneaux factoriels," in: Les Tendances Géométriques en Algèbre et Théorie des Nombres (1966), pp. 219-223.

351. P. Samuel, Modules réflexifs et anneaux factoriels. Sémin. Dubreil et Pisot. Fac. Sci., Paris, 17(1):8/01-8/04 (1967).

352. D. F. Sanderson, A generalization of divisibility and injectivity in modules. Canad. Math. Bull., 8(4):505-513 (1965).

353. F. L. Sandomierski, "Relative injectivity and projectivity," Doctoral dissertation, Pennsylvania State Univ. (1964) 67 pp.; Dissert. Abstr., 25(11):6664 (1965).

354. F. L. Sandomierski, Nonsingular rings. Proc. Amer. Math. Soc., 19(1):225-230 (1968).

355. D. A. Sankowsky, "Relative homological dimensions of rings and modules," Doctoral dissertation, Univ. California, Berkeley (1967), 44 pp. Dissert. Abstr., B28(10):4206 (1968).

356. M. Satyanarayana, Rings with primary ideals as maximal ideals. Math. Scand., 20(1):52-54 (1967).

357. M. Satyanarayana, Semisimple rings. Amer. Math. Monthly, 74(19):1086 (1967).

358. M. Satyanarayana, Characterization of local rings. Tohoku Math. J., 19(4):411-416 (1967).

359. J.-P. Serre, Sur les modules projectifs, Sémin P. Dubreil, M-L Dubreil-Jacotin et C. Pisot; Fac. sci. Paris, 1960-1961, 14 année, fasc. 1, Paris 2/01-2/16 (1963).

360. C. S. Seshadri, Triviality of vector bundles over the affine space K^2. Proc. Nat. Acad. Sci. USA, 44(5):456-458 (1958).

361. C. S. Seshadri, Algebraic vector bundles over the product of an affine curve and the line. Proc. Amer. Math. Soc., 10(5):670-673 (1959).

362. L. Silver, Noncommutative localizations and applications. J. Algebra, 7(1):44-76 (1967).

363. L. W. Small, An example in Noetherian rings. Proc. Nat. Acad. Sci. USA, 54(4): 1035-1036 (1965).

364. L. W. Small, "Some remarks on the homological dimensions of a quotient field," Mimeographed notes, Univ. of California, Berkeley, Calif. (1966).

365. L. W. Small, Hereditary rings. Proc. Nat. Acad. Sci. USA, 55:25-27 (1966).

366. L. W. Small, On some questions in Noetherian rings. Bull. Amer. Math. Soc., 72(5):853-857 (1966).

367. L. W. Small, Orders in Artinian rings. J. Algebra, 4(1):13-41 (1966).

368. L. W. Small, Semihereditary rings. Bull. Amer. Math. Soc., 73(5):656-658 (1967).

369. L. W. Small, Correction and addendum: Orders in Artinian rings. J. Algebra, 4:505-507 (1966).

370. L. W. Small, Orders in Artinian rings. II. J. Algebra, 9(3):266-273 (1968).

371. L. W. Small, A change of rings theorem. Proc. Amer. Math. Soc., 19(3):662-666 (1968).

372. D. A. Smith, Chevalley bases for Lie modules. Trans. Amer. Math. Soc., 115(3): 283-299 (1965).

373. E. Snapper, Injective modules under change of rings. Proc. Amer. Math. Soc., 16(4):788-790 (1965).

374. A. Solian, Characterization of some functors of modules. Rev. roumaine math. pures et appl., 11(3):283-285 (1966).

375. A. Solian, Foncteurs qui transforment les épimorphismes locaux en monomorphismes locaux. Rev. roumaine math. pures et appl., 11(4):401-410 (1966).

376. R. Sridharan, Homology of non-commutative polynomial rings. Nagoya Math. J., 26(1):53-59 (1966).

377. O. Stănașilă, Citeva observații asupra topologizării modulelor. Studii și cercetări mat. Acad. RSR, 18(5):681-686 (1966).

378. A. Steger, Diagonability of idempotent matrices. Pacif. J. Math., 19(3):535-542 (1966).

379. B. T. Stenström, Pure submodules. Arkiv mat., 7(2):159-171 (1967).

380. B. T. Stenström, High submodules and purity. Arkiv mat., 7(2):173-176 (1967).

381. R. W. Stringall, Endomorphism rings of primary Abelian groups. Pacif. J. Math., 20(3):535-557 (1967).

382. J. R. Strooker, Lifting projectives. Nagoya Math. J., 27(2):747-751 (1966).

383. A. Sugano, A note on Azumaya's theorem. Osaka J. Math., 4(1):157-160 (1967); correction ibid., 5:153 (1968).

384. A. Sugano, Note on semisimple extensions and separable extensions. Osaka J. Math., 4(2):265-270 (1967).

385. S. Suzuki, Note on formally projective modules. J. Math. Kyoto Univ., 5(3): 193-196 (1966).

386. Y. Suzuki, On automorphisms of an injective module. Proc. Japan Acad., 44(3): 120-124 (1968).

387. R. G. Swan, Induced representations and projective modules. Ann. Math., 71(3): 552-578 (1960).

388. R. G. Swan, The number of generators of a module. Math. Z., 102(4):318-322 (1967).

389. F. Szász, Einige Kriterien für die Existenz des Einselementes in einem Ring. Acta scient. math., 28(1-2):31-37 (1967).

390. H. Tachikawa, On dominant dimensions of QF-3 algebras. Trans. Amer. Math. Soc., 112(2):249-266 (1964).

391. G. Thierrin, Anneaux métaprimitifs. Canad. J. Math., 17(1):199-205 (1965).

392. K. Tiwari, On complete ring of quotients. J. Indian Math. Soc., 31(2):95-99 (1967).

393. A. K. Tiwary, On the quotients of indecomposable injective modules. Canad. Math. Bull., 9(2):187-190 (1966).

394. Chi-te Tsai, Report on injective modules. Queen's Papers Pure and Appl. Math., No. 6, 243 pp. (1966).

395. K. Uchida, A note on the projective modules over group rings. Proc. Japan Acad., 43(1):13-16 (1967).

396. Y. Utumi, On continuous rings and self injective rings. Trans. Amer. Math. Soc., 118(6):158-173 (1965).

397. Y. Utumi, On the continuity and self-injectivity of a complete regular ring. Canad. J. Math., 18(2):404-412 (1966).

398. Y. Utumi, Self-injective rings. J. Algebra, 6(1):56-64 (1967).

399. P. Vámos, On ring classes defined by modules. Publs. math., 14(1-4):1-8 (1967).

400. P. Vámos, The dual of the notion of "finitely generated". J. London Math. Soc., 43:643-646 (1968).

401. B. J. Varela, Une note sur des modules élastiques. Rev. colomb. mat., 1(3):26-28 (1967).

402. W. V. Vasconcelos, On local and stable cancellation. An. Acad. brasil. ciênc., 37(3-4):389-393 (1965).

403. W. V. Vasconcelos, R-sequences and the homology of Macaulay rings. An. Acad. brasil ciênc., 38(2):249-252 (1966).

404. W. V. Vasconcelos, Ideals and cancellation. Math. Z., 102(5):353-355 (1967).

405. O. E. Villamayor, Sur une representation matricielle de l'anneaux d'endomorphismes d'un module quelconque. Publs. Fac. cienc. fisicomat. Univ. nac. La Plata, No. 213, pp. 185-190 (1956).

406. C. P. Walker, Relative homological algebra and Abelian groups. III. J. Math., 10(2):186-209 (1966).

407. C. L. Walker and E. A. Walker "Quotient categories of modules," Proc. conf. categor. algebra (1966), pp. 404-420.

408. Y. Watanabe, Simple algebras over a complete local ring. Osaka J. Math., 3(1):13-20 (1966).

409. R. H. Wenger, "Semigroups having quasi-Frobenius algebras," Doctoral dissertation, Michigan State Univ. (1965), 70 pp.; Dissert. Abstr., 26(8):4703 (1966).

410. E. R. Willard, "Functors between categories of modules," Doctoral dissertation, Pennsylvania State Univ. (1964), 81 pp.; Dissert. Abstr. 25(11):6668-6669 (1965).

411. E. R. Willard, Properties of projective generators. Math. Ann., 158(5):352-364 (1965).

412. Ling-Erl Eileen Ting Wu, A characterization of self-injective rings. Ill. J. Math., 10(1):61-65 (1966).

413. Ling-Erl Eileen Ting Wu and J. P. Jans, On quasi projectives. Ill. J. Math., 11(3): 439-448 (1967).

414. Ling-Erl Eileen Ting Wu, H. Y. Mochizuki, and J. P. Jans, A characterization of QF-3 rings. Nagoya Math. J., 27(1):7-13 (1967).

415. F. Wuytack and J. Depunt, Operators over I-collections of modules. Bull. Soc. math. Belg., 17(1):37-54 (1965).

416. F. Wuytack and J. Depunt, Note on I-collections of modules. Bull Soc. math. Belg., 17(4):484-494 (1965).

417. C. R. Yohe, Triangular and diagonal forms for matrices over commutative noetherian rings. J. Algebra, 6(3):335-368 (1967).

418. A. Zaka, Residue rings of semi-primary hereditary rings. Nagoya Math. J., 30:279-283 (Aug. 1967).

419. A. Zaks, Global dimension of Artinian rings. Proc. Amer. Math. Soc., 18(6): 1102-1106 (1967).

420. A. Zaka, A note on semi-primary hereditary rings. Pacif. J. Math., 23:627-628 (1967).

421. H. Zassenhaus, Orders as endomorphism rings of modules of the same rank. J. London Math. Soc., 42:180-182 (1967).

422. J. M. Zelmanowitz, Endomorphism rings of torsionless modules. 5(3):325-341 (1967).

Lattice Theory

M. M. Glukhov, I. V. Stelletskii, and T. S. Fofanova

The present survey is devoted to results in the papers on lattice theory reviewed in Referativnyi Zhurnal (Mat.) during 1965-1968, and is a natural continuation of Skornyakov's survey article [38]. Papers on vector lattices, (partially, lattice) ordered algebraic systems (semi-groups, groups, rings), as well as a large circle of papers on projective geometry, reviewed in the section on "Lattices" but bearing only a slight relation to the main theme of the present article, are not considered here.

In conclusion we mention certain monographs. The books by Morgado [339], Rutherford [420-422], and Donnellan [146] are devoted to general lattice theory. The monographs of Adelfio and Nolan [67], Casanova [120], Flegg [173], Carvallo [119], Dwinger [153], and Halmos [208] are devoted to the theory of Boolean algebras. We mention also the books by Liber [26] and Gillie [190]. In his monograph [293] Lehman considers congruence relations given on a set with one binary and one unary operations, on which a real-valued norm has been defined, such that the corresponding equivalence classes form a normed Boolean algebra. Analogously, a normed Dedekind lattice is constructed for a set with two binary operations.

§1. Boolean Algebras

1. Definition of Boolean Algebras. Diego and Suárez [143] showed that a Boolean algebra can be defined as a universal algebra $\langle B, \rightarrow, 0 \rangle$ with the identities $(a \rightarrow a) \rightarrow b = b$ and $a \rightarrow ((c \rightarrow 0) \rightarrow (b \rightarrow 0)) = (a \rightarrow b) \rightarrow (a \rightarrow c)$, or $\langle B, \rightarrow, ^{-} \rangle$ with the identities $(a \rightarrow a) \rightarrow \bar{\bar{b}} = b$ and $a \rightarrow \overline{\overline{c \rightarrow \bar{b}}} = (a \rightarrow \bar{\bar{b}}) \rightarrow (a \rightarrow \bar{\bar{c}})$. Petcu [379] defined a Boolean algebra with the aid of the operations $\cup$ and $^-$ and of the identities $x \cup x = x$; $((x \cup y) \cup z) \cup (\overline{\overline{t \cup u} \cup \overline{t \cup u}}) = t \cup (x \cup (y \cup z))$. He has also shown that by replacing $\overline{\overline{t \cup u} \cup \overline{t \cup u}}$ by t in the second axiom we obtain an axiom system for semilattices. Siosan [452-454] has devoted a number of papers to the investigation of the axiomatics of Boolean and Newman algebras. He has found several new independent systems of identities defining Newman algebras (for example $x \cap (y \cup z) = (x \cap y) \cup (x \cap z)$, $x \cap (y \cup \overline{y}) = x$, $x \cap y = y \cap x$, $x \cap x = x$, $x \cap \overline{x} = y \cap \overline{y}$), the addition of the identity $x \cup x = x$ to which leads to an independent system of axioms for a Boolean algebra. Somewhat different identity systems defining Newman algebras were found by Wooyenaka [494]. By adding the identities $x \cup x = x$, $\overline{(x \cup x)} \cup x = \overline{x}$, $(x \cap x) \cap y = x \cap (x \cap y)$ to the systems he found, he obtained an independent system of identities for quasi-Boolean algebras, quasi-Boolean rings, and Boolean rings, respectively. The relation between the axiomatics of Newman and Boolean algebras has been considered also by Albada [69]. Let us remark that a very detailed survey of the results on the various definitions of Boolean algebras is contained in Rudeanu's book [410] devoted to the axiomatics of various classes of partially ordered sets. The book contains a number of the author's own new results (see [35, 409] and states certain open questions concerning mainly the independence of axiom systems.

2. Imbedding in Complete Algebras. Free Extensions and Free Products of Boolean Algebras.

In [458] Solovay gives simpler proofs of Gaifman and Hales' results on the existence of a countably-generated complete Boolean algebra and of a complete ∞-distributive Boolean algebra with k generators or arbitrarily high cardinalities. Kripke [283], using Solovay's methods, proves a stronger assertion: every Boolean algebra can be imbedded in a countably-generated complete Boolean algebra. Monteiro [335] generalized to semihomomorphisms (relative to union) Sikorski's theorem of the existence of the extension of the homomorphism of any subalgebra of a Boolean algebra A into a complete Boolean algebra B upto a homomorphism of A into B (proved with the use of Zorn's lemma). From Sikorski's theorem we can derive a theorem on prime ideals: any proper ideal of a Boolean algebra can be imbedded in a maximal

ideal. Luxemburg [299] investigated the logical relations between
Sikorski's theorem, the theorem on prime ideals, and the axiom of
choice in a specific axiomatic system of set theory.

Let α be any infinite cardinal number. A Boolean algebra
B_α is called an α-complete extension of algebra A if B_α is α-com-
plete, the algebra A is α-generated, and there exists an α-isomor-
phism of A into B (an isomorphism preserving the faces existing
in A of sets of cardinality $\leq \alpha$); an α-complete extension B_α is
called free if every homomorphism of A into an α-complete Boolean
algebra B can be extended upto an α-homomorphism of B_α into B.
The papers by Rieger, Sikorski, Yaqub, and Dwinger, devoted to the
investigation of α-complete and complete free extensions of Boolean
algebras, are well known. In connection with the results of Gaifman and
Hales on the absence of a complete extension for any infinite free Boolean algebra there arose the problem of describing Boolean algebras
having free complete extensions. Using the above-mentioned results of Gaifman and Hales, Day [134] proved that a Boolean algebra has a free complete extension if and only if it is superatomic.
Day's paper [135] was devoted to the study of superatomic Boolean
algebras. In [158] Dwinger and Yaqub generalize the majority of
results of free α-complete and complete extensions of Boolean algebras to sets $\{B_t, t \in T\}$ of algebras.

A free α-complete extension of any set $\{B_t, t \in T\}$ of Boolean
algebras always exists, and it is isomorphic to a free α-complete
extension of the free product of the algebras B_t. It is natural that
just as in the case of one algebra the free complete extension of
the set of Boolean algebras does not always exist. In this regard
the following results generalizing Day's results are of interest.
If $\{B_t, t \in T\}$ is a set of Boolean algebras and each of the algebras
B_t contains more than two elements, then the free complete extension of the set $\{B_t, t \in T\}$ exists if and only if each algebra B_t is
superatomic. In 1963 Dwinger and Yaqub investigated the generalized products of Boolean algebras with an amalgamated subalgebra.
In particular, it was proven that a generalized product always
exists and is unique to within isomorphism. Mączyński [301] generalized this result to families of α-distributive α-complete
Boolean algebras containing an α-regular subalgebra isomorphic
to a given Boolean algebra. The investigations of Mączyński were
continued in his joint paper with Traczyk [303]. A free product and

a free amalgamated product of Boolean algebras are special cases
of the so-called direct limit of a partially-ordered system of
Boolean algebras. Dwinger's paper [157] was devoted to the study
of this construction in the categories $\mathfrak{B}$ of Boolean algebras, $\mathfrak{B}_\alpha$
of α-complete Boolean algebras, and $\mathfrak{B}_\infty$ of complete Boolean algebras. In another of his papers [154] Dwinger considered spaces
dual to the inverse limit systems of Boolean algebras. Hewitt
[225], by studying the direct limits of Boolean algebras, shows that
the concepts of direct and inverse limits can be defined in any class
of algebras of the same type, closed under homomorphisms and
under the formation of subalgebras, free unions, and free products.

In [311] Mangani determines the free extension of a Boolean
algebra in a sense other than that considered above. He calls a
Boolean algebra B a free extension of a Boolean algebra A if there
exist a subalgebra $B^* \cong A$ and a transfinite sequence $\{a_\alpha\}$ such
that by setting $A_\alpha = \{xa_\alpha \cup y\bar{a}_\alpha \mid x,\ y \in \bigcup_{\beta < \alpha} A_\beta\}$, we shall have $B = \bigcup A_\alpha$.
It turns out that B is uniquely defined by the algebra A and the
cardinality of the sequence. Every homomorphism of algebra A
into C can be extended upto a homomorphism of B into C. Hence
it is concluded that A is a retract of algebra B, and certain other
properties of B are derived.

A Boolean algebra U_α of power (cardinality) $\leq \alpha$ is called
α-universal if any Boolean algebra of power $\leq \alpha$ is isomorphically
imbedded in U_α, and is called α-homogeneous if for any subalgebra
$A \subset U_\alpha$ of power $< \alpha$ every isomorphism of A into U_α can be extended upto an automorphism of U_α. These concepts are special
forms of the uniform and homogeneous systems introduced earlier
by Jónsson for a class of algebraic systems. By assuming the
fulfillment of the continuum hypothesis (the cardinal number α^+
closest to α equals 2^α), Jónsson pointed out the sufficient conditions for the existence of α^+-universal and α^+-homogeneous systems. After Dwinger and Yaqub investigated generalized free
products of Boolean algebras it became clear that the Jónsson conditions are fulfilled for Boolean algebras (before this it was not
clear that the so-called amalgamation property was fulfilled (see
[38, 101]). Hence, for any infinite cardinal number α there exists
(and to within isomorphism is unique) a α^+-universal and α^+-homo-
geneous Boolean algebra U_{α^+}. Keisler [272] proves that when the
continuum hypothesis is fulfilled $S(\omega)/S_\omega(\omega)$ is an ω^+-universal

ω^+-homogeneous Boolean algebra, where $S(\omega)$ is the Boolean algebra of all subsets of the set of positive integers and $S_\omega(\omega)$ is its ideal consisting of all finite subsets. Mączyński [302] considered retracts of the Boolean algebra U_{α^+}. In particular, he showed that the class of complete Boolean algebras of power $\le \alpha$ coincides with the set of retracts of algebra U_{α^+}. Monk proves that for α there exists a complete Boolean algebra $\mathfrak{A}$ such that any nontrivial α-homomorphic image of $\mathfrak{A}$ has cardinality $\ge \alpha$. On the basis of this lemma he shows that nontrivial α-injective Boolean algebras do not exist [333].

3. Certain Special Classes and Generalizations of Boolean Algebras.

We point out a number of papers devoted to normed Boolean algebras and spaces. Climescu [122] proves that for any n > 0 there exist Boolean algebras admitting of a measure of n values. Using n maximal ideals of a Boolean algebra, he determined a measure having not more than 2^n values. A conjecture on the existence of a Boolean algebra of 2^n elements with $1, 2, \ldots, 2^n$-valued measure is stated. Lehman [293, 294] considers the algebra $\langle S, \cup, ^- \rangle$ with a real norm [x] such that: 1) $[a \cup b] + [a \cup \bar{b}] = [a] + [b \cup \bar{b}]$; 2) $[a \cup b] \ge [a], [b]$; 3) $[a \cup (b \cup c)] = [(a \cup b \cup c]$, and considers various congruences of it, under which the equivalence classes form a Boolean algebra with norm [x]. The weakest and the strongest congruences are indicated. Boolean metric spaces and normed rings were studied by Melter [325]. An associative ring R is called a p-ring if px - 0 and $x^p = x$, x $\in$ R. The ring R can be looked upon as a normed Boolean space with a metric $|x| = x^{p-1}$ over its idempotent algebra (as a B_p-space). Melter [326] proves that every p-ring is a subdirect sum (and is a direct sum when p = 2, see Smithson [457]) of the fields GF (p), and that every Boolean metric space is imbeddable into a B_p-space. Subrahmanyam [471] studies Boolean vector spaces of finite dimension k and their linear (affine and metric) transformations. In particular, it is proved that the isometry group is isomorphic to the B-extension of a symmetric group of degree k + 1. This theorem was generalized to the case of affine transformations by Jagannadham [250]. In [251] Jagannadham continued the investigations, initiated by Foster and subsequently developed by Subrahmanyam [471], on B-extensions of universal algebras with finitary operations. We note in passing certain papers which study linear transformations of Boolean spaces, or Boolean matrices. A matrix

A over a Boolean algebra B is called orthogonal if $AA^T = A^T A = E$.
In 1934 Wedderburn proved that orthogonal Boolean matrices coincide with invertible matrices. Rutherford [424] indicated a canonic representation of an orthogonal matrix in the form of a prod-
uct of elementary ones and proved that in case $|B| = 2^m$ the group
of orthogonal matrices is isomorphic with the direct sum of m
symmetric groups S_n. In [423] he showed that the eigenvalues of
one eigenvector form a sublattice of the basic Boolean algebra,
while the eigenvectors corresponding to one eigenvalue form a
Boolean vector space. An analog of the characteristic equation of
a matrix was derived. Blyth [108] studies the so-called distributive
matrices (in connection with the fact that the multiplication opera-
tion of Boolean matrices: $(a_{ij})(b_{ij}) = (\bigcup_{t}(a_{it} \cap b_{tj}))$, is not distributive
relative to the operation of intersection: $(a_{ij}) \cap (b_{ij}) = (a_{ij} \cap b_{ij})$).
Certain aspects of the reduction of the multiplication operation of
Boolean matrices to the multiplication of simpler (elementary)
matrices are considered by Scognamiglio [441]. Fischer [170]
studied a Boolean vector space with a Hamming metric, paying
fundamental attention to the classification of the vectors relative
to the Hamming distance. Kirsch [277] investigated isotone mappings onto the set of positive integers of a Boolean algebra with a
linear order $\preceq$.

If in the usual definition of a real algebra of finite rank the
scalar field is replaced by a Boolean algebra B, then we obtain a
Frobenius-Boole algebra (an FB-algebra). If $B = \{0, 1\}$, the FB-
algebra is termed primitive. Andreoli [72, 73] proves that if B
is complete and atomic, then the FB-algebra is a direct sum of
primitive FB-algebras. Scognamiglio [437] and Fadini [163] were
occupied with FB-algebras of rank 2. Servi [449, 450] constructed
a topological theory of Boolean algebras with operators k satisfying the conditions $k0 = 0$, $x \le kx$. Flachsmeyer studies representations of the so-called separable complete Boolean algebras by
regular open subsets to topological spaces [172]. Abbott [59] defines a semi-Boolean algebra as a join semilattice in which every
principal filter is a Boolean algebra. The connection with impli-
cation algebras is established and various examples of semi-
Boolean algebras are cited. Wyler [497] introduces the notion of
a clan (a lattice with a subtraction operation satisfying specified
conditions) and shows that an l-group and a Boolean ring will be clans

if we set $b-a - b + (-a)$ in them. By the same token, Birkhoff's
Problem 105 receives an affirmative answer. A very broad generalization of Boolean algebras, covering Newman algebras,
Brouwer algebras, autometrized lattices, commutative l-groups,
etc., is the concept of a metric algebra which is defined as a commutative algebra $\langle A, +, 0 \rangle$ with an antisymmetric reflexive relation $\le$, provided with a metric. Swamy's large paper (also see
[397, 399]) is devoted to the study of autometrized algebras. We
mention a number of papers related to logic. In 1955-1957 Halmos,
with the aim of constructing an algebraic interpretation for the
calculus of first-stage one-place predicates and for the calculus
of first-stage predicates (with and without equality), published a
series of papers in which he defined and investigated the so-called
monadic and polyadic Boolean algebras. The theory of polyadic
algebras found its own development in the works of Daigneault
[131, 132], Fenstad [165], Monk [332], and Jurie [265]. In all these
papers the question of the representation of the algebras occupies
an important place. Magari [310] studied the representations of
monadic algebras. Marrona [312] gave a new definition (as the algebra
$\langle M; \cap, ^- \rangle$ with the identities $a = a \cap \overline{\overline{a} \cap \overline{b}}, \ \overline{\overline{c \cap a} \cap \overline{b} \cap \overline{a}} = a \cap \overline{\overline{b} \cap \overline{c}})$ of
the Morgan lattice introduced previously by Moisil. Ruzsa's paper
[425] is devoted to a formal logical study of Boolean algebras and
certain of their generalizations. Normal forms are analyzed and
attention is paid to the completeness of the axiomatics. McCall
[323] also has considered the questions of completeness and of the
solvability problem for a Boolean algebra as a deductive calculus.
de Jongh and Troelstra [263] investigated U-representable lattices
and the pseudo-Boolean algebras defined by them, having a connection with intuitionistic propositional logic. See Ponasse [385]
for the algebrization of the propositional calculus.

Doctor [145] refined, from the category-theoretic point of
view, the meaning of the well-known Stone's assertion on the mathematical equivalence of the theories of Boolean rings and spaces.
In particular, it was proved that the category of Boolean algebras
is equivalent to the category of Boolean rings. The equivalence or
the dual equivalence of certain subcategories of the above-mentioned categories and the categories of Boolean spaces was established. Analogous questions were treated by Preller [388]. In

another paper [389] she derived also the well-known properties of Boolean algebras from general category-theoretic facts.

4. **Boolean Functions.** In the survey of papers (appearing during 1965-1968) on functions over Boolean algebras and their generalizations we shall not concern ourselves with the numerous articles and books on functions over a two-element Boolean algebra B_2. They refer largely to cybernetics or to the algebra of logic. The majority of papers on Boolean functions deal with the questions of the representation and classification of the functions. There are various approaches to the definition of Boolean functions. By a Boolean function, or more precisely, by a function of n variables over a Boolean algebra B, we shall mean the single-valued mapping $B^n \to B$. Rudeanu [35] proves that any Boolean function can be represented by a polynomial composed of variables and constants from B with the aid of a finite number of Boolean operations $(\cup, \cap, ^-)$ only in the case when $B = B_2$. In this same paper he describes all the Sheffer functions of two variables over a Boolean algebra B and all the polynomials in two variables with whose aid we can define a ring on the algebra B. It is proved that all rings defined in this manner are isomorphic to the usual Boolean ring corresponding to the algebra B. Analogous questions were treated by Grätzer [198]. In particular, he proved that the class of Boolean functions representable by polynomials coincides with the class of functions possessing the property (sP): $a_i \equiv b_i (\theta)$ implies $f(a_1, ..., a_n) \equiv f(b_1, ..., b_n) (\theta)$ for any congruence θ on B. In this connection, in another paper [200] he called an (sP)-function over a distributive lattice L with zero and unit a Boolean function. It turns out that every Boolean function $f: L_n \to L$ is uniquely defined by its own characteristic function $\varphi = f \,|B_2$ (the restriction of f onto the Boolean algebra B_2, where B_2 is a two-element subalgebra in L).

In [35] Rudeanu shows that every polynomial over a Boolean algebra B can be uniquely represented in the disjunctive normal form

$$f(x_1, ..., x_n) = \bigcup_{a_1, ..., a_n} \lambda_{a_1, ..., a_n} x_1^{a_1} ... x_n^{a_n}, \quad a_i = 0,1,$$

with coefficients $\lambda_{a_1, ..., a_n} \in B$. If, however, we consider Boolean

polynomials in variables only (without using constants) then to them
there will correspond representations in which $\lambda_{a_1, \ldots, a_n} = 0,1$. Pinter [383] considers the question of obtaining the minimal conjunc-
tive forms of Boolean functions.

Every function of the algebra of logic can be represented in
the form $f(x_1, \ldots, x_n) = f(x_1, \ldots, x_{n-1}, 1) x_n \cup f(x_1, \ldots, x_{n-1}, 0) \bar{x}_n$.
Goodrich [193] generalizes this representation to a function over a
similar Boolean ring. Scognamiglio indicates the canonic form of
functions obtainable from one function by means of the operations
of superposition and complementing [438]. Righi [400] studies
selfdual functions. Scognamiglio [436] divides Boolean functions
into two classes – algebraic and transcendental.

A set of Boolean functions is called a family if it is closed
under the operations of superposition and mapping of variables.
By supplementing Kuntzmann's investigations, Benzaken found
[102] all the maximal subfamilies of the family of growing Boolean
functions with the condition $f(x_1, \ldots, x_n) \geq f^*(x_1, \ldots, x_n)$, where the
function f^* is the dual of f. In [104] he studied certain other
families of growing functions. In particular, he derived a criterion for a function to belong to a given family. By continuing the researches of Slepian, Sagalovich, Gaşpar, and others,
Constantinescu and Gaşpar [22] analyzed functions in the $\sum_{i=1}^{s} p_i$ variables $x_1, \ldots, x_{p_1}, x_{p_1+1}, \ldots, x_{p_1+p_2}, \ldots$; they consider functions to be
equivalent if one of them is obtained from the other by a variable
transformation from the group $H = H_{p_1} \times H_{p_2} \times \ldots \times H_{p_s}$, where
H_{p_i} is the group of statements on the set of variables $x_{p_1+\ldots+p_{i-1}+1}$,
$\ldots, x_{p_1+\ldots+p_i}$ (symmetric or alternating or hyperoctahedral). A
method is indicated for computing the number of equivalence
classes. Bassi [93, 94] discussed the theory of polynomials in a
Boolean algebra equipped with an additional unary operation (closure).
Using this theory the author attempts to cover from a single point
of view the various dualities encountered in mathematics.

We mention further the series of papers by Rudeanu [405,
407, 408, 411, 414-416] dealing with the solution of Boolean equations, and the paper by Ivanescu [248] on the solution of a system
of pseudo-Boolean equations.

§ 2 . Identity and Defining Relations in Lattices

Szász [473, 474] examined the lattice identities 1) $(x \cap y) \cap z = x \cap (y \cap z)$; 3) $x \cap y = y \cap x$; 5) $x \cap (x \cup y) = x$; 7) $x \cap x = x$; 9) $x \cap (y \cap z) = (x \cap y) \cap (x \cap z)$; 11) $x \cap (y \cap z) = (y \cap x) \cap (z \cap x)$; and, respectively, the dual identities 2, 4, 6, 8, 10, 12 and the quasi-identity system 13) $x \cap y = y \Leftrightarrow y \cup x = x$. It was proved that $S_I = \{1 - 6\}$; $S_{II} = \{1 - 4,7,13\}$, $S_{III} = \{3 - 6,9,10\}$; $S_{IV} = \{5,6,11,12\}$ are independent axiom systems for a lattice. Here, all the examples of algebras, proving the independence of the axioms, contain the smallest possible number of elements. Petcu started from the following identities, fulfilled in any lattice: (1) ten variants of the associative law, obtained from 1), 2) by permutations of the letters in the left and right hand sides; (2) eight variants of the absorption law (for example, $(y \cap x) \cup x = x$); and (3) eighty absorption-associative laws (for example, $z \cup (y \cup x) = (x \cup (y \cup z)) \cap ((x \cup (y \cup z)) \cup t))$. In [377, 378] he found 32 independent axiom systems each of which consists of two variants of the associative law and two variants of the absorption law, and 96 independent systems containing at least two absorption-associative identities and at least one idempotent one. In summary, Petcu answered a number of the questions posed by Sorkin (Uch. Zap. Mosk. Gos. Ped. Inst., 1:61-66 (1962)). However, the question of the existence of a system of two identities defining a lattice was answered affirmatively only 1968 by Kalman [266]. This question has been answered affirmatively also for a semilattice. Padmanabhan [370] showed that semilattices may be given by the system of identities $\{1, 7\}$. On the other hand, Potts [387] proved that it is impossible to give a semilattice by less than two identities. Iqbalunnisa [240] delineated lattice classes by means of the identities: I. $\bigcap_{i=1}^{n} (x \cup y_i) = x \cup \left(\bigcup_{i=1}^{n} \left(\bigcap_{j=1}^{n} z_{ij} \right) \right)$, where n ≥ 3, $z_{ij} = x \cup y_i$ when i = j and $z_{ij} = y_j$ when i ≠ j; II. the same as in I, but n ≥ 2, $z_{ij} = y_j$ when i = j and $z_{ij} = x \cup y_j$ when i ≠ j, as well as by means of the duals to I and II. Lattices in which identities I and II are fulfilled, he called, respectively, supermodular and almost-modular lattices of order n. Since every supermodular lattice is modular, by the same token this solves Grätzer and Schmidt's Problem 10 on the existence of classes of weakly modular lattices defined by identities and different from one-element, distributive, and modular lattices. As n increasce, tho olasses of supermodular and almost-modular lattices of power n approach, respectively, the classes of distributive and modular lattices. Zelinka [498] proved the independence of the well-known system of seven axioms of Birkhoff and Birkhoff for distributive lattices with unit. The axiomatics of lattices are touched on also by Ruedin in a paper on distributive groupoids [419] (see [418] for the proof). A survey of the various approaches to the definition of lattices, distributive lattices, and modular lattices is given in Rudeanu's book [410] (see § 1).

By $L(n_1 + ... + n_k)$ and $FM(n_1 + ... + n_k)$ we denote, respectively, a lattice and a modular lattice freely generated by a cardinal sum of k chains of lengths $n_1, ..., n_k$. It is well known that $FM(n_1 + n_2)$ (for any n_1, n_2), $FM(1 + 1 + 1)$, $FM(2 + 1 + 1)$ are finite, while $FM(1 + 1 + 1 + 1)$ is infinite. Rolf [402] proves that the lattice $FM(2 + 2 + 2)$ contains an infinite chain. The lattice $L(n_1 + ... + n_k)$ is a special case of the so-called almost free lattices, i.e., lattices given by generating elements and by defining relations of the form $a_i * a_j = a$, where $*$ is $\cup$ or $\cap$ and k = i or j. Chebotareva [47], using the properties of canonic words, solved positively the belonging problem for almost free lattices. (Note that the belonging problem in the general case of lattices remains open although the word problem and the isomorphism problem have been solved psitively.) Glukhov [8] solved positively the word problem and the isomorphism problem for almost free modular lattices by proving that every partial almost free modular lattice is imbeddable in a modular lattice. We remark that all the fundamental algorithmic problems for modular lattices remain open. The lattices $L(n_1 + n_2)$ with the additional binary normality relation $\trianglelefteq$ were considered by Dean and Kruse [136]. An element a of a lattice with the relation $\trianglelefteq$ and with unit is called normal (subnormal) if $a \trianglelefteq 1$ (respectively, $a \trianglelefteq ... \trianglelefteq 1$). This concept of a normal element is a generalization of the normal divisor in a subgroup lattice of a group and is different from the well-known concept of Kurosh-normality. Let $L'(n + m)$ be a lattice freely generated by the chains $\{a_n \leq a_{n-1} \leq ... \leq a_1\}$ and $\{b_m \leq b_{m-1} \leq ... \leq b_1\}$ with the normality relations $a_i \trianglelefteq a_{i-1}$ and $b_i \trianglelefteq b_{i-1}$. It was proved that $L'(1 + m)$ and $L'(n + 1)$ are distributive, $L'(2 + 2)$ and $L'(3 + 2)$ are finite and all their elements are subnormal, and $L'(3 + 3)$ is infinite and not all its elements are subnormal. An example was constructed of a lattice of subgroups, in which the join of two subnormal elements is not subnormal, although such a situation is impossible in a modular lattice.

In [412, 413] Rudeanu considers nine different constraints on the chain of a lattice (modularity of the lattice, semimodularity, finiteness of the chains, the Jordan-Dedekind conditions, etc.) and describes all possible implications of the form A ⇒ B, where A, B are logical functions of the nine conditions indicated (constructed from these conditions with the aid of conjunction, disjunction, and negation).

Let us mention the results on lattices obtained by Shevrin in [50] devoted to the lattice properties of semigroups. In this paper he points out arithmetically closed and not closed classes S($\mathfrak{U}$) of all possible subgroup lattices of semigroups of class $\mathfrak{U}$ and proves the unsolvability of the elementary theory of the lattice class S($\mathfrak{U}$) for the semigroup class $\mathfrak{U}$ containing all torsion-free Abelian groups. Also of interest are the results of Khisamiev [45] on the unsolvability of the elementary theory of a free lattice with n generators when n ≥ 3, and also those of Taitslin [40] on the unsolvability of the elementary theories of lattices of ideals of certain rings.

Grätzer [202] continued the study, started in 1963 by Jónsson, of the lattice of all equational classes of lattices. Jónsson had shown that the zero element of this lattice is the class of one element lattices and that this class is covered uniquely by the class D of distributive lattices which itself is covered only by the classes M_5 and N_5 generated, respectively, by a 5-element modular nondistributive lattice and by a 5-element nonmodular lattice. By solving the problem posed by Jónsson of describing the set of elements covering M_5, Grätzer proved that a finite modular subdirectly irreducible lattice generates an equational class covering M_5 if and only if it is isomorphic to one of the two lattices with the diagrams shown in Fig. 1.

§ 3.   Distributive Lattices

Let M be the set of subsets of elements of a complete lattice and let ΠM be the set of all mappings t: M → L such that t(X) ∈ X for each X ∈ M. Lowig [298] proved that the distributive identity

$$\bigcap_{X \in M} \left(\bigcup_{x \in X} x \right) = \bigcup_{t \in \Pi M} \left(\bigcup_{x \in M} t(X) \right)$$ is equivalent to the dual one.

The set $Q^M = \{x \in R \,|\, x \leq q$ for some $q \in Q\}$ is called the M-closure of the subset Q of a partially-ordered set R. Let the set P_T of join-irreducible elements of a lattice L satisfy a descending chain condition and a certain further condition. Avann [81] proved that in this case the set $L_M (P_T)$ of M-closures of all finite crowns of the set P_T is a distributive lattice also with a descending chain condition; the correspondence $x \to x^M$ establishes an isomorphism between the set P_T and the set of join-irreducible elements of the lattice $L_M (P_T)$. Avann studied the logical implications between the following properties of a distributive lattice: A) L satisfies a descending chain condition; B) L is compactly generated; C) $a \cap (\bigcup_{s \in S} s) = \bigcup_{s \in S} (a \cap s)$ for any $a \in L$, $S \subseteq L$; D) $x \leq \bigcup_{s \in S} s \Rightarrow \exists s_0 \in S$, $x \leq s_0$; for every join-irreducible element $x \in L$; E) L is atomic; F) the set P_T satisfies a descending chain condition; G) if $a \in L$ and $a \neq 1$, then there exists at least one element $b \in L$ such that b covers a; A') $-$ G') are dual conditions. (L is almost everywhere assumed complete and weakly atomic.)

In the last chapter of an extensive article devoted to a systematic study of pseudocomplemented lattices, Varlet [481] considered pseudo-complemented distributive lattices identically satisfying the condition (e): $a^* \cup a^{**} = 1$, namely, Stone lattices (S.1.). He derived a number of conditions equivalent to (e). The properties of ideals picking out S.1. from the class of pseudocomplemented distributive lattices were studied; it was proved that the cardinal product, the ordinal sum, and the lattice of ideals of S.1. are once again S.1. Grätzer and Schmidt had earlier proved that a pseudocomplemented distributive lattice is a S.1. if and only if it satisfies the condition (GS): the set-theoretic union of two different minimal prime ideals of a lattice L coincides with L (Birkhoff's Problem 70). Other formulations of this result were given by Katriňák [18] and Varlet [485] (the latter in the language of filters).

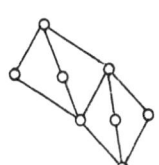 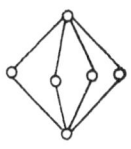

Fig. 1

Necessary and sufficient conditions for a lattice of ideals to be a S.1. were given in [18]. Katriňák [19] also obtained the necessary and sufficient conditions under which a distributive lattice with zero and unit possesses the property (GS).

Grätzer [199] proved that every S.1. is isomorphic to a *-sublat-
tice (the subalgebra relative to $\cup$, $\cap$, *) of the lattice of all ideals
of a Boolean algebra of all subsets of a certain set, and showed that
the converse assertion is true. The proof was carried out by the
methods of mathematical logic. An elementary lattice-theoretic
proof of this result was given by Bruns [112]. In [155, 156] Dwinger presents many previously well known results on Post algebras,
often with new approaches and simplified proofs. Traczyk [478]
gave a definition of a Post algebra as a universal algebra. He also
established [477] several equivalent axiom systems and properties
of Post algebras, and proved a theorem on the topological representation of such an algebra [479].

A lattice L is said to be conditionally implicative (c.i.) if for
any x, y $\in$ L, x $\nleq$ y, the set $\{z \mid z \in L, xz \leq y\}$ has a largest element. (Such a lattice is always distributive.) Properties of c.i.
lattices have been studied by Balbes and Horn [86]. Let $\{L_\alpha \mid \alpha \in P\}$ be a set of distributive lattices indexed by the elements of a
poset P. A distributive lattice L is called the order sum of the lat-
tices L_α $\left(L = \sum_{\alpha \in P} L_\alpha\right)$, if for all $\alpha \in P$ there exist monomorphisms
φ_α: $L_\alpha \to L$ satisfying the conditions: 1) $\bigcup_{\alpha \in P} \varphi_\alpha (L_\alpha)$ generates the
lattice L; 2) if x $\in$ L_α, y $\in$ L_β, and $\alpha < \beta$, then $\varphi_\alpha (x) < \varphi_\beta (y)$; 3)
if M is some distributive lattice and f_α: $L_\alpha \to$ M, $\alpha \in$ P, is a
family of homomorphisms such that when $\alpha < \beta$, $f_\alpha (x) \leq f_\beta (y)$ for
all x $\in$ L_α, y $\in$ L_β, then the homomorphism f: L $\to$ M exists and,
moreover, $f \varphi_\alpha = f_\alpha$ for all $\alpha \in$ P. Balbes and Horn, who introduced this concept, proved that $\sum_{\alpha \in P} L_\alpha$ exists and is unique to within
an isomorphism, and derived a number of properties of this lattice.
The construction of an order sum, generalizing the free product
and the ordinal sum of lattices can obviously be used not solely for
the distributive case.

Certain results on the connections between the set of join-
irreducible elements of a finite distributive lattice L and its imbeddings were obtained by Khisamiev [46]; he also proved a number of
theorems on the solvability of the $\forall \exists$ -theory and of the universal
theory for certain concrete classes of lattices.

In a set M let there be defined an operation τ by which an
ordered pair of elements a, b $\in$ M is related to a nonempty subset

aτb of set M. Let $\Lambda\tau B - \{x \mid x \in a_\tau b$ for some a $\in$ A, h $\in$ B$\}$.
Axioms: 1) (aτt) τc = aτ (bτc); 2) aτb = bτa; 3) c $\in$ aτb $\Rightarrow$ a $\in$ bτc.
A set M with operation τ satisfying these axioms was called a mul-
timodule by Nakano [355]. By assuming that besides τ an associative binary operation is given on the set M and by connecting
these operations by natural axioms, he introduced the concept of a
multiring. The usual modules and rings are special cases of multimodules and multirings, and a number of known theorems ensue
from the more general results on these new systems. The concepts of ordered multirings and multimodules are introduced in a
natural way. Characteristics are given of modular lattices as
multimodules and of relatively-complemented distributive lattices
as multirings.

Give'on [192] considered the set L_n of nth-order matrices
with elements from a distributive lattice with zero and unit, (which
also is a distributive lattice with zero and unit). Conditions of
nilpotency, invertibility, and orthogonality for L_n-matrices and
their other properties were ascertained. See Avann [79, 80], Ky
Fan [164], and Goodstein [194] for certain functions on a distributive
lattice. Padmanabhan [369] has given the characteristics of distributive lattices, as well as of chains and modular lattices, in
terms of ternary relations. One special case of distributive lattices, of interest in logical applications, was considered by Belnap
and Spencer [99]. See [384] for the structure of distributive quasi-
lattices. Necessary and sufficient conditions for the distributivity
of skew (noncommutative) lattices were pointed out by Gerhardts
[188].

§4. Geometrical Aspects and the Related
Investigations

A series of papers in this direction have been devoted to the
investigation of the concept of parallelism (which is introduced
analogously by Hsu [230]) and of affine lattices. Maeda [304] considered a weakly modular (a $\cap$ b $\neq$ 0 $\Rightarrow$ (a, b) M) matroid lattice L
of length $\geq$ 4, which he called affine if the weak parallelism axiom
is fulfilled in it: if p is a point, l is a straight line p $\nleq l$, then
there exists not more than one straight line q such that q $\parallel l$, p $\leq$ q
(in the strong axiom the words "not more than one straight line"
are replaced by "one and only one straight line"). In an affine

matroid lattice (a.m.l.) the straight line l is called anticomplete if for any point $p \nleq l$ there exists a straight line k such that $l \parallel$ k, $p \leq$ k. An element a $\in$ L is anticomplete if any straight line l, $l \leq$ a, is anticomplete. By I (p) we denote the maximal anticomplete element containing the point $p \in$ L. If I (p) $\neq$ 1, then for any point q either I (p) $\parallel$ I (q) or I (p) = I (q). An element a $\in$ L is $\parallel$-closed if $r \leq a \Rightarrow$ I (r) $\leq$ a. Maeda proved that the set $\Omega_0 = \{$I(p) $|p \in$ L$\}$ of a nonmodular a.m.l. is an irreducible projective space and the set M of all $\parallel$-closed elements is an irreducible modular matroid sublattice isomorphic to the lattice of linear subsets of the set Ω_0. The sublattice M is called the modular center of the a.m.l. L. A nonmodular a.m.l. is modularly irreducible (M = $\{0, 1\}$) if and only if it satisfies the strong parallelism axiom. In [305] Maeda proved certain analogs of theorems known for continuous geometries, for affine lattices. By continuing the investigation of the connection between affine and projective lattices, he introduced the concept of the concept of the modular contraction R = [p, I (p)] for a.m.l. If Λ is a complemented modular lattice and $S \subset \Lambda$ is an ideal, then the set L = $\Lambda \backslash (S \backslash 0)$ with the same order relation turns out to be a weakly modular lattice called the Wilcox lattice. Let L be an a.m.l. and let a $\parallel\parallel$ b $\Leftrightarrow$ a $\parallel$ b or a = b. The relation $\parallel\parallel$ is an equivalence. The set of cosets S = $\{$[a] | a $\in$ L, a is anticomplete$\}$ is isomorphic to the modular contraction R of lattice L and L can be treated as a Wilcox lattice, Λ = L $\cup$ S. A definition of parallelism not using the notion of a point and certain results obtained previously in [305], generalized to arbitrary Wilcox lattices, are given in [306]. Fujiwara [182, 183] considered parallelism in left complemented lattices. By using a construction generalizing a Wilcox lattice he succeeded in obtaining results analogous to Maeda's theorems [305]. Sachs [426] proved that if in a matroid lattice L there exists a set M of elements satisfying certain conditions, then the set $\overline{M}$ dual to M is a matroid lattice of the same length as L. In this case $\overline{M}$ is said to be inverse to L. Sachs investigated the conditions for the existence of more than one inverse and the conditions for the determinability of lattice L by the lattice $\overline{M}$. The concept of perspectivity in a matroid lattice and its connection with modularity and parallelism was studied by Maeda [307]. Using the point perspectivity relation introduced by him, we can extend dimension theory to matroid lattices.

A complete complemented weakly-modular lattice is called a locally modular lattice (l.m.l.) if it contains a modular ideal I such

that a = sup $\{I \cap [0, a]\}$ for any a $\in$ L. Ramsay [393] proved that
a l.m.l. is semimodular and derived a number of conditions equivalent to local modularity. He proved that from an orthocomplemented weakly-modular lattice L we can pick out a maximal local-
ly-modular direct summand and described the remaining terms of
this expansion. Necessary and sufficient conditions were obtained
for L to be a dimension lattice in the sense of Loomis [297] and it
was shown that every l.m.l. is a locally-finite dimension lattice. In an
orthocomplemented lattice L an element a is called modular if the lattice [0, a] is modular and (a, x) M† for all x $\in$ L. A lattice in
which every element is a sum of modular elements is called nearly
modular. MacLaren has studied the properties of such lattices
[300]. He established that a complete orthocomplemented lattice
(c.o.l.) is locally finite if and only if it is nearly modular and he
derived the necessary and sufficient conditions for a c.o.l. to be
a dimension lattice of type I.

A lattice L is called: a) an IC-lattice if for any e $\in$ L there
exists a complement f $\in$ L such that (f, e) M and the pair (e, f) is
dually modular; b) a quasiorthomodular lattice (q.-o.-m.l.) if L
and the lattice dual to it are IC-lattices. These lattices were
studied by Janowitz [260] who proved, in particular, that the properties of being an IC-lattice and a q.-o.-m.l. are hereditary under
transition to an interval. He also investigated [255] the mappings
φ: L $\rightarrow$ L of q.-o.-m.l. called quantors: 1) $0\varphi = 0$; 2) $e \leqslant e\varphi$, $\forall e \in L$;
3) $(e \wedge f\varphi)\varphi = e\varphi \wedge f\varphi$, $\forall e$, $f \in L$. The characteristics of quantors on the
class of local orthomodular lattices were obtained. In this same
paper it was proven that a dimension lattice of type I with certain
additional conditions can be coordinated by a Baer *-semigroup.
Janowitz [254] also studied the semigroup S (P) of all transformations of a poset P, possessing a certain property. If S (P) is a
Baer semigroup, the P is a lattice coordinatizable by this semigroup. The semigroup S (P) was characterized for the case when
the set P admits of involution and constraints on S (P) were indicated which are necessary and sufficient for P to be a q.-o.-m.l.
Also see [258]. Foulis [177] established interesting connections
between the lattice-theoretic properties of an orthomodular lattice
and the semigroup properties of the Baer *-semigroup coordinati-
zing it. Schreiner [435] studied the properties of the relation

---

†Translator's Note: This is the authors' notation for the phrase "the ordered
pair (a, x) is a modular pair." The more customary usage is M (a, x).

(a, b) M in a orthomodular lattice and the conditions equivalent to its symmetry. Maeda [308] proved that in a orthocomplemented relatively-atomic lattice the following conditions are equivalent: a) (a, b) M $\Rightarrow$ (b, a) M; b) (a, b) M and (a $\cap$ b) = 0 $\Rightarrow$ (b, a) M; c) for any atom p $\in$ L, p $\nleq$ a, the element a $\cup$ p covers a. In [309], using the distributivity and orthogonality of certain elements, he introduced D-relations and D-implications in orthocomplemented lattices and in terms of these he described conditions equivalent to orthomodularity, thus generalizing the analogous results of Foulis, Nakamura, and Piron.

Artmann [74] considered questions connected with the coordinatization by a ternary ring of an arbitrary complemented modular lattice possessing a homogeneous basis of degree n = 3 (regular rings are applicable for n $\geq$ 4). He called such a lattice plane. He obtained the properties characterizing the connection of a certain class of automorphisms of plane lattices with their ternary rings and pointed out the condition under which a plane lattice is imbedded in a direct product of a plane lattice constructed over a regular ring and of a certain family of projective planes coordinatizable by distributive quasidivision rings. Paper [75] is devoted to conditions equivalent to a plane lattice being a subdirect product of projective planes. Occupying himself with Skornyakov's 26th problem (see [455], p. 167), Artmann [76] indicated the properties of the lattice L (R) necessary and sufficient for the commutativity of ring R.

Various generalizations of the usual "betweenness" relations and their geometric properties in certain classes of lattices were considered by Bumcrot [114, 115]. Jakubik and Kolibiar [252] studied α-Euclidean lattices (α is a cardinal number) and showed that a lattice is α-Euclidean if and only if it is representable as a α-product of chains. By introducing the concepts of an R-extension (R is a regular ring) of an Abelian group G and of an R-vector space, generalizing the Boolean extension and the Boolean space over a Boolean algebra, Subrahmanyam ([38], § 1) and Gopala [196] proved that every R-extension of an Abelian group G is an R-vector space and also proved the converse.

Wille [491] considered a certain geometric system, which he called an n-grade geometry, and derived the necessary and sufficient conditions for the isomorphism of an arbitrary lattice L and the lattice of subspaces of this geometry.

Let L be a complete modular $\cap$-continuous lattice with an ascending or descending chain condition. With a certain set Γ of isomorphisms between the factors of lattice L Fort [175] connects an equivalence relation on the set of co-irreducible factors (a factor A/B is co-irreducible if the element B is $\cap$-irreducible in the lattice A/B), the cosets with respect to which are called Γ-types. To every element $x \in$ L there is associated a certain set Mx of Γ-types; x is called Γ-isotypic if Mx contains one Γ-type. It was proven that every element of lattice L has an irreducible representation as the intersection of a finite number of Γ-isotypic elements, uniquely to within Γ-equivalence of the component; certain properties and applications to the lattice of submodules were pointed out. Ewald [162] has considered the properties of extensions (not modular but possessing relative complements) for complete atomic $\cap$-continuous complemented modular lattices.

See Ramsay [394] on the subdirect decompositions of continuous geometries. Brickman and Fillmore [110] showed that the lattice L (A) of invariant subspaces of a linear transformation A is distributive if and only if the transformation is cyclic, and also obtained more properties of the lattice L (A) and of the lattice L (Q) of a nilpotent transformation Q. Crapo [126] studied the properties of the mappings of complete atomic continuous semimodular lattices. Also see p. 146 for modular lattices.

§ 5.   Homological Aspects

In recent years there has been a sharp increase in interest in the application of homological methods in various areas of general algebra, which is reflected also in the researches on lattice theory. To tell the truth, the number of papers devoted to these aspects is comparatively small. This is explained, possibly, by the general difficulty of posing and solving homological problems in non-Abelian categories.

The concepts of projection and injection in the category of Boolean algebras and in the category of complete Boolean algebras were studied by Halmos ([38, § 1). The papers by Keisler [273] Mączyński [302] on universal homogeneous Boolean algebras, and also the paper by Linton [296], may prove to be useful in the study of morphisms of Boolean algebras. Isbell's paper [246] deals with and solves all but one of 24 problems: for what cardinal number m is the class of all (Boolean algebras/fields of sets) that have no infinite (complete/freeprojective) (sub/quotient) algebra or field closed under the formation of (free sums/direct products) of families of cardinality m?

By analyzing the category of distributive lattices Balbes [85] proved that here complete Boolean algebras, and these only, are injective objects. A distributive lattice is projective if and only if it is a retract of a free distributive lattice. A projective distributive lattice does not contain uncountable chains.

The fundamental technical tools used by Balbes to study the projection of distributive lattices were E-free sequences.

An expression (e) of the form $x_{i_1} \cdot \ldots x_{i_n} \leqslant x_{j_1} + \ldots + x_{j_m}$. is called an inequality in the set $\{x_i\}$, $i \in I$. We say that a sequence $\{a_i\}$, $i \in J \supseteq I$, of elements of a distributive lattice L satisfies inequality (e) if $a_{i_1} \cdot \ldots \cdot a_{i_n} \leqslant a_{j_1} + \ldots + a_{j_m}$. Let E be some set of inequalities in the set $\{x\}$, $i \in I$. The sequence $\{a\}$, $i \in J \supseteq I$, is called E-free if it satisfies all of the inequalities in E and their consequences and does not satisfy any other inequality whatsoever. It was proven that a distributive lattice L generated by an E-free sequence $\{a_i\}$, $i \in I$, is projective if and only if for each $i \in I$ there exist finite nonempty subsets $S_{i,1}, \ldots, S_{i,p(i)} \subseteq I$ satisfying the conditions: (1) $a_i = \sum_{k=1}^{p(i)} \left(\prod_{l \in S_{i,k}} a_l \right)$; (2) let $x_{i_1} \cdot \ldots \cdot x_{i_n} \leqslant x_{j_1} + \ldots + x_{j_m} \in E$; then for any $f(i_t)$, $0 < f(i_t) \leq p(i_t)$, $t = 1, \ldots, n$, there exist $q \in \{1, \ldots, m\}$ and $r \in \{1, \ldots, p(j_q)\}$ such that $S_{j_q, r} \subseteq \bigcup_{t=1}^{n} S_{i_t, f(i_t)}$. In spite of the cumbersomeness of this criterion it allows us to obtain examples of projective lattices and, in a number of concrete cases, to solve successfully the question of whether this or the other lattice is projective. A set G of elements of a distributive lattice L is called a-disjointed if xy = a for all x, y $\in$ G. Any a-disjointed set of elements of a projective distributive lattice L is finite. Among the further results in [85] we note the following: 1) Boolean algebras and Boolean rings are projective if and only if they are finite; 2) a chain is projective if and only if it is countable; 3) a finite distributive lattice is projective if and only if the sum of any two meet-irreducible elements is meet-irreducible; every finite projective distributive lattice is isomorphic with the family of all nonempty proper hereditary subsets of a finite lattice, and, moreover, the converse also is true; 5) the direct product $\prod\limits_{i \in I} L_i$ of finite distributive lattices is projective if and only if the lattices L_i are projective for all $i \in I$ and $|L_i| = 1$ for all but finitely many $i \in I$.

Balbes and Horn [86] introduced the concept of a conditionally implicative lattice and proved that every projective distributive lattice is conditionally implicative. They also investigated the conditions under which the order sum $\sum\limits_{\alpha \in P} L_\alpha$ of projective distributive lattices L_α is projective (see § 3 of the present survey for the definition of a conditionally implicative lattice and for the description of the construction of $\sum\limits_{\alpha \in P} L_\alpha$. A number of results were obtained in this direction, of which we note two: 1) the order sum $\sum\limits_{\alpha \in P} L_\alpha$ of distributive lattice L_α with zero and unit is projective if and only if all the summands and the lattice $P^{*\dagger}$ are projective; 2) the ordinal sum $L_0 \oplus L_1$ of free distributive lattices is projective if and only if one of the following conditions is fulfilled: a) the lattices L_0 and L_1 are countable; b) at least one of the lattices L_0 and L_1 is finite.

By investigating Lukasiewicz's three-valued algebras (i.e., distributive lattices L with 0 and 1, in which certain operations N and $\mu: L \to L$, called negation and possibility, are given), Monteiro [336] proved that such an algebra is injective if and only if: 1) lattice L is complete and 2) there exists an element c of L such that $Nc = c$.

Topping's paper [476] is devoted to the solution of analogous problems in the category of vector lattices. It turned out that this category does not contain nontrivial injective objects. A vector lattice A containing a positive orthogonal family $\{a_i\}$, $i \in I$, of

---

$\dagger$ $P^* = \sum\limits_{\alpha \in P} L_\alpha$, where L_α is a one-element lattice for all α.

elements generating it linearly is projective if and only if it is countable. Thus, this category is found to have traits in common with the categories of Boolean algebras and of complete Boolean algebras. In vector lattices Topping introduced an analog of the Jacobson radical and studied its structure. He also obtained a number of interesting results on semisimple, radical, and local vector lattices. If A is a vector lattice, then there exists a vector lattice E in which A is a maximal ideal and, moreover, $\text{Rad } E = A$. Every linearly ordered vector lattice is either radical or local. A vector lattice A, projective with respect to the module of its own radical, splits (as a vector space) into the direct sum $A = N \oplus S$, $N = \text{Rad } A$ and S is semisimple, $S = A/N$.

An object of a category is called exactly injective if it is injective relative to exact monomorphisms in the sense of Grothendieck. In the category of posets the exact monomorphisms coincide with imbeddings. Banaschewski and Bruns [88] derived certain conditions equivalent to the exact injection of a poset and also showed that an essential exactly-injective extension in this category coincides with completion by sections. Fort [176] introduced the concept of essential extension for an element of a complete modular ∩-continuous lattice and proved a number of properties analogous to the properties of essential extensions of modules. An element of a lattice, not having proper essential extensions, was called injective by him. The existence of an injective envelope for every element was proven, as well as certain other properties for the concepts introduced.

A category G of lattices with 0 and 1 (and with homomorphisms preserving 0 and 1), as well as its complete subcategory Gd generated by distributive lattices, were considered by Diaconescu [142]. He proved that in the category G there exist the cokernel of any pair of morphisms with common beginning and end and the direct limit of any upward-directed diagram, and proved that the category Gd is right-complete. Here the free product of two distributive lattices (in the category Gd) is their tensor product.

Another direction was taken by Folkman [174], Rota [404], and Mather [314]. A set C of elements of a lattice L with 0 and 1 is called a cross-cut if: 1) $0, 1 \notin C$; 2) $x, y \in C \Rightarrow x$ and y are incomparable; 3) any finite chain of elements of lattice L can be extended upto a chain containing elements of set C. Rota determined the homology groups H_n (C) for a finite lattice L and showed that they do not depend on the choice of C. Folkman generalized his results to infinite lattices. Let K (C) be an abstract simplicial complex whose vertices are the elements of cross-cut C, while the simplexes are finite subsets $\{x_1, ..., x_n\} \subset$ C such that either $x_1 \wedge ... \wedge x_n \neq 0$, or $x_1 \vee ... \vee x_n \neq 1$. We define H_i (C) = H_i (K (C)). Now let K (L) be an abstract simplicial complex whose vertices are the elements of the set $L - \{0, 1\}$ and let the simplexes be finite linearly-ordered subsets of this set. We set H_i (L) = H_i (K (L)).

Folkman proved that H_i (C) $\simeq$ H_i (L). He also computed the homology groups for a certain special case of finite lattices which he termed geometric. Mather showed that the homotopic type of complex K (C) does not depend on the choice of the cross-cut (for finite lattices).

§ 6. Lattices of Congruences and of Ideals of a Lattice

Schmidt [430] showed that for a finite lattice L in which $(x \cap y) \cup z = x \cup z = y \cup z$, there exists a finite universal algebra with a congruence lattice isomorphic to L. Hence, Birkhoff's Problem 48 receives an affirmative answer for lattices of the form indicated. Papert [372] studies congruence lattices of similattices. A Boolean algebra is representable as a congruence lattice of a semilattice if and only if it is completely distributive. Paine's paper [371] is devoted to lattices of equivalence relations. Kapolinskaya [16] described lattices representable by congruence lattices of distributive lattices. The class of lattices whose congruence lattices are isomorphic to the congruence lattices of distributive lattices also has been described (Kapolinskaya and Stelletskii [17]).

Using semicongruences Permutti [374] characterizes semimodular lattices, while Fujiwara [185] characterizes semimodular lattices, distributive lattices, and relatively-complemented lattices.

Morgado has devoted a series of papers [342, 344, 345] to the study of α-complete congruences and of the ideals of α-complete lattices, where α is some cardinal number.

In a large series of papers Iqbalunnisa [233-239, 241-244] continued the study of lattices of congruences and of ideals in lattices. In particular, [237] and [239] are devoted to the solution of several problems of Grätzer and Schmidt (see [38]). By continuing the researches of Grätzer and Schmidt, Kinugawa and Hashimoto [276] study relative maximal and prime ideals in lattices. They have proven that a lattice is distributive if and only if each of its relative maximal ideals is prime. An analog of Cohn's ring-theoretic theorem also has been obtained. Janowitz [256] studies various ideals of lattices and solves some of the problems of Grätzer and Schmidt (also see [429]).

In another paper [257] Janowitz considered the normal ideals and the center of the ideal lattice of a given lattice. Schmidt [431] constructed an example of a lattice containing an ideal which is not a homomorphism kernel of this lattice but is the homomorphism kernel of some sublattice of it.

Congruences and ideals in pseudocomplemented lattices have been examined also by Varlet [482, 483]. In another paper [484] he studied congruency in a semilattice L. He showed, in particular, that the congruence lattice of L is Boolean if and only if every closed interval in L is linearly ordered and has a finite length. Benzaken [105] studies the congruence lattice of a distributive pseudolattice and of its ideals. (A distributive pseudolattice is a semilattice relative to the operation + with associative and distributive (but not commutative) multiplication, for which the condition: $a + ab = a(b \cdot a) = a$ is valid,)

Venkatanarasimhan [487, 486] studied ideals in semilattices and in modular lattices. He showed that the lattice of all ideals of a pseudo-complemented semilattice is a pseudocomplemented lattice and every normal ideal decomposes into a sum of principal normal ideals and into the intersection of certain principal ideals.

Mayer-Kalkschmidt and Steiner [319] obtained certain results concerning meet-closed and inductive (i.e., join-closed directed systems of subsets) families of subsets of an arbitrary set and families of ideals of a poset.

§ 7.   Lattices of Subsets, of Subalgebras, etc.

The papers relevant here can be divided into two groups. In the first group belong the papers which study lattices representable by sets satisfying some conditions or others. For example,

the lattices of subgroups, of subrings, etc. In the second group
belong the papers which study algebraic systems (groups, rings,
etc.) by drawing on the lattice of their subsystems.

Let us look at papers in the first group. In 1946 Birkhoff
proved that for any group G there exists a universal algebra whose
automorphism group is isomorphic with G. Also well known is the
result of Birkhoff and Frink that compactly-generated lattices of
all subalgebras of universal algebras. Grätzer and Schmidt showed
that every compactly-generated lattice is isomorphic to the lattices
of all congruences of some universal algebra ([38], § 4). Continuing
these investigations Schmidt proved that for any group G and for
any compactly-generated lattice L we can find a universal algebra
whose automorphism group is isomorphic to G, and we can find a
congruence lattice isomorphic to L (see [428]). Furthermore, we
can find a universal algebra whose automorphism group is isomorphic to G, while the subalgebra lattice is isomorphic to L (see
[427]).

A binary relation whose diagram is a subalgebra in $A \times A$ is
called a correspondence of a universal algebra A. Iskander [14]
(also see [203]) showed that the necessary and sufficient condition
for a lattice L to be isomorphic to the lattice of all correspondences of some universal algebra, and for its automorphism α, whose
square is an identity ($\alpha^2 = \varepsilon$), to be realized by passing to the opposite correspondence, is that the lattice L be compactly generated.
This result generalizes the Birkhoff-Frink theorem. In another paper [15] Iskander found the necessary and sufficient conditions that
for the given lattices $\mathfrak{Q}_1$, $\mathfrak{Q}_2$, $\mathfrak{Q}_3$, for a given sublattice L_i, and for
the automorphism α_i of the lattice $\mathfrak{Q}_i$ ($\alpha_i^2 = \varepsilon$, i = 1, 2) there exist
partial universal algebras A_1 and A_2 such that L_i be isomorphic to
the lattice of all subalgebras, that $\mathfrak{Q}_i$ be isomorphic to the lattice
of all correspondences of algebra A_i, that the automorphism α_i be
realized by passing to the opposite correspondence, and that the
lattice $\mathfrak{Q}_3$ be isomorphic to the lattice of the correspondences of
algebra A_1 with algebra A_2, i.e., to the lattice of binary relations
of set A_1 with set A_2, preserving the operations defined in A_1 and A_2.

Grätzer [201] characterized lattices of subalgebras with infinitary operations and proved that every complete lattice is isomorphic to the lattice of all congruences of some universal algebra
with infinitary operations.

Two problems, posed by Valutse [5], are connected with the questions being considered: 1. Will every sublattice of the lattice of all subsets of a set M, containing M, be a lattice of subalgebras of some algebra with the unary operations (M, Ω)? 2. Will every sublattice of equivalences of set M, containing an equality and a universal relation, be a lattice of congruences of some algebra with the unary operations (M, Ω)?

As yet unsolved is the question of under what conditions will an abstract lattice be isomorphic to the lattice of subgroups of some group. Anishchenko [3] has examined a special case of this problem. Namely, he proved that for a lattice L to be isomorphic to the lattice of subgroups of some Abelian p-group with more than two generators, it is necessary and sufficient to fulfill the following conditions: a) L is a modular lattice; b) if an element a $\in$ L is not join-reducible and a $\geq$ b, then b also is join-irreducible; c) the condition dual to b); d) each element other than 0 covers either exactly one or exactly p + 1 elements.

We now note the papers in the second group. A number of papers relating to the theory of semigroups draw on lattices of congruences [197] or of subsemigroups [11-13, 49, 52, 386]. Many group-theoretic papers consider lattices of subgroups and of normal divisors [1, 15, 23, 56, 57, 101, 117, 127-130, 160, 221, 223, 356]. Paper [31] is devoted to the theory of topological (locally-compact) groups. The papers [44, 90, 96, 109, 261, 262, 278, 281, 401] on ring theory also relate to this direction. Paper [386] is devoted to a semilattice of partitionings of an arbitrary set.

§ 8.   Closure Operators

In a series of papers Morgado has continued the study of the lattice Φ (L) of closure operators of a complete lattice L. In particular, necessary and sufficient conditions have been obtained for a closure operator to be a distributive (upper distributive), completely distributive (completely upper distributive) element of lattice Φ (L) (see [341]). An element $\Delta'\in \Phi$ (L) is central if and only if there exists an element $\Delta' \in \Phi$ (L) such that the equality

$$(\Delta \cap \varphi_x) \cup (\Delta' \cap \varphi_x) = (\Delta \cup \varphi_x) \cap (\Delta' \cup \varphi_x)$$

holds for every dual atom $\varphi_x \in \Phi(L)$ (see [343]). Furthermore, necessary and sufficient conditions have been obtained for the center of lattice $\Phi(L)$ to be a complete lattice [340].

Continuing the earlier investigations on quasi-isomorphisms of lattices ([38], §6), Morgado [347] constructed an example showing that from the homomorphism $L \rightarrow M$ of a complete lattice L onto a complete lattice M there does not ensue the homomorphism of the corresponding lattices of closure operators. Namely, let L be a five-element nonmodular lattice and let M be a free lattice with two generators. Then there does not exist a homomorphism of $\Phi(L)$ onto $\Phi(M)$. The necessary and sufficient conditions for the lattice $\Phi(M)$ to be a complete homomorphic image of lattice $\Phi(L)$ have also been established. In [350, 352] Morgado studies lattices of residuated closure operators, i.e., of closure operators φ for which there exists a closure operator φ^* such that $\varphi\varphi^* \leq \varepsilon \leq \varphi^*\varphi$, where ε is the identity operator. In particular, Janowitz's problem on the conditions for the lattices of residuated closure operators of two complete lattices to be isomorphic is studied. Notice also that any component of a direct decomposition of lattice $\Phi(L)$ is isomorphic to the lattice of closure operators of some complete lattice [35].

Davies [133] constructed the least extension of a closure operator given on a complete sublattice M of a complete lattice L, upto a closure operator on L.

Banaschewski [87] studies regular closure operators, i.e., closure operators satisfying the additional conditions: $\lambda(x \cup y) = \lambda x \cup \lambda y$ and, if $x \cap \lambda y = \emptyset$, then $\lambda x \cap \lambda y = \emptyset$, given on a sublattice of the lattice of all closed sets of a T_1-space, whose elements form a basis of closed sets of the given space. The connection between these operators and certain equivalence relations in the original T_1-space was established.

§9. Topological Aspects

Many papers have been devoted to the study and comparison of the various topologies defined by order. Using the modern approach Sekanina and Sekanina [444] (also see [447, 448]) introduce the concept of a topology compatible with ordering. Namely, we say that a topology is compatible with ordering if it is a T_1-topology and if for any two elements a < b there exist neighborhoods O_1 and O_2 of them such that the neighborhood O_1 does not contain elements x ≥ b while the neighborhood O_2 does not contain elements y ≤ a. A T_1-topology is called strongly compatible with ordering if for any two elements a < b there exist neighborhoods O_1 and O_2 of them such that the condition x ≱ y is valid for any x ∈ O_1, y ∈ O_2. It was shown that a large part of the known topologies (interval, order-convergence, topology of ideals, topology of topological lattices) are compatible with ordering (if it is a T_1-topology). It was ascertained also which topologies are strongly compatible with ordering. Conditions were found under which an interval topology is the least element of a poset of all order-compatible topologies.

Atsumi [78] indicated the necessary and sufficient condition for a lattice to be Hausdorf and bicompact in its own interval topology (every directed set should have a directed subset convergent in the order topology). Matsushima [316] studies the B^*-topology in distributive lattices. (A B^*-topology is defined as the topology whose pseudobase of closed sets is formed by the sets $B^*(a, b) = \{x \mid b \in B(a, x)\}$, where $B(a, x) = \{y \mid ay + xy = (a + y)(x + y)\}$). He examines the connection between the B^* and interval topologies. In particular, certain conditions for them to be coincident are established. Alo and Frink [70] study the topology of ideals, the interval topology, and a new Birkhoff interval topology on lattice products. Completely irreducible ideals and dual ideals form the pseudobase of open sets for the topology of ideals. The base of closed sets for the Birkhoff interval topology is formed by sets whose intersections with any closed intervals are intersections of finite unions of closed intervals. The question: For which of the listed topologies does the topology of lattice products coincide with the product of the cofactor topologies? was considered. This question has been answered positively for the topology of ideals of a finite lattice product, for the interval topology of the product of an infinite set of lattices with 0 and 1, and for the new interval topology of the product of a finite number of chains. The topology of ideals was generalized in a natural way to the $\mathfrak{m}$-ideals of a poset (see Mayer and Novotný [318, 367]). By an $\mathfrak{m}$-ideal of a poset P we mean a subset I of it such that $M^{*+} \subset I$ for any $M \subset I$ whose cardinality $|M| < \mathfrak{m}$. It was established that the topology of $\mathfrak{m}$-

ideals of the cardinal product of $\mathfrak{m}$-directed sets P_ν ($\nu \in N$, $|N| <$ $\mathfrak{m}$) coincides with the product of the topologies of $\mathfrak{m}$-ideals of the cofactors. (A set is called $\mathfrak{m}$-directed if each of its subsets of cardinality less than $\mathfrak{m}$ is bounded).

By taking the set $V_i = \{ (x, y) \mid x\theta_i y \}$, where $\{ \theta_i \} = C$ is a downward directed system of congruences of lattice L, as a system of entourages, Rema [398] converted L into a uniform space (L, C). If every congruence of lattice L satisfies the condition: $x\theta y \Leftrightarrow x + t = y + t$ for some $t \in \theta$ (0), then (L, C) is called a C^*-lattice. C^* lattices were studied. If (L, C) is a compact Hausdorff complemented C^*-lattice, then L is a continuous geometry. McCartan [320] examined several topologies, including the Dedekind one (see [38]). Hansell [209], continuing the investigations of Wolk and Naito (see [38], § 7), studies posets not containing infinite subsets of pairwise-incompatible elements.

Kent [275] compares the convergence of filters relative to the interval topology with order convergence. These convergences turn out to be dual to each other in a certain sense. Papers [9, 315, 171] also are devoted to a comparison of various convergences. Filters in lattices are considered in [30, 274].

Various topologies connected with ordering in lattices are studied also in [21, 29, 41, 42, 111, 118, 140, 220, 227, 232, 269, 270, 317, 321, 337, 368, 375, 396, 417, 456, 469, 480, 489].

Drake and Throm [149, 150] continue the study of the representations of a lattice Γ by lattices of closed subsets of a T_0-space. Estimates were obtained for the cardinality of the family of such representations. The ordering of the set of T_0-spaces representing a given lattice Γ was considered (also see [167]).

A number of papers have been devoted to the study of the lattice of topologies on a fixed set. Steiner [464, 465] showed that locally connected topologies form a complete lattice and a sublattice of the lattice of all linear topologies of a linear space (also see [466]). Pervin and Biesterfeldt [376] studied the lattice of uniform convergence spaces.

The problem of the existence of complements in the lattice of all topologies on a fixed set is treated by Gaifman [186] and Berri [106]. Their results overlap those of Steiner [459, 460, 462].

It was shown that the lattice of all topologies on any set X is a complemented lattice. It is distributive only in the case when $|X| \leq 2$, and is autodual only when $|X| \leq 3$. The concept of a principal topology was introduced. It was proved that a principal complement exists for each T_1-topology. Schnare [433] showed that every complement to a T_1-topology possesses a majorizing complement of the same topology. An ultratopology (i.e., a topology majorized only by the discrete topology) possesses a maximal complement if and only if it is principal. Every nontrivial topology on a finite set X has at least $|X| - 1$ complements (also see [434]). Steiner [461] found the sufficient existence conditions for a compliment in the lattice of T_1-topologies, which in general, is absent. The paper [467] by Steiner and Steiner is devoted to this same question. In another paper [463] they constructed a complement of the usual real line topology in the lattice of all T_1-topologies on the line.

A number of papers have looked at lattice generalizations of the theory of topological spaces. Relevant here is the paper [373] by Papert, which studies the category of complete Brouwer lattices (corresponding to the lattices of open sets) with complete join-homomorphisms as morphisms. Holsztyński [228] generalizes the concept of topological dimension to the language of distributive lattices with 0 and 1, corresponding to the lattice of closed sets of a topological space. Dowker and Papert [148] obtained the lattice-theoretic generalization of the Urysohn lemma. Papers [178, 204, 205, 224, 391] are along this same line.

The concept of proximity in Boolean algebras was examined by Meenakshi [324]. Abbott [58] studies implicative algebras (i.e., meet-semilattices whose principal ideals are Boolean). The technique evolved by him permits us to specify algebraically a topology on a given set and to carry over the fundamental topological concepts to implicative algebras. Bănică and Stănăşilă [89] study the spectrum of a complete lattice L(i.e., a certain topological space of quasisimple elements of L) and consider an application to ring theory.

Metric lattices are studies by Previale [391] and Rema [395]. Riečan's paper [32] touches on the theory of metric lattices.

§ 10. Partially-Ordered Sets

The investigation of isotonic mappings plays a major part here. Let P be a partially-ordered set (poset) and let the mapping f: P → P be isotonic. Kolibiar [279] showed that if P has a zero and if every completely ordered chain has a sup in P, then the set P (f) of fixed points (f.p.) of this mapping is not empty and has a least element; furthermore, sup C (in P (f)) exists for every chain C ⊂ P (f) and, if C is maximal, sup C (in P) = sup C (in P (f)). A poset P is called left- (right-) complete if for any non-empty subset X ⊂ P there exists inf X ∈ P (sup X ∈ P). Kurepa [284] established that if the set P is left- (right-) complete and if the mapping f is isotonic (antitonic), and, moreover, if $\{x \mid x \in P, f (x) \geq x\} \neq \emptyset(\{x \mid x \in P, f (x) \geq x\} \neq \emptyset)$, then P (f) is a left- (right-) complete nonempty set. If f is an antitonic mapping, then P (f) is an antichain (empty or nonempty); if, further, the set P is left- or right-complete, then P (f) can be of arbitrary cardinality. In [285] Kurepa studied the set ωP of completely ordered subsets of set P, equipped with a certain order relation. He showed that a strictly increasing mapping f : ωP → P does not exist. Hence it follows that such a mapping does not exist for the set of all subsets of set P, ordered by inclusion. Some analogous results were proven for the case when P is a subset of the set of real numbers. Demarr [138] proved a number of theorems on the existence and properties of common f.p. for a certain nonempty commutative family of isotonic mappings and obtained the following theorems for a right-complete set and for the pair of isotonic mappings f, g: P → P: 1) if f and g are commutative and f g has f.p., then f and g have a common f.p.; 2) let x ≤ f (x) ⇒ f (x) ≤ g (x) and f (x) ≤ x ⇒ g (x) ≤ f (x), then if f or g has f.p., they have a common f.p. Wong's paper [492] is devoted to solving analogous problems.

A poset is called linearly complete if each of its maximal chains is complete. Kogalovskii [20] proved that for f.p. to exist of every isotonic transformation of a poset (of a semilattice) it is necessary and sufficient that this set be linearly complete. The study of posets is equivalent to the study of groupoinds possessing certain properties. If G is such a groupoid, then the order† rela-

---

†Everywhere in what follows, by "order" we mean a partial ordering.

tion π: aπb $\Leftrightarrow$ ab = a is called natural. Let ρ and ρ' be some order
relations on the groupoids G and G', respectively, and let π and π'
be natural order relations. Fiala and Novak [166] studied the conditions under which the set of isotonic mappings (G, ρ) $\to$ (G', ρ')
coincides with the set of isotonic mappings (G, π) $\to$ (G', π') (the
latter coincide with homomorphisms of the groupoids).

Also see Hsu [231] and Sturm [470] on isotonic mappings of
posets.

An isotonic mapping φ: M $\to$ $M_0 = \{0, 1\}$ such that if a maximal element u $\in$ M exists, then $\varphi(u) = 1$, is called the character
of the chain M. Kowalski and Pondělíček [282] proved the theorems:
1) the chain M is antiisomorphic to the ordered set M^Λ of its own
characters if and only if it is a completely ordered set with largest
element; 2) the chain M is isomorphic to $M^{\Lambda\Lambda}$ if and only if M is a
finite set. Certain problems on permutation chains were considered
by Jullien [264]. Bruns [113] proved an assertion which in certain
applications is interchangeable with Iwamura's lemma on the fact
that every infinite directed set is the join of chains of directed sets
of lower cardinality. Let k, l be positive integers. There exists a
positive integer h (k, l) satisfying the condition: if a set S consists
of h elements and if $f(X) \subseteq X$ for each $X \subseteq S$ and, moreover, if
$f(X)$ consists of not more than l elements, then we can find subsets
$X_0, ..., X_k$ of set S such that $X_0 \subset X_1 \subset ... \subset X_k$ and $f(X_0) = ...f(X_k)$;
Rado [392] proved this generalized theorem of Harzheim on chains
of finite sets. Ivanescu [249] pointed out various methods for reducing the problem of seeking all minimal decompositions of a finite
poset in chains to the problem of seeking the minimal value of a
pseudo-Boolean function with certain conditions (pseudo-Boolean
programming). See Havel [217] on the refinements of chains. Let
r and s be two binary relations on a set X, defined by the sets
R, S $\subseteq$ X $\times$ X. They are said to be chain-equivalent if $(R \cup R^{-1})\backslash\Delta =$
$(S \cup S^{-1})\backslash\Delta$ (Δ is the diagonal in X $\times$ X). Bednarek [98] proved that
that for every quasiorder there exists a partial order chain-equiv-
alent to it. He indicated [97] several equivalent formulations of two
theorems by Mickle and Rado on binary reflexive and symmetric
relations on abstract sets, each of which is equivalent to Zorn's
lemma. Fine [168] derived certain conditions for the existence of
a maximal element in a poset and gave an application of it to groups.

Novoa [366] gave the name n-ordered set to the pair (X, φ_n), where X is an arbitrary set and φ_n is a function, defined on all possible oriented n-simplexes of set X, taking the value 1, 0, -1 and possessing two properties which are analogs of antisymmetry and transitivity (the source of this concept was a certain function φ_n defined on ordered sets of n points in a Euclidean space). It turns out to be possible to carry over to such sets certain concepts of posets (the Dedekind cuts, completeness, the topology defined by the n-order, etc.) and the results in the theory of ordered sets and lattices.

Let T be a finite poset with 0 and 1, satisfying the condition: if x, y $\in$ T, x $\leq$ y, then any maximal chain connecting x and 1 has at least one element in common with any maximal chain connecting 0 and y. Benzaken [103] called such sets SP-sets (series-parallel). He proved that every SP-set is a lattice and established some further properties. Duda [152] called a poset X a stem if it possesses the least element and if for each x $\in$ X the set of predecessor elements is finite and totally ordered. The structure of a stem is studied and the concepts of proper topology and dimension are introduced. It is proved that every stem X can be converted into a metric space (X, ρ) with an integral metric ρ satisfying the condition: for any x, y $\in$ X and any integer k, $0 < k < \rho(x, y)$, there exists one and only one element z $\in$ X such that $\rho(x, z) = k$ and $\rho(z, y) = \rho(x, y) - k$. The property of being representable by such a space is a characteristic property of a stem.

Let us give certain definitions pertaining to Ginsburg and Isbell [191]. Let P be a poset. A subset $L \subset P$ is called: a) residual if $\{y \mid y \leq x$ for some $x \in L\} \subseteq L$; b) cofinal if it intersects every nonempty residual subset of set P. c) total if it contains a residual cofinal subset. A covering $\{V_\alpha\}$ of set P is called normal if the join of the set $\{R \mid R \subseteq P, R$ is residual and $R \subseteq V_\alpha$ for some $\alpha\}$ is cofinal. The mapping f: $P \to P'$ is called convergent if the preimages of total sets are total and the preimages of normal coverings are normal. The convergent mappings f, g: $P \to Q$ are said to be equivalent (f $\sim$ g) if for any residual $R \subset Q$ we can find a total $T \subset P$ such that $f^{-1}(R) \cap T = g^{-1}(R) \cap T$. The sets P and Q are cofinally similar if they can be imbedded as cofinal subsets in some poset; this is equivalent to the presence of a convergent mapping f: P → Q for which there exists a convergent mapping g: Q → P such that fg ~ 1 and gf ~ 1, i.e., to the type equivalency of sets P and Q. Paper [191] is devoted to the investigation of the category $\mathfrak{A}$ of ordered sets and convergent mappings, of the category $\mathfrak{B}$ of ordered sets and equivalence classes of convergent mappings (it is called the category of cofinal types), and also of their complete subcategories $\mathfrak{J}$ and $\mathfrak{J}^*$ called the category of canonic sets and convergent mappings and the category of canonic cofinal types, respectively.

It was proven that the category $\mathfrak{B}$ is imbeddable in the category of sets. A number of interesting functors were constructed, in particular, a functor from category $\mathfrak{J}$ into the category $\mathfrak{B}^*$ of complete Boolean algebras and complete homomorphisms inducing duality between category $\mathfrak{J}^*$ and category $\mathfrak{B}^*$. A number of other results were obtained. In [245] Isbell studies directed sets with the relation ≤: D ≤ E ⇔ the existence of a convergent function f: D → E. Conditions are investigated for the existence of sup and inf for families of such sets. A number of theorems are obtained on standard sets $\Delta_{\lambda\mu}$ (the direct set of all subsets of cardinality $\aleph_\lambda$ of an arbitrary set of cardinality $\aleph_\mu$). These investigations are a continuation of those of Tukey and of Schidt [432]. See Isbell [247] on the existence of sup for every directed subset of cardinality ≤ $\aleph_\alpha$ of a given poset.

Let H be a nonempty poset and let $\{G_\alpha \mid \alpha \in H\}$ be a system of posets. A lexicographic product is the cartesian product $G = \prod_{\alpha \in H} G_\alpha$ with a binary relation ≤ defined by the rule: $f \le g \Leftrightarrow$, if there exists $\alpha_0 \in H$ such that $f(\alpha_0) \ne g(\alpha_0)$, then there exists $\alpha_1 \le \alpha_0$ such that $f(\alpha_1) < g(\alpha_1)$. The papers by Novák [364], Rosenberg [403], and Harzheim [213] are devoted to the investigation of the properties of the lexicographic product. The last-named considered the imbedding of totally ordered sets in certain lexicographically ordered sets. In general, a lexicographic product is not a poset. The necessary and sufficient condition that $\prod_{\alpha \in H} G_\alpha$ be a poset was pointed out by Novak. Here, the conditions on the set H can introduce a new order $\prec$ such that H will satisfy a decreasing chain condition and $\prod_{\alpha \in H} G_\alpha = \prod_{\alpha \in H, \prec} G_\alpha$. Novák also gives a certain estimate of the dimension of the set $\prod\limits_{\alpha \in H} G_\alpha$. His papers [361-363] are devoted to various generalizations of the dimension of a poset.

By examining arbitrary posets A and B, Novák [360] found the necessary and sufficient conditions for the existence of a poset X such that $A \oplus X \simeq B$ ($\oplus$ is the ordinal sum).

Let n be an arbitrary cardinal number and let m (n) be the minimal number such that every poset of cardinality n is isomorphically imbedded in a lattice of cardinality $\leq$ m (n). Kurepa [287] proved that m (n) $\leq 2^n$ when n > 1. If n $\geq \aleph_0$, then m (n) = n. Hence it follows that any infinite poset can be imbedded in a lattice of the same cardinality.

A series of papers [62-65] by Abian and Deever has been devoted to the representation of a poset P by the set S of sequences of one and the same finite or transfinite type, composed of 0 and 1. Conditions under which such a representation exists, the properties of set S, the least order type of the sequences, etc., are ascertained. The representation of a set P by sequences composed of 0, 1, and the letter u are studied in [66, 61]. The results of Sierpiński and Popruženko were generalized. It was shown that every poset P is isomorphic to a set S of sequences of the type of P, formed from 0, 1, u (partially ordered in some manner). Harzheim's paper [212] is related to this theme.

Pretzel [390] (representation in the form of a join of chains) and Adnadević [68] occupied themselves with quostions on the representation of posets.

For any nonempty set S we shall denote by T (S) and L (S) the sets of partitions in S and on S, respectively, ordered in a natural way. Havel [216, 219] studied the properties of a certain mapping T (S) $\times$ T (R) $\to$ T (S $\times$ R) and showed that its restriction onto L (S) $\times$ L (R) is a lattice monomorphism. Also see his paper [215] as well as [161, 229] on a certain partition set. The subsets A, B of a finite poset K are said to be related (A$\bar{\rho}$B) if every element a $\in$ A is less than some b $\in$ B and every element b $\in$ B is larger than some a $\in$ A. Leininger [295] showed that if $\overline{K}$ is a partition of set K into nonintersecting classes, then $\overline{K}$ is partially ordered by the relation $\bar{\rho}$. He studied the properties of a partition $\overline{K}$ satisfying the condition: for any subchain C $\subset$ K no two elements are contained in one subset of $\overline{K}$. Let S be a simply ordered set of order type ξ. Milner [330] defined $\xi \in P(\alpha, \beta)$ if for any A, B$\subset$S such that S = A$\cup$B and a $\leq$ b for all a $\in$ A and b $\in$ B, there exists either A' $\subset$ A of order type α or B' $\in$ B of order type β. This paper, set-theoretic in nature, investigated the properties of the set P (α, β). Certain set-theoretic results on ordered sets are to be found also in Novák [365] and Harzheim [214].

See Frucht [181] on Birkhoff's Problem 6. Gedeonová [187] proved a certain theorem on posets, from which ensues a generalized Jordan-Hölder theorem for groups. Let G (S) be an automorphism group of poset S and let τ (S) be its order type. Morel [338] pointed out the conditions for τ (S), equivalent to the nontriviality of the group G (S), and conditions equivalent to G (S) not being Abelian. In the latter case its cardinality $\geqslant 2^{\aleph_0}$

Let 2p be the set of all subsets of an arbitrary set P. Certain theorems on the structure of the set U of all partial orders on 2p, invariant relative to the permutations of set P, were proved by Sekanina [445].

In conclusion we remark that semigroups of directed transformations and semigroups of endomorphisms of ordered sets were studied by Lyapin [27, 28], Aizenshtat [2], Devadze [10], and Skryago [39]. See the survey on semigroups [7] for a more complete presentation in this direction and also for a bibliography.

Posets are investigated also in [24, 25, 123, 231, 338, 442, 446, 488].

§ 11. Other Questions

Here we should note a series of papers devoted to modular lattices. Hájek [206, 207] gave certain numerical characteristics of modular lattices of finite length and their characteristics in the language of semisimple decompositions. Delany [137] delineated two important classes of $\cup$-continuous modular lattices in which every element has an irreducible representation in the form of meets of meet-irreducible elements: lattices with a maximality condition and atomic lattices. He also obtained some more interesting theorems on $\cup$-continuous modular lattices. Monk [334] obtained a theorem on the representation of one class of modular

lattices by the submodules of a module over a ring whose left
ideals form a chain. Draškovičová [151] gave the characteristics of
modular lattices in the language of a ternary relation which is a
natural generalization of the "betweenness" relation in metric lattices. A subset A of a lattice L is called a line if there exists a
one-to-one mapping of it onto a chain C possessing certain properties. Kolibiar [280] studied the properties of lines and showed
that an analog of the Jordan-Hölder theorem is valid for lines in
modular lattices. Lattices with a normality relation and the properties of normal series were examined by Janoš [253] and by
Dean and Kruse [136]. Ascoli and Teppati [77] obtained a lattice-
theoretic analog of the Schur lemma. A number of decomposition
theorems, generalizing certain well-known results in the theory of
ideals to the language of multiplicative lattices, were obtained by
McCarthy [322], Hino [226], and Steinfeld [468]. Kappos and Papangelou [268] showed that every $\cap$-continuous lattice can be imbedded
with preservation of all joins and meets in a $\cap$-continuous and $\cap$-
complete lattice. An example refuting one of Dappos' results [267]
was obtained.

Edel'man [54] pointed out a sufficient condition for the isomorphism of the intervals $[a \cap b, b]$ and $[a, a \cup b]$.

Andreoli [71] found the necessary and sufficient conditions
for an unoriented graph to be realized as the graph of a modular
lattice of finite length. In compactly generated modular lattices
Head [222] introduced the concepts of pure and almost locally pure
elements (being generalizations, respectively, of the concepts for
subgroups of an Abelian group), and he studied their properties.
Also see Mihalek [327, 328], Miyashita [331], Milić [329], and Bumcrot [116] on modular and semimodular lattices. Janowitz [259]
proved that the center of a complete relatively complemented lattice
is a complete sublattice. Wille studied certain properties of semi-
complements and various classes of semicomplemented lattices.
Goodstein's paper [195] was devoted to the property of weak complements, being a generalization of relative complements. Pickett
[381] gave a generalization of an equivalence relation in the language of n-place predicates. Another generalization of an equivalence relation, connected with partitions, is due to Frontera [180].

Benado [100] continued the study of multilattices (see [38] for
the definition). Bravtsev [4] derives an example of an algebra with

a modular congruence lattice which, however, does not commute, which gives a negative answer to Hashimoto's question (see [38]).

Let us note the papers on semilattices. Nemitz [358, 359] studies the structure of implicative semilattices (i.e., of relatively pseudocomplemented semilattices). Every homomorphism of such a lattice is uniquely defined by the kernel and its structure is very similar to the structure of relatively complemented lattices. Salii [37] considers the question of the isomorphic representations (in particular, by semilattices of partial mappings) of semilattices with a binary relation ξ satisfying the conditions: 1) $(a, a)\in\xi$, 2) $a_1\cap a_2 = a_1, \Rightarrow (a_1, a_2)\in\xi$, 3) $a_1\cap a_1 = a_1, a_2\cap a_2 = a_2,, (a_1, a_2)\in\xi \Rightarrow (a_1, a_2)\in\xi$.

Munn [353] proves that any semilattice is isomorphically imbeddable in a uniform (i.e., with isomorphic principal ideals) semilattice with unit. It is also established that a semilattice is uniform if and only it is isomorphic to the semilattice of all idempotents of some bisimple inverse semigroup. In [354] he studies the congruence lattices of bisimple ω-semigroups. Shevrin [49, 50] continues the investigation of questions on the lattice definability of semilattices [51]. Analogous questions are considered for lattices by Filippov [43]. (In particular, for lattices of modular, distributive, relatively complemented, and Boolean algebras.)

Frontera [179] analyzes, on a class of lattices, a certain numerical function F (R) whose equality to zero is necessary (but not sufficient) for the reducibility of lattice R. The question remains open: Will this condition not be sufficient for modular complemented lattices? Crapo [125] studies the Möbius function on posets and on lattices. An analog of one number-theoretic formula was obtained for lattices by Pic [380].

BIBLIOGRAPHY

1. I. N. Abramovskii, "Groups in which the subgroup lattices is a relatively complemented lattice," Algebra Logika. Seminar, 6(1):5-8 (1967).
2. A. Ya. Aizenshtat, "Regular endomorphism semigroups of ordered sets," Uch. Zap. Leningrad. Gos. Ped. Inst. im. A. I. Gertsena, 387:3-11 (1968).
3. S. A. Anishchenko, "Representation of certain modular lattices by subgroup lattices," Mat. Zap. Krasnoyarsk. Ped. Inst., 1:1-21 (1965).
4. A. F. Bravtsev, "Congruences of the direct product of algebras," Algebra Logika. Seminar, 6(1):39-43 (1967).

5. I. I. Valulse, "Some aspects of the theory of unary algebras," in: Abstracts of the Reports of the Second Scientific Engineering Conference of the Kishniev Polytechnic Institute [in Russian], Kishniev (1966), 229 pp.

6. V. G. Vinokurov, "A lattice method for defining dimension," Dokl. Akad. Nauk SSSR, 168(3):504-507 (1966).

7. L. M. Gluskin, "Semigroups," in: Algebra. 1964 [in Russian], Science Summaries, VINITI, Akad. Nauk SSSR, Moscow (1966), pp. 161-202.

8. M. M. Glukhov, "Algorithmic solvability of the word problem for completely free modular lattices," Sibirsk. Mat. Zh., 5(5):1027-1034 (1964).

9. E. E. Gurevich and G. Ya. Rotkovich, "Comparison of various definitions of 0-convergence in lattices," Uch. Zap. Leningrad. Gos. Ped. Inst. im. A. I. Gertsena, 274:52-58 (1968).

10. Kh. M. Devadze, "Generating set of the endomorphism semigroup of a finite linearly ordered set," Uch. Zap. Leningrad. Gos. Ped. Inst. im. A. I. Gertsena, 387:101-111 (1968).

11. A. E. Evseev, "Semigroups with ordinally decomposable semilattices of sub-semigroups," Izv. Vyssh. Uchebn. Zavedenii. Matematika, No. 6, 74-84, (1965).

12. A. E. Evseev, "Certain classes of semigroups with ordinally decomposable semi-lattices of subsemigroups," Uch. Zap. Leningrad. Gos. Ped. Inst. im. A. I. Gertsena, 302:63-69 (1967).

13. A. E. Evseev, "Lattice isomorphisms of free nilpotent semigroups," Uch. Zap. Leningrad. Gos. Ped. Inst. im. A. I. Gertsena, 387:112-125 (1968).

14. A. A. Iskander, "Lattice of correspondences of a universal algebra," Izv. Akad. Nauk SSSR, Ser. Mat., 29(6):1357-1372 (1965).

15. A. A. Iskander, "Partial universal algebras with given subalgebra lattices and correspondences," Mat. Sb., 70(3):438-456 (1966).

16. L. N. Karolinskaya, "Congruence lattices of distributive lattices," Izv. Akad. Nauk, SSSR, Ser. Mat., 28(5):1037-1054 (1964).

17. L. N. Karolinskaya and I. V. Stelletskii, "Congruences on weakly modular lattices," Sibirsk. Mat. Zh., 7(5):1033-1038 (1966).

18. T. Katriňák, "A remark on Stone lattices. I," Math.-Fyz. Časop., 16(2):128-142 (1966).

19. T. Katriňák, "A remark on Stone lattices. II," Mat.-Fyz. Časop., 17(1):20-37 (1967).

20. S. P. Kogalovskii, "Linearly complete ordered sets," Uspekhi Mat. Nauk, 19(2): 147-150 (1964).

21. S. P. Kogalovskii, "On Frink's theorem," Uspekhi Mat. Nauk, 19(2):143-145 (1964).

22. P. Constantinescu and T. Gaşpar, "Number of types of Boolean functions with respect to certain direct products of subgroups of a hyperoctahedral group," Bull. Math. Soc. Sci. Math. Phys. RPR, 7(3):189-198 (1963).

23. P. G. Kontorovich, S. G. Ivanov, and G. P. Kondrashov, "Distributive pairs of elements in a lattice," Dokl. Akad. Nauk SSSR, 160(5):1001-1003 (1965).

24. I. Korets, "Construction of all partially ordered sets homomorphically mappable onto a given partially ordered set," Acta Fac. Rerum Natur. Univ. Comenianae, Math., 10(5):23-36 (1966).

25.   Z. Ladzianska, "Characterization of a directed partially ordered set by means of a "betweenness" relation," Mat.-Fyz. Časop., 15(2):162-167 (1965).

26.   A. E. Liber, A Binary Boolean Algebra and Its Application [in Russian], Saratov Univ., Saratov (1966), 80 pp.

27.   E. S. Lyapin, "Semigroups of directed transformations of ordered sets," Mat. Sb., 74(1):39-46 (1967).

28.   E. S. Lyapin, "Types of elements and the zeros of semigroups of directed transformations," Uch. Zap. Leningrad. Gos. Ped. Inst. im. A. I. Gertsena, 387:171-180 (1968).

29.   L. L. Maksimova, "Topological spaces and quasi-ordered sets," Algebra i Logika. Seminar, 6(4):51-59 (1967).

30.   B. Makhkamov, "Filters of a distributive lattice," UzSSR Fanlar Akad. Akhboroti Fiz.-Mat. Fanlari Ser., Izv. Akad. Nauk UzSSR, Ser. Fiz.-Mat. Nauk, No. 4, 29-32 (1965).

31.   Yu. N. Mukhin, "Locally compact groups with a distributive lattice of closed subgroups," Sibirsk. Mat. Zh., 8(2):366-375 (1967).

32.   B. Riečan, "Continuous extension of monotone functionals of a certain type," Mat.-Fyz. Časop., 15(2):116-125 (1965).

33.   V. V. Rogov, "Complemented groups," Mat. Zap. Krasnoyarsk. Ped. Inst., 1:54-58 (1965).

34.   S. Rudeanu, "The definition of Boolean algebras by means of binary operations," Rev. Math. Pures et Appl., 6(1):171-183 (1961).

35.   S. Rudeanu, "Boolean functions and Sheffer functions," Rev. Math. Pures et Appl., 6(4):747-759 (1961).

36.   L. E. Sadovskii, "Certain lattice-theoretic aspects of group theory," Uspekhi Mat. Nauk, 23(3):123-157 (1968).

37.   V. N. Salii, "Lower lattices," Reports Third Siberian Conf. Math. Mech. 1964 [in Russian], Tomsk Univ., Tomsk (1964), pp. 237-238.

38.   L. A. Skornyakov, "Lattice theory," in: Algebra. 1964 [in Russian], Science Summaries, VINITI, Akad. Naúk SSSR, Moscow (1966), pp. 237-274.

39.   A. M. Skryago, "The ideals of the endomorphism semigroup of one type of partially ordered sets," Nauchn. Tr. Krasnodarsk. Gos. Ped. Inst., No. 41, 9-10 (1965).

40.   M. A. Taitslin, "The elementary theories of lattices of ideals in polynomial rings," Algebra i Logika. Seminar, 7(2):94-97 (1968).

41.   V. V. Fedorchuk, "Ordered sets and the product of topological spaces," Vestn. Mosk. Univ., Mat., Mekh., No. 4, 66-71 (1966).

42.   V. V. Fedorchuk, "Ordered spaces," Dokl. Akad. Nauk SSSR, 169(4):777-780 (1966).

43.   N. D. Filippov, "Projection of lattices," Mat. Sb., 70(1):36-54 (1966).

44.   P. A. Freidman, "Rings with a distributive lattice of subrings," Mat. Sb., 73(4):513-534 (1967).

45.   N. G. Hisamiev, "Unsolvability of an elementary theory of a free lattice," Algebra i Logika. Seminar, 6(5):45-48 (1967).

46.   N. G. Hismiev, "Several remarks on distributive lattices," Algebra i Logika. Seminar, 6(4):107-110 (1967).

47. L. K. Chebotareva, "The belonging problem for free and almost free lattices," Sibirsk. Mat. Zh., 8(2):399-405 (1967).

48. L. N. Shevrin, "Fundamental question in the theory of projection of semilattices," Mat. Sb., 66(4):568-597 (1967).

49. L. N. Shevrin, "Lattice isomorphisms of commutative holoidal semigroups," Izv. Vyssh. Uchebn. Zavedenii. Matematika, No. 1, 153-160 (1966).

50. L. N. Shevrin, "Elementary lattice properties of semigroups," Dokl. Akad. Nauk SSSR, 167(2):305-308 (1966).

51. L. N. Shevrin, "Fundamental questions in the theory of projection of semilattices. II," Mat. Zap. Ural'skii Univ., 5(3):107-122 (1966).

52. L. N. Shevrin, "Semi-isomorphisms and lattice isomorphisms of semigroups with a cancellation law," Dokl. Akad. Nauk SSSR, 171(2):296-298 (1966).

53. M. F. Shirokhov, "Decomposition sequences of a Boolean algebra," An. Ştiint. Univ. Iaşi, Sec. Ia, 11B:75-87 (1965).

54. S. L. Edel'man, "Certain properties of intervals in a lattice," Mat. Zap. Krasnoyarsk. Ped. Inst., 1:69-75 (1965).

55. I. M. Yaglom, "Boolean algebras," in: Certain Aspects of Modern Mathematics and Cybernetics [in Russian], Prosveshchenie, Moscow (1965), pp. 230-324.

56. B. V. Yakovlev, "Lattice isomorphisms of metabelian groups of degree P," Mat. Zap. Krasnoyarsk. Ped. Inst., 1:59-68 (1965).

57. B. V. Yakovlev, "Lattice isomorphisms of periodic nilpotent groups," Mat. Zap. Ural'skii Univ., 5(1):106-116 (1965).

58. J. C. Abbott, "Remarks on an algebraic structure for a topology," General Topology and Related Modern Analysis and Algebra, Vol. 2, Prague, (1967), pp. 17-21.

59. J. C. Abbott, Semi-boolean algebra. Mat. Vesn., 4(2):177-198 (1967).

60. A. Abian, The Stone space of a Boolean ring. Enseign. math., 11(2-3):194-198 (1965).

61. A. Abian and D. Deever, Representation of partially and simply ordered sets by terminating sequences. Boll. Unione mat. ital., 21(4):371-376 (1966).

62. A. Abian and D. Deever, On representation of partially ordered sets. Math. Z., 92(5):353-355 (1966).

63. A. Abian and D. Deever, On the minimal length of sequences representing simply ordered set. Z. math. Logik und Grundl. Math., 13(1):21-23 (1967).

64. A. Abian and D. Deever, Representation of simply ordered sets and the generalized continuum hypothesis. Roczn. Polsk. towarz. mat., Ser. 1, 11(1):183-188 (1967).

65. A. Abian and D. Deever, On the bounds of the minimal length of sequences representing simply ordered sets. Arch. math. Logik und Grundlagenforsch., s.a., Nos. 1-2, pp. 3-5 (1968).

66. A. Abian and D. Deever, On representation of partially ordered sets. Math. Ann., 169(2):328-330 (1967).

67. S. A. Adelfio, Jr. and C. F. Nolan, Principles and applications of Boolean algebra. New York, Hayden (1964), ix, 319 pp.

68. D. Adnadević, O predstavljunju konačnich delimično uzedenih skupova. Mat. Vesn., 3(1):17-21 (1966).

69. P. Albada, Axiomatique des algebres de Boole. Bull. Soc. Math. Belg., 18:260-272 (1966).

70. R. A. Alo and O. Frink, Topologies of lattice products. Canad. J. Math., 18(5):1004-1014 (1966).

71. L. R. Alvarez, Undirected graphs realizable as graphs of modular lattices. Canad. J. Math., 17(6):923-932 (1965).

72. G. Andreoli, Algoritmi booleani subordinati ad algoritmi numerici conversi. Algebre di Frobenius—Boole. Ricerca, 14:3-9 (Jan.-Apr., 1963).

73. G. Andreoli, Algebre booleane con operazioni quasi-lineari, e loro interpretazione con matrici booleane. Giorn. mat. Battaglini, 91(1):5-23 (1963).

74. B. Artmann, Automorphismen und Koordinaten bei ebenen Verbänden. Mitt. Math. sem. Giessen, No. 66 (1965), 58 pp.

75. B. Artmann, Ein Struktursatz für ebene Verbände. Arch. Math., 18(1):23-32 (1967).

76. B. Artmann, Bemerkungen zu einem verbandstheoretischen Problem von Skornjakov. Arch. Math., 18(3):226-229 (1967).

77. R. Ascoli and G. Teppati, A lattice theoretical analogue of the Schur lemma. Boll. Unione mat. ital., 22(2):200-204 (1967).

78. K. Atsumi, On complete lattices having the Hausdorff interval topology. Proc. Amer. Math. Soc., 17(1):197-199 (1966).

79. S. P. Avann, Application of the joinirreducible excess function to semimodular lattices. Math. Ann., 142(4):345-354 (1961).

80. S. P. Avann, Increases in the join-excess function in a lattice. Math. Ann., 154(5):420-426 (1964).

81. S. P. Avann, Dependence of finiteness conditions in distributive lattices. Math. Z., 85(3):245-256 (1964).

82. J. Badida, O asociativnej operácii na určitej triede sväzov. Acta Fac. rerum. natur. Univ. Comenianae. Math., 9(2):71-74 (1964).

83. J. Badida, O asociativnej operacii Δ na triede szäzov Ω_0. Acta Fac. rerum natur. Univ. Comenianae. Math., 10(5):37-41 (1966).

84. J. Badida, O asociativnych operáciach na určitých triedach sväsov. Acta Fac. rerum natur. Univ. Comenianae. Math., 10(7):31-41 (1966).

85. R. Balbes, Projective and injective distributive lattices. Pacif. J. Math., 21(3):405-420 (1967).

86. R. Balbes and A. Horn, Order sums of distributive lattices. Pacif. J. Math., 21(3):421-435 (1967).

87. B. Banaschewski, Regular closure operators. Arch. Math., 14(4-5):271-274 (1963).

88. B. Banaschewski and G. Bruns, Categorical characterization of the MacNeille completion. Arch. Math., 18(4):369-377 (1967).

89. C. Bănică and O. Stănăsilă, Spectrul unei latici. Studii şi certări mat. Acad. RSR, 19(1):137-145 (1967).

90. D. W. Barnes, Lattice isomorphisms of associative algebras. J. Austral Math. Soc., 6(1):106-121 (1966).

91. D. L. Barnett, A condition for a ring to be Boolean. Amer. Math. Monthly, 74(1, Part I):44-45 (1967).

92. F. Bartolozii, Immersione di un reticolo distributivo finito nel reticolo degli interi rispetto alla divisione. Boli. Unione mat. ital., 21(2):178-185 (1966).

93. A. Bassi, Sui polinomi in un'algebra del Boole con topologia. Atti Accad. naz. Lincei. Rend. Cl. sci. fis., mat. e natur., 40(1):29-34 (1966).

94. A. Bassi, Polinomi e dualitá in un'algebra del Boole con topologia. Rend. mat. e applic., 25(1-2):94-112 (1967).

95. F.-W. Bauer, Der Homotopietyp und der Verband der zulässigen Untergruppen. Math. Ann., 164(4):336-343 (1966).

96. K. Baumgartner, Skalarbereich und Unterraumverband eines Vektorraumes. Math. Z., 93(1):60-68 (1966).

97. A. R. Bednarek, On the Mickle—Rado covering theorems. Fundam. math., 53(1): 93-97 (1963).

98. A. R. Bednarek, On chain-equivalent orders. Nieuw arch. wiskunde, 12(3):137-138 (1964).

99. N. D. Belnap and J. H. Spencer, Intensionally complemented distributive lattices. Portug. Math., 25(1-2):99-104 (1966).

100. M. Benado, Remarques sur la théorie des multitreillis. VI. (Contributions a la théorie des structures algébriques ordonnées). Mat.-fyz. časop., 14(3):163-207 (1964).

101. S. Benedetto, Isomorfismi tra reticoli di sottogruppi e reticoli di sottogruppi normali. Atti. Ist. veneto sci. lettere ed arti. Cl. sci. mat. e natur., 123:227-235 (1965).

102. C. Benzaken, Définition et propriétés de certaines familles de fonctions booléennes croissantes. C. r. Acad. sci., 259(7):1369-1371 (1964).

103. C. Benzaken, Treillis série-paralléle. C. r. Acad. sci., 260(21):5431-5434 (1965).

104. C. Benzaken, Les familles de fonctions booléennes déduites de certaines familles de fonctions booléennes croissantes. Critères de détermination de l'indice d'une fonction croissante. C. r. Acad. sci., 260(6):1528-1531 (1965).

105. C. Benzaken, Pseudo-treillis distributifs et applications. Note I. Pseudo-treillis distributifs. Bul. Inst. politehn. Iaşi, 12(3-4):13-18 (1966).

106. M. P. Berri, The complement of a topology for some topological groups. Fundam. math., 58(2):159-162 (1966).

107. H. J. Biesterfeldt, Jr., Uniformization of convergence spaces. Part 1. Definitions and fundamental constructions. Math. Ann., 177(1):31-42 (1968).

108. T. S. Blyth, Λ-distributive Boolean matrices. Proc. Glasgow Math. Assoc., 7(2): 93-100 (1965).

109. M.-P. Brameret, Treillis d'idéaux et structure d'anneaux. Sémin. Dubreil et Pisot. Fac. Sci. Paris, 16(1):1/01-1/12 (1967).

110. L. Brickman and P. A. Fillmore, The invariant subspace lattice of a linear transformation. Canad. J. Math., 19(4):810-822 (1967).

111. D. R. Brown, Topological semilattices on the two-cell. Pacif. J. Math., 15(1): 35-46 (1965).

112. G. Bruns, Ideal-representations of Stone lattices. Duke Math. J., 32(3):555-556 (1965).

113. G. Bruns, A lemma on directed sets and chains. Arch. Math., 18(6):561-563 (1967).

114. R. J. Bumcrot, "A comparative study of betweenness relations in lattices," Doctoral dissertation, Univ. of Missouri (1962), 111 pp.; Dissert. Abstr., 23(8): 2921-2922 (1963).

115. R. J. Bumcrot, Betweenness geometry in lattices. Rend. Circolo mat. Palermo, 13(1):11-28 (1964).

116. R. J. Bumcrot, On lattice complements. Proc. Glasgow. Math. Assoc., 7(1):22-23 (1965).

117. M. C. R. Butler, A class of torsion-free abelian groups of finite rank. Proc. London Math. Soc., 15(4):680-698 (1965).

118. J. J. Carruth, A note on partially ordered compacts. Pacif. J. Math., 24(2):229-231 (1968).

119. M. Carvallo, Monographie des treillis et algèbre de Boole, Gauthier-Villars, Paris (1966), 130 pp. (second edition).

120. G. Casanova, L'algébre de Boole, Press. Univ. France, Paris (1967), 128 pp.

121. A. C. Choudhury, Application of lattice theory to geometry. Math. Student, 33(1):23-26 (1965).

122. A. Climescu, Măsuri numerica cu un număr finit de valori in algebrele Boole. Bul. Inst. politehn. Iaşi, 9(1-2):1-6 (1963).

123. J. L. V. Cordoba, Limite inductivo generalizado de conjuntos. Rev. mat. hisp.-amer., 25(4-5):202-206 (1965).

124. N. C. A. da Costa, Opérations non monotones dans les treillis. C. r. Acad. sci., 263(14):A429-A432 (1966).

125. H. H. Crapo, The Möbius function of a lattice. J. Combin. Theory, 1(1):126-131 (1966).

126. H. H. Crapo, Structure theory for geometric lattices. Rend. Semin. mat. Univ. Padova, 38:14-22 (1967).

127. M. Curzio, Elementi U-quasi-distributivi in alcuni reticoli di gruppi. Ricerche mat., 13(1):70-79 (1964).

128. M. Curzio, Sui gruppi per cui sono riducibili alcuni reticoli di sottogruppi notevoli-Matematiche, 19(1):1-10 (1964).

129. M. Curzio, Una caratterizzazione reticolare dei gruppi abeliani. Rend. mat. e applic., 24(1-2):1-10 (1965).

130. M. Curzio and M. Giordano, Una proprieta reticolare di alcuni p-gruppi localmente finiti. Mathematiche, 21(1):18-22 (1966).

131. A. Daigneault, On automorphisms of polyadic algebras. Trans. Amer. Math. Soc., 12(1):84-130 (1964).

132. A. Daigneault, Freedom in polyadic algebras and two theorems of Beth and Craig. Michigan. Math. J., 11(2):129-135 (1964).

133. R. O. Davies, The least extension of a closure operator on a sublattice. Proc. Cambridge Philos. Soc., 63(1):9-10 (1967).

134. G. W. Day, Free complete extensions of Boolean algebras. Pacif. J. Math., 15(4): 1145-1151 (1965).

135. G. W. Day, Superatomic Boolean algebras. Pacif. J. Math., 23(3):479-489 (1967).

136. R. A. Dean and R. L. Kruse, A normality relation for lattices. J. Algebra, 3(3): 277-290 (1966).

137. J. E. Delany, "Meet representations in upper continuous modular lattices," Doctoral dissertation, Iowa State Univ. Sci. and Technol. (1966), 61 pp.; Dissert. Abstr.. B27(4):1213 (1966).

138. R. E. Demarr, Common fixed points for isotone mappings. Colloq. math., 13(1): 45-48 (1964).

139. R. E. Demarr, Partially ordered spaces and metric spaces. Amer. Math. Monthly, 72(6):628-631 (1965).

140. R. E. Demarr, Order convergence and topological convergence. Proc. Amer. Math. Soc., 16(4):588-590 (1965).

141. J.-C. Derdérin, Residuated mappings. Pacif. J. Math., 20(1):35-43 (1967).

142. R. Diaconescu, Sume directe in categoria laticilor. Studii şi cercetări mat. Acad. RSR, 19(6):817-824 (1967).

143. A. Diego and A. Suárez, Two sets of axioms for Boolean algebras. Portug. math., 23(3-4):139-145 (1964).

144. V. Dlab, General algebraic dependence structures and some applications. Colloq. math., 14:265-273 (1964).

145. H. P. Doctor, The categories of Boolean lattices, Boolean rings and Boolean spaces. Canad. Math. Bull., 7(2):245-252 (1964).

146. T. Donnellan, Lattice Theory, Pergamon Press, Oxford (1968), 283 pp.

147. C. H. Dowker and D. Papert, Quotient frames and subspaces. Proc. London Math. Soc., 16(2):275-296 (1966).

148. C. H. Dowker and D. Papert, "On Urysohn's lemma," General Topology and Related Modern Analyses and Algebra, Vol. 2, Prague (1967), pp. 111-114.

149. D. Drake, "On the representations of an abstract lattice as the family of closed sets of a topological space," Doctoral dissertation, Univ. Colorado (1964), 69 pp.; Dissert. Abstr., 25(10):5955-5956 (1965).

150. D. Drake and W. J. Throm, On the representations of an abstract lattice as the family of closed sets of a topological space. Trans. Amer. Math. Soc., 120(1): 57-71 (1965).

151. H. Draškovičová, Über die Relation "zwischen" in Verbänden. Mat.-fyz. časop., 16(1):13-20 (1966).

152. R. Duda, On stems. Colloq. math., 15(2):253-261 (1966).

153. P. Dwinger, Introduction to Boolean Algebras, Würzburg, Physica-Verl. (1961), 61 pp.

154. P. Dwinger, The dual space of the inverse limit of an inverse limit system of Boolean algebras. Proc. koninkl. nederl. akad. wet., A67(2):164-172 (1964); Indagationes math., 26(2):164-172 (1964).

155. P. Dwinger, Notes on Post algebras. I. Proc. Koninkl. nederl. akad. wet., A69(4): 462-468 (1966); Indagationes math., 28(4):462-468 (1966).

156. P. Dwinger, Notes on Post algebras. II. Proc. Koninkl. nederl. akad. wet., A69(4): 469-478 (1966); Indagationes math., 28(4):469-478 (1966).

157. P. Dwinger, Direct limits of partially ordered systems of Boolean algebras. Proc. Koninkl. nederl. akad. wet., A70(3):317-325 (1967); Indagationes math., 29(2): 317-325 (1967).

158. P. Dwinger and F. M. Yaqub, Free extensions of sets of Boolean algebras. Proc. Koninkl. nederl. akad. wet., A67(5):567-577 (1964); Indagationes math., 26(5): 567-577 (1964).

159. P. Eligio Di Vizio, Omomorfismi ed isomorfismi connessi alle operazioni param-
etrizzate nelle strutture algebriche booliane. Ann. pontif. Ist. super. sci. e
lettere "S. Chiara", No. 13, pp. 431-466 (1963).

160. M. Emaldi and G. Zaches, I gruppi risolubili relativamente complementati.
Ricerche mat., 14(1):3-8 (1965).

161. P. Erdös and R. Rado, Partition relations and transitivity domains of binary relations. J. London Math. Soc., 42(4):624-633 (1967).

162. G. Ewald, Erweiterung projektive geometrischer Verbände. Math. Z., 87(1):101-
114 (1965).

163. A. Fadini, Teoria degli elementi complessi nelle algebre di Boole. Ann. pontif.
ist. super. sci. et lett. "S. Chiara", No. 12, pp. 223-243 (1962).

164. Ky Fan, Subadditive functiones on a distributive lattice and an extension of
Szász's inequality. J. Math. Analysis and Applic., 18(2):262-268 (1967).

165. J. E. Fenstad, On representation of polyadic algebras. Kg. norske vid. selskabs
forhandl., 27(8):36-41 (1964).

166. F. Fiala and V. Novák, On isotone and homomorphic mappings. Arch. mat.,
2(1):27-32 (1966).

167. P. D. Finch, On the lattice equivalence of topological spaces. J. Austral. Math.
Soc., 6(4):495-511 (1966).

168. A. Fine. Abstract covering theorems. Fundam. Math., 52(2):205-207 (1963).

169. W. L. Fischer, Die Boolesche Algebra. Archimedes, 19(1):2-4 (1967).

170. W. L. Fischer, Anwendung der Boole−Ringe. Archimedes, 19(3):39-43 (1967).

171. J. Flachsmeyer, Einige topologische Fragen in der Theorie der Booleschen
Algebren. Arch. Math., 16(1):25-33 (1965).

172. J. Flachsmeyer, "Uber die Realisierung von Boole-Algebren als Boole-Algebren
regulär offener Mengen," General Topology and Related Modern Analysis and
Algebra, Vol. 2, Prague (1967), pp. 133-139.

173. H. G. Flegg, L'algebre de Boole et son utilisation, Dunod, Trad. Paris (1967),
245 pp.

174. J. Folkman, The homology groups of a lattice. J. Math. and Mech., 15(4):631-
636 (1966).

175. J. Fort, Eléments Γ-isotypiques dans les (ℑ) algébres modulaires. C. r. Acad.
sci., 258(6):1676-1678 (1964).

176. J. Fort, Contribution a l'étude des éléments tertiaires et isotypiques dans les modules et les (τ)-algebres. Bull. Soc. math. France, 92(1):Supp., 99 pp.

177. D. J. Foulis, Semigroups co-ordinatizing orthomodular geometries. Canad. J.
Math., 17(1):40-51 (1965).

178. Z. Frolik, Fixed points of maps of extremally disconnected spaces and complete
Boolean algebras. Bull. Acad. polon. sci. Sér. sci. math. astron. et phys., 16(4):
269-275 (1968).

179. M. B. Frontera, Una functión numérica en los reticulos finitos que se anula para
los reticulos reducibles. Publs. semin. mat. Fac. cienc. Zaragoza, No. 3, pp.
103-111 (1962).

180. M. B. Frontera, "Generalizacion del concepto de relacion de equivalencia en un
conjunto," Actas 5 Reun. anual. mat. esp., Valencia, 1964, Madrid (1967), pp.
203-220.

181. W. R. Frucht, Sobre el problema 6 de Birkhoff. Scientia (Chile), 31(134):5-21 (1964).

182. S. Fugiwara, A note on left complemented lattices. Res. Bull Fac. Liber. Arts. (Natur. Sci.), 2(3):19-22 (1963).

183. S. Fugiwara, On the parallelism in some left complemented lattice. Res. Bull. Fac. Liber. Arts, Oita Univ. (Natur. Sci.), 2(4):1-8) (1964).

184. S. Fujiwara, The generalization of Wilcox lattices. Res. Bull. Fac. Liberal Arts Oita Univ., 2(5):1-6 (1965).

185. T. Fugiwara, Permutability of semilattice congruences on lattices, Canad. J. Math., 19(2):370-375 (1967).

186. H. Gaifman, Remarks on complementation in the lattice of all topologies. Canad. J. Math., 18(1):83-88 (1966).

187. E. Gedeonová, Der Jordan—Höldersche Satz für unendliche Ketten in teilweise geordneten Mengen. Acta Fac. rerum natur. Univ. Comenianae. Math., 10(3): 41-50 (1965).

188. M. D. Gerhardts, Zur Characterisierung distributiver Schiefverbände. Math. Ann., 161(3):231-240 (1965).

189. M. D. Gerhardts, Über die Zerlegbarkeit von nichtkommutativen verbänden in kommutative Teilverbände. Proc. Japan Acad., 41(10):883-888 (1965).

190. A. C. Gillie, Binary Arithmetic and Boolean Algebra, McGraw-Hill, New York (1965), 248 pp.

191. S. Ginsburg and J. R. Isbell, The category of cofinal types. I. Trans. Amer. Math. Soc., 116(4):386-393 (1965).

192. Y. Give'on, Lattice matrices. Inform. and Control, 7(4):477-484 (1964).

193. E. D. Goodrich, Boolean-like functions. Amer. Math. Monthly, 73(1):40-44 (1966).

194. R. L. Goodstein, The solution of equations in a lattice. Proc. Roy. Soc. Edinburgh, A67(3):231-242 (1967).

195. R. L. Goodstein, On weak complements in a lattice. Rev. roumaine math. pures et appl., 12(8):1059-1063 (1967).

196. R. N. R. Gopala, Vector spaces over a regular ring. Math. Ann., 167(4):280-291 (1966).

197. J. Grappy, Demi-groupes dont le treillis des congruences est un treillis complémenté. Sémin. Dubreil et Pisot. Fac. Sci. Paris, 17(2):21/01-21/21 (1967).

198. G. Grätzer, On Boolean functiones. (Notes on lattice theory II). Rev. math. pures. et appl. (RPR), 7(4):693-697 (1962).

199. G. Grätzer, A generalization of Stone's representation theorem for Boolean algebras. Duke math. J., 30(3):469-474 (1963).

200. G. Grätzer, Boolean functions on distributive lattices. Acta math. Acad. scient. hung., 15(1-2):195-201 (1964).

201. G. Grätzer, On the family of certain subalgebras of a universal algebra. Proc. Koninkl. nederl. akad. wet., A68(5):790-802 (1965); Indagationes math., 27(5): 790-802 (1965).

202. G. Grätzer, Equational classes of lattices. Duke Math. J., 33(3):613-622 (1966).

203. G. Grätzer and W. A. Lampe, On subalgebra lattices of universal algebras. J. Algebra, 7(2):263-270 (1967).

204. G. Grimeisen, Zum Produkt R-topologischer Verbäde, Spisy přirodověd. fak. univ. Brně, No. 9, pp. 463-468 (1964).

205. G. Grimeisen, Ein Produkt R-topologischer Verbände. Math. Z., 88(4):309-319 (1965).

206. O. Hájek, Representation of finite-length modular lattices. Czech. Math. J., 15(4):503-520 (1965).

207. O. Hájek, Characteristics of modular finite-length lattices. Czech. Math. J., 15(4):521-525 (1965).

208. P. R. Halmos, Lectures on Boolean Algebras, Van Nostrand, New York (1963), 152 pp.

209. R. W. Hansell, Monotone subnets in partially ordered sets. Proc. Amer. Math. Soc., 18(5):854-858 (1967).

210. D. J. Hansen, A functional characterization of a Boolean determinant. J. London Math. Soc., 41(4):723-727 (1966).

211. Z. S. Harris, Decomposition lattices. Univ. of Pennsylvania Transform. and Discourse Analysis Paper, No. 70, 33 pp. (1967).

212. E. Harzheim, Einbettungssätze für totalgeordnete Mengen. Math. Ann., 158(2): 90-108 (1965).

213. E. Harzheim, Einbettung totalgeordneter Mengen in lexikographische Produkte. Math. Ann., 170(4):245-252 (1967).

214. E. Harzheim, Kombinatorische Betrachtungen über die Struktur der Potenzmenge. Math. Nachr., 34(3-4):123-141 (1967).

215. V. Havel, Zerlegungen in Kartesischen Funktionen. (Vorläuf. Mitt.) Comment. math. Univ. carolinae, 6(1):43-47 (1965).

216. V. Havel, Kartesisch assoziierte Zerlegungen. (Vorläuf. Mitt.). Comment. math. Univ. carolinae, 6(1):49-51 (1965).

217. V. Havel, Unessential Zassenhaus refinements. Arch. mat., 1(1):35-38 (1965).

218. V. Havel, Eigenschaften des Halbverbandes der Zerlegungen in einer Menge. Arch. mat., 1(4):213-215 (1965).

219. V. Havel, On associated partitions. Časop. pěstov. mat., 91(3):241-245 (1966).

220. I. Hayashi, The normality of the product of two linearly ordered spaces. Proc. Japan Acad., 43(4):300-304 (1967).

221. T. J. Head, Note on groups and lattices. Nieuwe arch. wiskunde, 13(2):110-112 (1965).

222. T. J. Head, Purity in compactly generated modular lattices. Acta math. Acad. scient. hung., 17(1-2):55-59 (1966).

223. H. Heineken, Über die Charakterisierung von Gruppen durch gewisse Untergruppenverbände. J. reine und angew. Math., 220(1-2):30-36 (1965).

224. C. Heuchenne, La saturation dans les treillis. Bull. Soc. roy. sci. Liege, 32(9): 632-652 (1963).

225. G. C. Hewitt, Limits in certain classes of abstract algebras. Pacif. J. Math., 22(1): 109-115 (1967).

226. K. Hino, A decomposition theorem in a multiplicative system. Osaka J. Math., 2(1):147-152 (1965).

227. J. Holec and J. Mařik, Continuous additive mappings. Czech. Math. J., 15(2): 237-243 (1965).

228. W. Holsztyński, Topological dimension of lattices. Bull. Acad. polon. sci. Sér. sci. math., astron. et phys., 14(2):63-69 (1966).

229. J. Horějš, On decompositions in sets. Spisy přírodověd fak. univ. Brně, No. 1, pp. 15-22 (1964).

230. Chen-Jung Hsu, On lattice theoretic characterization of the parallelism of affine geometry. Ann. Math., 50(2):1-7 (1949).

231. N. C. Hsu, Elementary proof of Hu's theorem on isotone mappings. Proc. Amer. Math. Soc., 17(1):111-114 (1966).

232. A. J. Insel, A relationship between the complete topology and the order topology of a lattice. Proc. Amer. Math. Soc., 15(6):847-850 (1964).

233. Iqbalunnisa, Maximal congruences on a lattice. J. Madras Univ., B33(2):113-128 (1963).

234. Iqbalunnisa, On neutral elements in a lattice J. Indian Math. Soc., 28(1):25-31 (1964).

235. Iqbalunnisa, Normal, simple congruences and weakly modular lattices. Current Sci., 33(3):78 (1964).

236. Iqbalunnisa, Weak modularity and permutability of congruences in lattices. Current Sci., 33(4):108 (1964).

237. Iqbalunnisa, Standard ideals in lattices. Current Sci., 33(2):44-45 (1964).

238. Iqbalunnisa, Neutrality in weakly modular lattices. Acta math. Acad. scient. hung., 16(3-4):325-326 (1965).

239. Iqbalunnisa, On some problems of G. Gratzer and E. T. Schmidt. Fundam. math., 57(2):181-185 (1965).

240. Iqbalunnisa, On types of lattices. Fundam. math., 59(1):97-102 (1966).

241. Iqbalunnisa, Permutable congruence in a lattice. Ill. J. Math., 10(2):235-239 (1966).

242. Iqbalunnisa, Characterizations of weakly modular lattices. Fundam. math., 59(3):301-305 (1966).

243. Iqbalunnisa, On a Galois correspondence between the lattice of ideals and the lattice of congruences on a lattice L. Fundam. math., 59(1):103-108 (1966).

244. Iqbalunnisa, Normal, simple and neutral congruences on lattices. Ill. J. Math., 10(2):227-234 (1966).

245. J. R. Isbell, The category of cofinal types. II. Trans. Amer. Math. Soc., 116(4): 394-416 (1965).

246. J. R. Isbell, Spaces without large projective subspaces. Math. Scand., 17:89-105 (1965).

247. J. R. Isbell, Directed unions and chains. Proc. Amer. Math. Soc., 17(6):1467-1468 (1966).

248. P. L. Ivănescu, Systems of pseudo-Boolean equations and inequalities. Bull. Acad. polon. sci., Sér. sci. math. astron et phys., 12(11):673-680 (1964).

249. P. L. Ivănescu, On the minimal decomposition of finite partially ordered sets in chains. Rev. roumaine math. pures et appl., 9(10):897-903 (1964).

250. P. V. Jagannadham, Linear transformations in a boolean vector space. Math. Ann., 167(3):240-247 (1966).

251. P. V. Jagannadham, A characterisation of Boolean extension of universal algebras. Math. Ann., 172(2):119-123 (1967).

252. J. Jakubik and M. Kolibiar, Über euklidische Verbände. Math. Ann., 155(4):334-342 (1964).

253. L. Janoš, Vlastnosti Zassenhausovy konstrukce v grupách a svazech. Časop. pěstow. mat., 88(2):246-249 (1963).

254. M. F. Janowitz, Baer semigroups. Duke Math. J., 32(1):85-95 (1965).

255. M. F. Janowitz, Quantifier theory on quasiorthomodular lattices. Ill. J. Math., 9(4):660-676 (1965).

256. M. F. Janowitz, A characterisation of standard ideals. Acta math. Acad. scient. hung., 16(3-4):289-301 (1965).

257. M. F. Janowitz, A note on normal ideals. J. Sci. Hiroshima Univ., Ser. A, Div. 1, 30(1):1-9 (1966).

258. M. F. Janowitz, A semigroup approach to lattices. Canad. J. Math., 18(6):1212-1223 (1966).

259. M. F. Janowitz, The center of a complete relatively complemented lattice is a complete sublattice. Proc. Amer. Math. Soc., 18(1):189-190 (1967).

260. M. F. Janowitz, IC-lattices. Portug math., 24(3-4):115-122 (1965).

261. R. E. Johnson, Distinguished rings of linear transformations. Trans. Amer. Math. Soc., 111(3):400-412 (1964).

262. R. E. Johnson, Potent rings. Trans. Amer. Math. Soc., 119(3):524-534 (1965).

263. D. H. J. de Jongh and A. S. Troelstra, On the connection of partially ordered sets with some pseudo-Boolean algebras. Proc. Koninkl. nederl. akad. wet., A69(3): 317-329 (1966); Indagationes math., 28(3):317-329 (1966).

264. P. Julien, Sur quelques problèmes de la théorie des chaînes permutées. C. r. Acad. sci., AB262(4): A205-A208 (1966).

265. P.-F. Jurie, Notion de quasi-somme amalgamée: premiéres applications l'algébre booléienne polyadique. C. r. Acad. sci., 264(24):A1033-A1036 (1967).

266. J. A. Kalman, A two-axiom definition of lattices. Rev. roumaine math. pures et appl., 13(5):669-670 (1968).

267. D. A. Kappos, "Einbettung eines beliebigen Verbandes in einem o-topologischen Verband," Proc. Internat. Congr. Math. (1954), 2, Amsterdam (1954), pp. 30-31.

268. D. A. Kappos and F. Papangelou, Remarks on the extension of continuous lattices. Math. Ann., 166(4):277-283 (1966).

269. M. Karlowicz and K. Kuratowski, On relations between some algebraic and topological properties of lattices. Fundam. math., 58(2):219-228 (1966).

270. R. Kaufman, Ordered sets and compact spaces. Colloq. math., 17(1):35-39 (1967).

271. H. J. Keisler, Good ideals in fields of sets. Ann. Math., 79(2):338-359 (1964).

272. H. J. Keisler, Universal homogeneous Boolean algebras. Mich. Math. J., 13(2): 129-132 (1966).

273. H. J. Keisler, Universal homogeneous Boolean algebras. Michigan Math. J., 13(2): 129-132 (1966).

274. D. C. Kent, On the order topology in a lattice. Ill. J. Math., 10(1):90-96 (1966).

275. D. C. Kent, The interval topology and order convergence as dual convergence structures. Amer. Math. Monthly, 74(4):426-427 (1967).

276. S. Kinugawa and J. Hashimoto, On relative maximal ideals in lattices. Proc. Japan Acad., 42(1):1-4 (1966).

277. A. Kirsch, Über lineare Ordnungen endlicher Boolescher Verbände. Arch. Math., 17(6):489-491 (1966).

278. K. Koh, On one-sided ideals which are prime rings. Amer. Math. Monthly, 73(2): 176-177 (1966).

279. M. Kolibiar, Über Fixpunktsätze in geordneten Mengen. Spisy přírodověd. fak. univ. Brně, No. 9, pp. 469-472 (1964).

280. M. Kolibiar, "Linien in Verbänden," An. stiint Univ. Iași, Sec. Ia, 11B:89-98 (1965).

281. B. Kolman, Semi-modular Lie algebras. J. Sci. Hiroshima Univ., Ser. A, Div. I, 29(2):149-163 (1961).

282. O. Kowalski and B. Pondělíček, O charaktereh řetězců. Časop. pěstov. mat., 91(1):1-3 (1966).

283. S. A. Kripke, An extension of a theorem of Gaifman–Hales–Solovay. Fundam. math., 61(1):29-32 (1967).

284. D. Kurepa, Fixpoints of monotone mappings of ordered sets. Glasnik mat.-fiz. i astron., 19(3-4):167-173 (1964).

285. D. Kurepa, Monotone mappings between some kinds of ordered sets. Glasnik math.-fiz. i astron., 19(3-4):175-186 (1964).

286. D. Kurepa, Remark on recent two results of Dilworth and Gleason. Publs. Inst. math., 4:107-108 (1964).

287. D. Kurepa, Imbedding of ordered sets in minimal lattices. Publs. Inst. math., 6:165-170 (1966).

288. R. H. La Grange, Jr., "Some problems concerning Boolean algebras," Doctoral dissertation, Univ. of Colorado (1966), 63 pp.; Dissert. Abstr., B28(4):1615-1616 (1967).

289. F. Lapscher, Introduction de variables surabondantes dans une fonction booléenne et applications. C. r. Acad. sci., 259(23):4206-4208 (1964).

290. J. D. Lawson, "Vietoris mappings and embeddings of topological semilattices," Doctoral dissertation, Univ. of Tennessee (1967), 99 pp.; Dissert. Abstr., B28(9): 3783 (1968).

291. S. Leader, Kimber's theorem on weighted sets. Amer. Math. Monthly, 74(10): 1226-1227 (1967).

292. A. Lebow, A Schroeder–Bernstein theorem for projections. Proc. Amer. Math. Soc., 19(1):144-145 (1968).

293. A. Lehman, "Postulants for a normed Boolean algebra," Math. Res. Center, Univ. of Wisconsin Techn. Summary Rept. No. 389, Madison (1963), 46 pp.

294. A. Lehman, Postulates for a normed Boolean algebra. SIAM Rev., 7(2):253-273 (1965).

295. C. W. Leininger, On chains of related sets. Bull. Math. Biophys., 28(1):103-106 (1966).

296. F. E. J. Linton, Injective Boolean σ-algebras. Arch. Math., 17(5):383-387 (1966).

297. L. Loomis, The lattice-theoretic background of the dimension theory of operator algebra. Mem. Amer. Math. Soc., No. 18 (1955), 36 pp.

298. H. F. J. Lowig, Note on the self-duality of the unrestricted distributive law in complete lattices. Israel J. Math., 2(3):170-173 (1964).

299. W. A. J. Luxemburg, A remark on Sikorski's extension theorem for homomorphisms in the theory of Boolean algebras. Fundam. math., 55(3):239-247 (1964).

300. M. D. MacLaren, Nearly modular orthocomplemented lattices. Trans. Amer. Math. Soc., 114(2):401-16 (1965).

301. M. J. Maczyński, Generalized free m-products of m-distributive Boolean algebras with an m-amalgamated subalgebra. Bull. Acad. polon. sci. Sér. sci. mat., astron. et phys., 14(10):539-542 (1966).

302. M. J. Maczyński, A property of retracts of α^+-homogeneous, α^+-universal Boolean algebras. Bull. Acad. polon. sci. Sér. sci. math., astron. et phys., 14(11): 607-608 (1966).

303. M. J. Maczyński and T. Traczyk, The m-amalgamation property for m-distributive Boolean algebras. Bull. Acad. Polon. Sci. Sér. sci. math., astron. et phys., 15(2):57-60 (1967).

304. F. Maeda, Modular centers of affine matroid lattices. J. Sci. Hiroshima Univ., Ser. A, Div. 1, 27(2):73-84 (1963).

305. F. Maeda, Parallel mappings and coparability theorem in affine matroid lattice. J. Sci. Hiroshima Univ., Ser. A, Div. 1, 17(2):85-96 (1963).

306. F. Maeda, Point-free parallelism in Wilcox lattices. J. Sci. Hiroshima Univ., Ser. A, Div. 1, 28(1):19-32 (1964).

307. F. Maeda, Perspectivity of points in matroid lattices. J. Sci. Hiroshima Univ., 1964, Ser. A, Div. 1, 28(2):101-112 (1964).

308. S. Maeda, On the symmetry of the modular relation in atomic lattices. J. Sci. Hiroshima Univ., Ser. A, Div. 1, 29(2):165-170 (1961).

309. S. Maeda, On conditions for the orthomodularity. Proc. Japan. Acad., 42(3):247-251 (1966).

310. R. Magari, Sulle connessioni fra i vari modi di rappresentare un'algebra. Matematiche, 20(1):22-40 (1965).

311. P. Mangani, Estensioni libere di un'algebra di Boole. Boll. Unione mat. ital., 20(2):210-219 (1965).

312. R. Maronna, A characterization of Morgan lattices. Portug. math., 23(3):169-171 (1964).

313. L. H. Martin, Two ternary algebras and their associated lattices. Amer. Math. Monthly, 72(10):1088-1091 (1965).

314. J. Mather, Invariance of the homology of a lattice. Proc. Amer. Math. Soc., 17(5):1120-1124 (1966).

315. J. C. Mathews and R. F. Anderson, A comparison of two modes of order convergence. Proc. Amer. Math. Soc., 18(1):100-104 (1967).

316. Y. Matsushima, Between topology for lattices. J. Math. Soc. Japan, 16(4):335-341 (1964).

317. I. G. Maurer and I. Purdea, Despre o topologie in multimea relaţiilor binare. Studia Univ. Babeş-Bolyai. Ser. math.-phys., 9(1):21-24 (1964).

318. J. Mayer and M. Novotný, On some topologies on products of ordered sets. Arch. mat., 1(4):251-257 (1965).

319. J. Mayer-Kalschmidt and E. Steiner, Some theorems in set theory and applications in the ideal theory of partially ordered sets. Duke Math. J., 31(2):287-289 (1964).

320. S. D. McCartan, On subsets of a partially ordered set. Proc. Cambridge Philos. Soc., 62(4):583-595 (1966).

321. S. D. McCartan, A quotient ordered space. Proc. Cambridge Philos. Soc., 64(2): 317-322 (1968).

322. P. J. McCarthy, Primary decomposition in multiplicative lattices. Math. Z., 90(3):185-189 (1965).

323. S. McCall, The completeness of Boolean algebra. Z. math. Logik und Grundl. Math., 13(4):367-376 (1967).

324. K. N. Meenakshi, Proximity structures in boolean algebras. Acta scient. math., 27(1):85-92 (1966).

325. R. A. Melter, Boolean valued rings and Boolean metric spaces. Arch. Math., 15(4-5):354-363 (1964).

326. R. A. Melter, Contributions to Boolean geometry of p-rings. Pacif. J. Math., 14(3):995-1017 (1964).

327. R. J. Mihalek, Modular extensions of point-modular lattices. Proc. Japan Acad., 42(3):225-230 (1966).

328. R. J. Mihalek, Skewness in special semimodular lattices. Proc. Japan Acad., 42(3):237-238 (1966).

329. S. D. Milić, On the isomorphism of modular lattices. Math. Vesn., 2(2):153-155 (1965).

330. E. C. Milner, On complementary initial and final sections of simply ordered sets. J. London Math. Soc., 42(2):269-280 (1967).

331. Y. Miyashita, Quasi-projective modules, perfect modules, and a theorem for modular lattices. J. Fac. Sci. Hokkaido Univ., Ser. 1, 19(2):86-110 (1966).

332. J. D. Monk, Singulary cylindric and polyadic equality algebras. Trans. Amer. Math. Soc., 112(2):185-205 (1964).

333. J. D. Monk, Nontrivial 𝔐-injective Boolean algebras do not exist. Bull. Amer. Math., Soc., 73(4):526-527 (1967).

334. G. S. Monk, "The representation of primary lattices," Doctoral dissertation, Univ. of Minnesota (1966), 154 pp.; Dissert. Abstr., B27(3):886 (1966).

335. A. Monteiro, Généralisation d'un théorème de R. Sikorski sur les algèbre de Boole. Bull. sci. math., 89(3-4):65-74 (1965).

336. L. Monteiro, Sur les algebres de Lukasiewicz injectives. Proc. Japan. Acad., 41(7):578-581 (1965).

337. R. Moors, Compactification des espaces topologiques. Bull. Soc. roy. sci. Liège, 35(1-2):40-56 (1965).

338. A. Morel, Ordering relations admitting automorphisms. Fundam. math., 54(3): 279-284 (1964).

339. J. Morgado, Introdução a teoria dos reticulados. T. 1, T. 2. Textos mat. Inst. fis. e mat. Univ. Recife, No. 10, pp. 1-244 (1962); No. 11, pp. 245-342 (1962).

340. J. Morgado, Note on the central closure operators of complete lattices. Proc. Koninkl. nederl. akad. wet., A67(4):467-476 (1964); Indagationes math., 26(4): 467-476 (1964).

341. J. Morgado, Note on the distributive closure operators of a complete lattice. Portug. math., 23(1-2):11-25 (1964).

342. J. Morgado, None on α-complete congruences of α-complete lattices. Portug. math., 23(3-4):131-138 (1964).

343. J. Morgado, A characterization of the central closure operators of complete lattices. Proc. Koninkl. nederl. akad. wet., A68(1):70-73 (1965); Indagationes math., 27(1):70-73 (1965).

344. J. Morgado, Extensão de alguns resultados de Ore sôbre homomorfismos de reticulados completos. Notas e comuns mat., No. 1 (1965), 5 pp.

345. J. Morgado, Um teorema sôbre conguências α-completas de reticulados α-completos. Notas e comuns mat., No. 3 (1965), 8 pp.

346. J. Morgado, Note on the system of closure operators of the ordinal product of partially ordered sets. Notas e comuns mat., No. 5 (1965), 8 pp.

347. J. Morgado, On the complete homomorphic images of the lattice of closure operators of a complete lattice. Notas e comuns mat., No. 8 (1965), 13 pp.

348. J. Morgado, Lattices in which $a < b \lor c$ implies $a \le b$ or $a \le c$ Portug. math., 24(1-2):13-19 (1965).

349. J. Morgado, Note on the system of closure operators of the ordinal product of partially ordered sets. Portug. math., 24(3-4):189-195 (1965).

350. J. Morgado, On the lattices of residuated closure operators of complete lattices. Bol. Soc. mat. São Paulo, 18(1-2):31-37 (1966).

351. J. Morgado, Note on the factorization of the lattice of closure operators of a complete lattice. Proc. Koninkl. nederl. akad. wet., A69(1):34-41 (1966); Indagationes math., 28(1):34-41 (1966).

352. J. Morgado, On the lattices of residuated closure operators of complete lattices. Notas e comuns mat., No. 12, 7 pp. (1966).

353. W. D. Munn, Uniform semilattices and bisimple inverse semigroups. Quart. J. Math., 17(66):151-159 (1966).

354. W. D. Munn, The lattice of congruences on a bisimple ω-semigroup. Proc. Roy. Soc. Edinburgh, 1965-1967, A67(3):175-184 (1967).

355. T. Nakano, Rings and partly ordered systems. Math. Z., 99(5):335-376 (1967).

356. F. Napolitani, Elementi U-quasi distributivi nel reticolo dei sottogruppi di un gruppo. Ricerche mat., 14(2):93-101 (1965).

357. O. T. Nelson, Jr., Subdirect products of finite lattices. Doct. diss. Vanderbilt Univ., (1965), 49 pp.; Dissert. Abstr., 26(4):2240 (1965).

358. W. C. Nemitz, "Implicative semi-lattices," Res. Participat. Coll. Teachers Math., Norman, Okla. (1963).

359. W. C. Nemitz, Implicative semi-lattices. Trans. Amer. Math. Soc., 117(5):128-142 (1965).

360. V. Novák, O jedné vlastnosti ordinálního součtu. Časopis pěstov. mat., 89:78-94 (1964).

361. V. Novák, On the lexicographic dimension of linearly ordered sets. Fundam math., 56:9-20 (1964).

362. V. Novák, On the ω_μ-dimension and ω_μ-pseudodimension of ordered sets. Z. math. Logik und Grundl. Math., 10(1):43-48 (1964).

363. V. Novák, Some generalizations of the dimension of ordered sets. Spisy přírodověd. fak. univ. Brňe, No. 9, pp. 477-478 (1964).

364. V. Novák, On the lexicographic product of ordered sets. Czech. Mat. J., 15(2):
     270-282 (1965).

365. V. Novák, On universal quasi-ordered sets. Czech. Mat. J., 15(4):589-595 (1965).

366. L. G. Novák, On n-ordered sets and order completeness. Pacif. J. Math., 15(4):
     1337-1345 (1965).

367. M. Novotný, "Idealtopologien auf geordneten Menger," General Topology and
     Related Modern Analysis and Algebra, Vol. 2, Prague (1967), pp. 274-275.

368. M. Novotný and L. Skula, Über gewisse Topologien auf geordneten Mengen.
     Fundam. Math., 56(3):313-324 (1965).

369. R. Padmanabhan, On some ternary relations in lattices. Colloq. math., 15(2):
     195-198 (1966).

370. R. Padmanabhan, On axioms for semi-lattices. Canad. Math. Bull., 9(3):357-358
     (1966).

371. D. M. Paine, "Some grouplike properties of partition systems," Doctoral disser-
     tation, Univ. of Wisconsin (1963), 103 pp.; Dissert. Abstr., 24(4):1637 (1963).

372. D. Papert, Congruence relations in semi-lattices. J. London Math. Soc., 39(4):
     723-729 (1964).

373. S. Papert, An abstract theory of topological subspaces. Proc. Cambridge Philos.
     Soc., 60(2):197-203 (1964).

374. R. Permutti, Sulle semicongruenze di un reticolo. Rend. Accad. scil fis. e mat.
     Soc. naz. sci lettere ed arti Napoli, 31:160-167 (1964).

375. W. Y. Pervin and H. J. Biesterfeldt, Jr., Regular topologies and posets. Proc.
     Amer. Math. Soc., 17(4):903-905 (1966).

376. W. Y. Pervin and H. J. Biesterfeldt, Jr., Uniformization of convergence spaces.
     Part II. Conjugate convergence structures and bi-structures. Math. Ann., 177(1):
     43-48 (1968).

377. A. Petcu, Definirea laticilor numai cu ajutorul legilor de associativitate și ab-
     sorbtie. Studii și cercetări mat. Acad. RPR, 16(10):1265-1280 (1964).

378. A. Petcu, Short definitions of lattices, using the associative and absorption laws.
     Rev. roumaine math. pures et appl., 16(10):339-355 (1964).

379. A. Petcu, Definirea algebrelor booleene și a semilaticilor cu ajutorual a două
     axiome-identităti. Studii și cercetări mat. Acad. RSR, 19(6):891-895 (1967).

380. G. Pic, Sur un théoreme de la théorie des nombres et ses applications a la théorie
     des treillis et des groupes. Studia Univ. Babes-Bolyai. Ser. math.-phys., 11(2):
     21-30 (1966).

381. H. E. Pickett, A note on generalized equivalence relations. Amer. Math. Month-
     ly, 73(8):860-861 (1966).

382. R. S. Pierce, The global dimension of Boolean rings. J. Algebra, 7(1):91-99 (1967).

383. C. Pinter, "Sur l'obtention des formes minimales de fonctions booleennes con-
     jonctives, Doctoral dissertation, Univ. of Paris (1964), 90 pp.

384. J. Plonka, Representation theorems for idempotent quasi-abelian semigroups and
     distributive quasi-lattices. Bull. Acad. polon. sci. sér. sci. math., astron. et
     phys., 14(7):359-360 (1966).

385. D. Ponasse, Algébrisation du calcul proportionnel au moyen d'anneaux booleiens
     universels. Ann. Fac. scil Univ. Clermont. Math., No. 4, pp. 33-36 (1967).

386. S. Pontier, Sur les groupes (demigroups) dont le treillis des sous-groups (sous-demi-groupes) est géométrique. C. r. Acad. sci., 262(26):A1435-A1437 (1966).

387. D. H. Potts, Axioms for semi-lattices. Canad. Math. Bull., 8(4):519 (1965).

388. A. Preller, Sur le probléme universel (liberté) des algébres de Boole et des espaces de Boole par rapport aux ensembles. Publ. Dep. Math. (Lyon), 3(1):17-25 (1966).

389. A. Preller, Une catégorie duale de la catégorie des anneaux idempotents. Publ. Dep. Math. (Lyon), 3(1):26-30 (1966).

390. O. Pretzel, A representation theorem for partial orders. J. London Math. Soc., 42(3):507-508 (1967).

391. F. Previale, Reticoli metrici. Boll. Unione mat. ital., 21(3):243-250 (1966).

392. R. Rado, A theorem on chains of finite sets. J. London Math. Soc., 42(1):101-106 (1967).

393. A. Ramsay, Dimension theory in complete orthocomplemented weakly modular lattices. Trans. Amer. Math. soc., 116(4):9-31 (1965).

394. A. Ramsay, Decompositions of von Neumann lattices. J. Math. Analysis and Applic., 20(3):480-506 (1967).

395. P. S. Rema, Generalized metric lattices. I. The Bm-lattices. J. Madras Univ., B33(3):281-288 (1963).

396. P. S. Rema, On compact topological lattices. Math. Japonicae, 9(2):93-98 (1964).

397. P. S. Rema, On sets lattices and groups with Boolean metrization. Math. Student, 33(1):29-31 (1965).

398. P. S. Rema, Linear-compact congruence topologies in *-lattices. Fundam. math., 57(1):9-24 (1965).

399. P. S. Rema, Auto-topologies in Boolean algebras. J. Indian Math. Soc., 30(4):221-243 (1967).

400. R. Righi, Nota sulle funzioni booleane autoduali. Note recens. e notiz., 12(4):321-338 (1963).

401. K. W. Roggenkamp, Gruppenringe von unendlichem Dartellungstyp. Math. Z., 95(5):393-398 (1967).

402. H. L. Rolf, The free modular lattice FM (2 + 2 + 2) is infinite. Proc. Amer. Math. Soc., 17(4):960-961 (1966).

403. I. Rosenberg, Die transitive Hülle des lexikographischen Produktes. Matemat. časop., 17(2):92-107 (1967).

404. G. C. Rota, On the foundations of combinatorial theory. I. Theory of Möbius functions. Z. Wahrsheinlichkeitstheor. und verw. Geb., 2(4):340-368 (1964).

405. S. Rudeanu, Boolean equations and their applications to the study of Bridge-Circuits. I. Bull. math. Soc. sci. math. et phys. RPR, 3(4):445-473 (1959).

406. S. Rudeanu, Independent systems of axioms in lattice theory. Bull. math. Soc. sci. math. et phys. RPR, 3(4):475-488 (1959).

407. S. Rudeanu, Ecuaţiile booleene şi aplicaţiile for la studiul schemelor în punte. II. Comun. Acad. RPR, 11(6):611-618 (1961).

408. S. Rudeanu, Asupra rezolvării ecuaţiilor booleene prin metoda lui Löwenheim. Studii si cercetari mat. Acad. RPR, 13(2):295-308 (1962).

409. S. Rudeanu, Asupra unor sisteme de postulate ale algebrelor booleene. Comun. Acad. RPR, 12(8):893-899 (1962).

410. S. Rudeanu, Axiomele laticilor și ale algebrelor booleene. București. Acad. R.P.R. (1963), 159 pp.

411. S. Rudeanu, Solutions non redondantes des équations booléennes. Bull. math. Soc. sci. math. et phys. RPR, 7(1-2):45-49 (1963).

412. S. Rudeanu, Logical dependence of certain chain conditions in lattice theory. Spisy přírodověd. fak. univ. Brně, No. 9, pp. 484-485 (1964).

413. S. Rudeanu, Logical dependence of certain chain conditions in lattice theory. Acta scient. math., 25(3-4):209-218 (1964).

414. S. Rudeanu, Remarks on Motinori Goto's papers on Boolean equations. Rev. Roumaine Math. pures et appl., 10(3):311-317 (1965).

415. S. Rudeanu, Irredundant solutions of Boolean and pseudo-Boolean equations. Rev. Roumaine math. pures et appl., 11(2):183-188 (1966).

416. S. Rudeanu, Sur les equations booleennes de S. Četkovič. Publs. Inst. Math., 6:95-97 (1966).

417. M. E. Rudin, Interval topology in subsets of totally orderable spaces. Trans. Amer. Math. Soc., 118(6):376-389 (1965).

418. J. Ruedin, Groupoides distributifs et axiomatique des treillis Sémin. Dubreil et Pisot. Fac. sci. Paris, 20(1):4/01-4/36 (1968).

419. J. Ruedin, Sur les groupoides distributifs agant un élément neutre d'un côté et sur l'axiomatique des treillis. C. r. Acad. sci., 263(17):A559-A562 (1966).

420. D. E. Rutherford, Introduction to Lattice Theory, Oliver and Boyd, Edinburgh (1965), 118 pp.

421. D. E. Rutherford, Introduction to Lattice Theory, Hafner, New York (1965), 117 pp.

422. D. E. Rutherford, Introduction to Lattice Theory, Oliver and Boyd, Edinburgh (1965), 117 pp.

423. D. E. Rutherford, The eigenvalue problem for Boolean matrices. Proc. Roy. Soc. Edinburgh, A67(1):25-38 (1965).

424. D. E. Rutherford, Orthogonal Boolean matrices. Proc. Roy. Soc., 67(2):126-135 (1966).

425. I. Ruzsa, Über einige Erweiterungen des formalen Systems der elementaren Booleschen Algebra. Ann. Univ. scient. Budapest. Ser. math., 8:163-180 (1965).

426. D. Sachs, Reciprocity in matroid lattices. Rend. Semin. mat. Univ. Padova, 36(1):66-79 (1966).

427. E. T. Schmidt, Universale Algebren mit gegebenen Automorphismengruppen und Unteralgebrenverbänden. Acta scient. math., 24(3-4):251-254 (1963).

428. E. T. Schmidt, Universale Algebren mit gegebenen Automorphismengruppen und Kongruenzverbänden. Acta math. Acad. scient. hung., 15(1-2):37-45 (1964).

429. E. T. Schmidt, Remark on a paper of M. F. Janowitz. Acta math. Acad. scient. hung., 16(3-4):435 (1965).

430. E. T. Schmidt, Über endliche Verbände, die in einen endlichen Zerlegungsverband einbettbar sind. Studia scient. math. hung., Vol. 1, Nos. 3-4 (1966).

431. E. T. Schmidt, On the definition of homomorphism kernels of lattices. Math. Nachr., 3(1-2):25-30 (1967).

432. J. Schmidt, Konfinalität. Z. math. Logik und Grundl. Math., 1(4):271-303 (1955).

433. P. S. Schnare, Multiple complementation in the lattice of topologies. Fundam. math., 62(1):53-59 (1968).

434. P. S. Schnare, "Infinite complementation in the lattice of topologies," Doctoral dissertation Tulane Univ. (1967), 62 pp.; Dissert. Abstr., B28(9):3788 (1968).

435. E. A. Schreiner, Modular pairs in orthomodular lattices. Pacif. J. Math., 19(3): 519-528 (1966).

436. G. Scognamiglio, Funzioni transcendenti nell'algebra di Boole. Ann. pontifist. super. sci. e lett. "S. Chiara" No. 12, pp. 245-260 (1962).

437. G. Scognamiglio, Estensione della nozione di algebra. Algebre di secondo ordine su un'algebra di Boole. Giorn. mat. Battaglini, 90(1):93-108 (1962).

438. G. Scognamiglio, Analisi funzionale neele algebre booleane. Giorn. mat. Battaglini, 90(1):109-137 (1962).

439. G. Scognamiglio, Filze di funzioni algebriche booleane e matrici associate. Ricerca, 14:22-35 (Jan.-Apr., 1963).

440. G. Scognamiglio, Transformazioni algebriche su variabili booleane (relazioni fra matrici, determinanti ed operatori). Giorn. mat. Battaglini, 91(1):24-55 (1963).

441. G. Scognamiglio, Un metodo di calcolo dei prodotti delle matrici booleane elementari. Ann. pontif. Ist. super. sci. e lettere "S. Chiara", No. 13, pp. 413-429 (1963).

442. V. Sedmak, Geordnete Mengen und zufälligkeitsmass einer Fölge. Spisy přiro dověd. fak. univ. Brně, No. 9, pp. 486-487 (1964).

443. B. Segre, Introduction to Galois geometries. Atti Accad. naz. Lincei. Mem. Cl. sci. fis. mat. e natur., Sez. 1, 8(5):133-236 (1967).

444. A. Sekanina and M. Sekanina, Topologies compatible with ordering. Arch. mat., 2(3):113-126 (1966).

445. M. Sekanina, On ordering of the system of all subsets of a given set. Z. math. Logik und Grundl. Math., 10(4):283-301 (1964).

446. M. Sekanina, On the power of ordered sets. Arch. mat., 1(2):75-82 (1965).

447. M. Sekanina, "Topologies compatible with ordering," General Topology and Related Modern Analysis and Algebra, Vol. 2. Prague (1967), pp. 326-329.

448. M. Sekanina and A. Sekanina, Topology compatable with the order. Časop. pěstov. mat., 91:95-96 (1966).

449. M. Servi, Sull'estensione degli operatori de tipo V_{DD}. Rend. Semin. mat. Univ. e Politecn. Torino, 24:125-136 (1966).

450. M. Servi, Algebre di Fréchet: una classe di algebre booleane con operatore. Rend. Circolo. mat. Palermo, 14(3):335-366 (1966).

451. R. Sikorski, Boolean Algebras, 2nd ed., Springer, Berlin (1964), 237 pp.; Dtsch. Bibliogr. No. 48, p. 4239 (1968).

452. F. M. Sioson, On Newman algebras. I, II. Proc. Japan Acad., 41(1):31-39 (1965).

453. F. M. Sioson, Natural equational bases for Newman and Boolean algebras. Compositio math., 17(3):299-310 (1966).

454. F. M. Sioson, Natural equational bases for Newman and Boolean algebras. Compositio math., 17(3):299-310 (1967).

455. L. A. Skornyakov (Skorniakov), Complemented Modular Lattices and Regular Rings, Oliver and Boyd, Edinburg (1964), 182 pp.

456. L. Skula, Über Systeme von stetigen und isotonen Abbildungen. Spisy přírodověd. fak. univ. Brně, No. 9, pp. 490-492 (1964).

457. R. E. Smithson, A note on finite Boolean rings. Math. Mag., 37(5):325-327 (1964).

458. R. M. Solovay, New proof of a theorem of Gaifman and Hales. Bull. Amer. Math. Soc., 72(2):282-284 (1966).

459. A. K. Steiner, The topological complementation problem. Bull. Amer. Math. Soc., (1, Part I):125-127 (1966).

460. A. K. Steiner, The lattice of topologies: structure and complementation. Trans. Amer. Math. Soc., 122(2):379-398 (1966).

461. A. K. Steiner, Complementation in the lattice of T_1-topologies. Proc. Amer. Math. Soc., 17(4):884-886 (1966).

462. A. K. Steiner, "Ultraspaces and topological complementation problem," Doctoral dissertation, Univ. of New Mexico (1965), 57 pp.; Dissert. Abstr., 26(11):6753-6754 (1966).

463. A. K. Steiner and E. F. Steiner A T_1-complement for the reals. Proc. Amer. Math. Soc., 19(1):177-179 (1968).

464. E. F. Steiner, "Lattices of topologies on linear spaces," Doctoral dissertation, Univ. of Missouri (1963), 85 pp.; Dissert. Abstr., 24(3):1193 (1963).

465. E. F. Steiner, Lattice of linear topologies. Port. math., 23(3-4):173-181 (1964).

466. E. F. Steiner, Lattices of normed topologies. Port. math., 24(1-2):7-12 (1965).

467. E. F. Steiner and A. K. Steiner, Topologies with T_1-complements. Fundam. math., 61(1):23-28 (1967).

468. O. Steinfeld, Über Zerlegungssätze für teilweise geordneten Halbgruppen mit bedingten Distributivitätsregeln. Magyar. tud. akad. Mat. Kutató int. közl., 9(3):313-330 (1965).

469. D. P. Strauss, Topological lattices. Proc. London Math. Soc., 18(2):217-230 (1968).

470. T. Sturm, Isotonii rozšíření isotonnich zobrazení. Časop. pěstov. mat., 92(3):318-331 (1967).

471. N. V. Subrahmanyam, Boolean vector spaces. II. Math. Z., 87(3):401-419 (1965).

472. K. L. N. Swamy, A general theory of autometrized algebras. Math. Ann., 157(1):65-74 (1964).

473. G. Szász, On independent systems of axioms for lattices. Publs. math., 10(1-4):108-115 (1963).

474. G. Szász, Über einige Axiomensysteme der Verbände. Spisy přírodověd. fak. univ. Brně, No. 9, p. 493 (1964).

475. T. Tamura and W. Etterbeek, The lattice congruences of locally cyclic semigroups. Proc. Japan Acad., 42(7):682-684 (1966).

476. D. M. Topping, Some homological pathology in vector lattices. Canad. J. Math., 17(3):411-428 (1965).

477. T. Traczyk, Axioms and some properties of Post algebras. Colloq. math., 10(2):193-209 (1963).

478. T. Traczyk, An equational definition of a class of Post algebras. Bull. Acad. polon. sci. Sér. sci. math., astron. et phys., 12(3):147-149 (1964).

479. T. Traczyk, A generalization of the Loomis—Sikorski theorem. Colloq. math., 12(2):155-161 (1964).

480. S. S. Van Domelen, The structure of bounded commutative semisimple cones. Doct. dissert. Univ. New Mex., 1965, 47 pp. Dissert. Abstr., 26(11):6755 (1966).

481. J. Varlet, Contribution à l'étude des treillies pseudo-complémentés et des treillis de Stone. Mém. Soc. roy. scil Liège, Vol. 8, No. 4 (1963), 71 pp.

482. J. Varlet, Congruences dans les treillis pseudo-compléments. Bull. Soc. roy. sci. Liège, 32(9):623-625 (1963).

483. J. Varlet, Indéaux dans les lattis pseudo-complémentés. Bull. Soc. roy. sci. Liège, 33(3-4):143-150 (1964).

484. J. Varlet, Congruences dans les demi-lattis. Bull. Soc. roy. sci. Liège, 34(5-6): 231-240 (1965).

485. J. Varlet, On the characterization of Stone lattices. Acta scient. math., 27(1-2): 81-84 (1966).

486. P. V. Venkatanarasimhan, A note on modular lattices. J. Indian Math. Soc., 30(1):55-59 (1966).

487. P. V. Venkatanarasimhan, Ideals in semi-lattices. J. Indian Math. Soc., 30(1): 47-53 (1966).

488. J.-L. Verdier, Equivalence essentielle des systèmes projectifs. C. r. Acad. sci., 261(23):4950-4953 (1965).

489. J. H. Weston "Some results concerning m-compactness," Doctoral dissertation, Lehigh Univ. (1967), 55 pp.; Dissert. Abstr., B28(10):4211 (1968).

490. R. Wille, Halbkomplementäre Verbände. Math. Z., 94(1):1-31 (1966).

491. R. Wille, Verbandstheoretische Charakterisierung n-stufiger Geometrien. Arch. Math., 18(5):465-468 (1967).

492. J. S. W. Wong, Common fixed points of commuting monotone mappings. Canad. J. Math., 19(3):617-620 (1967).

493. A. C. Woods, A note on dense subsets of lattices. J. London Math. Soc., 41(4): 742-744 (1966).

494. Y. Wooyenaka, On postulates sets for Newman algebra and Boolean algebra. I, II. Proc. Japan Acad., 40(2):76-87 (1964).

495. F. B. Wright, Boolean averages. Canad. J. Math., 15(3):440-455 (1963).

496. F. Wuytack, Boolean composition algebras. Simon Stevin, 37(3):97-125 (1964).

497. O. Wyler, Clans. Compositio math., 17(2):172-189 (1966).

498. B. Zelinka, Důkaz nezávislosti Birkhoffova systému postulátů pro distributivni svazy s jednotkovým prvkem. Časop. pěstov. mat., 91(4):472-477 (1966).

GEOMETRY

Integral Geometry

G. I. Drinfel'd

PREFACE

This survey is devoted to papers in the area of integral geometry published during the last ten years.

During these years Chern's method of finding a measure in various uniform spaces received a further development; a method was developed based on a systematic application of the theory of Poincaré-Cartan integral invariants; intersting results were obtained in a number of aspects connected with the concept of kinematic measure. A large part of the first chapter of this survey is devoted to papers on the topics mentioned.

This same chapter cites the representative papers on the abstract foundations of integral geometry and the papers which are clearly the beginning of a new direction in the investigations on integral geometry, namely, a direction which differs in the application of more modern concepts and methods of mathematical analysis and geometry.

The second chapter of the survey is devoted to concrete problems and to the applications of integral geometry (generalizations of the formulas of Crofton, Lebesgue, and Steiner; Rashevskii's problems; Minkowski integrals; lattices and coverings, etc.). Papers in the probability-theoretic direction also are considered and the noted connection of integral geometry with pattern recognition is discussed.

It is difficult for the compiler of a survey not to show his personal tastes and interests, but this circumstance does not explain why the individual sections of the survey occupy a larger place than they may possibly deserve. There are other reasons.

The compiler assumed that he should not set forth in detail the concepts and ideas already presented in Blaschke's "Lectures" and Santaló's "Introduction..." and that it was enough to point out the new results. In the compiler's opinion new concepts and ideas should be presented in detail if, of course, they possess a certain generality.

In one place (§ 2 of Chapter I) the author expanded the presentation, taking it as his duty to dwell on the question of priority because this aspect is connected with the name of the outstanding Soviet scientist N. G. Chebotarev.

The survey also examines certain comparatively recent papers on integral geometry for the reason that they have not been included in other surveys and the work in recent years has been related to them.

Very little space in the survey has been allotted to papers connected with isoperimetric inequalities and with convex bodies; in the compiler's opinion it is natural to consider these topics in a special survey. Papers on integral geometry in the sense of Gel'fand are not considered at all for the very same reason.

Throughout the whole survey the differential forms are exterior ones; the summation sign is omitted if the summation takes place over an index encountered twice—upper and lower. Not all the papers cited in the Bibliography are mentioned in the survey, in one case, because the fundamental results of an author are covered in his later papers, and in other cases, because the paper is a popular one or is devoted to a very special problem.

Chapter I

MEASURES IN UNIFORM SPACES
§ 1. Chern's Program

The main requirement on the measures considered in integral geometry is that they be unique and invariant relative to the group of transformations acting in the space being considered. Furthermore, the measure is sought in the form of a certain integral of an exterior differential form and, therefore, it makes no difference whether we speak of the measure or of the density (the analog in probability theory is the distribution function and the probability density).

In Blaschke's Hamburg seminar, where integral geometry was mainly worked out up to 1940, no sufficiently general method was created for finding invariant measures. Such an algorithmically general method was first pointed out by Chern [26] in 1942. This method is presented in sufficient detail by Santaló [11] who used it until recent times for finding invariant measures. Briefly, this method consists in finding the relative (Maurer-Cartan) components of the transformation group, in constructing an invariant density, if it exists and is unique, in the form of the outer product of these components, and, finally, in clarifying (which already is not so simple a matter) the metric sense of the invariant found, by means of pasing to the point coordinates and to the parameters of the set of geometric elements being considered.

Using Chern's program Gaeta [34] analyzed sufficiently simple uniform spaces of a transitive group, namely, the spaces of straight lines and of oriented straight lines on a plane. Santaló [69] found a measure, invariant relative to a unimodular group of affine transformations, of the set of pairs of hyperplanes $\sum_{i=1}^{n} l_i x_i = m, \sum_{i=1}^{n} l_i x_i = m + 1.$ He obtained the formula

$$dP = |p_2 - p_1|^{-(n+2)} d\sigma \wedge dp_1 \wedge dp_2,$$

where $d\sigma$ is an area element of the unit hypersphere described by a vector normal to the hyperplanes, p_1 and p_2 are the distances of the planes from the origin.

In a later paper [71] Santaló considered a more general problem and proved the theorem: Let $\mathfrak{U}$ be a unimodular group of affine transformations. In order that the set of elements $H(L_{h_1}, L_{h_2}, \ldots, L_{h_q})$, formed by parallel linear subspaces of dimensions h_1, h_2, $\ldots$, h_q, which transform transitively by group $\mathfrak{U}$, have an invariant measure with respect to $\mathfrak{U}$ it is necessary and sufficient that $h_1 = h_2 = \ldots = h_q = h$, $q = n + 1 - h$.

For the density of the set Santaló obtained the expression

$$[(n-h)! \, S]^{-2n+h-1} dP_0 \wedge dP_{h+1} \wedge \ldots \wedge dP_n \wedge dL_{n-h},$$

where P_0, P_{h+1}, ..., P_n are the intersection points of the given $n-h+1$ parallel h-subspaces with the orthogonal $(n-h)$-spaces L_{n-h} passing through the origin; S is the volume of the simplex with vertices P_0, P_{h+1}, ..., P_n; dP_i is the volume element of L_{n-h} in P_i; dL_{n-h} is the density of the $(n-h)$-spaces L_{n-h}.

In [69] Santaló also proved the assertion: Let $\bar{x} = Ax$, $\det A = 1$, be a group of projective transformations in an n-dimensional space and let $x'\Phi x = 0$ be second-order nonsingular hypersurfaces. The density of the set of such hypersurfaces, invariant relative to the transformations considered, exists and equals

$$q_{nn}^{-1} dq_{00} \wedge dq_{01} \wedge \ldots \wedge dq_{n-1,n},$$

where $(q_{hl}) = (A^{-1})'\Phi A^{-1}$.

Luccioni [49] showed that the condition of nonsingularity of the hypersurfaces in Santoló's theorem is unnecessary. As far as the set of pairs of second-order singular hypersurfaces is concerned, it has a measure if the vertex spaces are convex and the sum of their dimensions equals $n-1$.

Luccioni [51] also considered the question of the existence of the measure of the set of linear subspaces of an affine space, not requiring them to be parallel as in [71] but assuming that each subspace is specified by a collection of n linearly independent points. However, the author's result, reducing the question to the same one for a projective space, is trivial (in any case, it is difficult to take it as being definitive).

With the aim of generalizing Santaló's theorem ([11], p. 135) on the existence of a measure of the set of nonintersecting linear subspaces, invariant relative to the group of projective transformations, Luccioni [50] introduced the following concepts: the set of nonintersecting linear subspaces of an n-dimensional space S_n, defined by specifying $n+1$ linearly independent points, is called a system of type S if their sum coincides with S_n; two systems of subspaces are said to be equivalent if every subspace of one system can be constructed from the subspaces of the second system with the aid of the operations of intersection and union. Luccioni proved, in particular, the following generalizations of Santaló's theorem:

1. The set of linearly independent linear subspaces of space S_n, defined by $n + 1$ linearly independent points, has a measure (invariant relative to the group of projective transformations) if and only if it is equivalent to a system of type S defined by those same points.

Let $S_{h_1}, \ldots, S_{h_r}; S_{k_1}, \ldots, S_{k_m}$ be linear nonintersecting subspaces of S_n. The measure of the union of the sets $(S_{k_1}, \ldots, S_{k_m}), (S_{h_1}, \ldots, S_{h_r})$ exists if and only if the sum of all $r + m$ subspaces coincides with S_n.

In a recent paper, adhering precisely to Chern's algorithm, Santoló [75] examined projective groups on a plane, relative to which point sets and line sets admit of an invariant measure. Further, Lie listed all the projective groups on a plane, depending on m parameters, $2 \leq m \leq 8$, $m \neq 7$, by writing out all the infinitesimal group operators.

In the paper mentioned Santaló showed that out of 33 projective groups on a plane only seven two-parameter groups and five three-parameter groups admit of invariant measures of the point set and the line set. The remaining groups do not admit of an invariant measure even for one of these sets. But such groups do admit of invariant measures of sets of pairs: pairs of points, pairs of straight lines, pairs made up of a point and a straight line. Santaló [73] found all the cases of the existence of such measures. In both these papers [73, 75] the densities were computed in all cases when they existed.

Yurtova and Lutsenko [16] proved the validity of the assumption of the reviewer (in Referativnyi Zhurnal, Matematika) of paper [73] that by using the well-known infinitesimal operators of the groups being considered we can find the invariant measures (or establish their absence) without turning to Chern's theorem, and, moreover, we can do this right away in the variables needed.

Sulanke [96] worked out Chern's theory in an n-dimensional uniform Klein space G/H. According to Chern a point $x \in G/H$ and a plane $a \in G/H$ are called incident if their preimages in G — the cosets in G with respect to H and K — have a nonempty intersection. By analyzing p-dimensional surfaces $V \subset G/H$ whose p-dimensional tangent elements (on a given surface) transitively transform the group G, Sulanke found the necessary and sufficient condition for the exterior form constructed by Chern's scheme ([11], p. 116) to be the density of the invariant measure on a surface V. He generalized Chern's condition ([11], p. 118) on the existence of a measure in a uniform space and obtained formulas for the representation of an area on V by means of an integral of the product of the number of points of intersection of V with the incident planes by the density of these planes (Crofton's formula).

§ 2. Invariant Measure of a Point Set and Integral Invariants

The measure of a point set in integral geometry, by its very definition, is an integral invariant of the transformation group.

The integral of a pth-order exterior differential form

$$\int_{H_p} A_{a_1 a_2 \ldots a_p} dx^{a_1} \wedge dx^{a_2} \wedge \ldots \wedge dx^{a_p} \tag{2.1}$$

over a p-dimensional point manifold lying on the system trajectories is called a pth-order Poincaré-Cartan integral invariant of the system of differential equations

$$\frac{dx^1}{X_1} = \frac{dx^2}{X_2} = \ldots = \frac{dx^n}{X_n} = dt, \ X_i = X_i(x^1, x^2, \ldots, x^n) \tag{2.2}$$

if the value of the integral does not change under a displacement of the points of H_p along the system trajectories. The necessary and sufficient invariance conditions are

$$X(A_{a_1 \ldots a_p}) + A_{i a_2 \ldots a_p} \frac{\partial X_{a_1}}{\partial x^i} + A_{a_1 i \ldots a_p} \frac{\partial X_{a_2}}{\partial x^i} + \ldots$$

$$\ldots + A_{a_1 \ldots a_{p-1} i} \frac{\partial X_{a_p}}{\partial x^i} = 0, \tag{2.3}$$

$$a_1, a_2, \ldots, a_p = 1, 2, \ldots, n; \ 1 \leqslant i \leqslant n,$$

$$X(f) = X_k \frac{\partial f}{\partial x^k}.$$

In particular, when p = n, integral (2.1) and conditions (2.3) take the form

$$\int\limits_{\dot{H}_n} M dx^1 \wedge dx^2 \wedge \ldots \wedge dx^n, \tag{2.4}$$

$$X(M) + M \sum_{i=1}^{n} \frac{\partial X_i}{\partial x^i} = \sum_{i=1}^{n} \frac{\partial (MX_i)}{\partial x^i} = 0.$$

The concept of an integral invariant of system (2.2) and the invariance conditions can be immediately generalized, at first to the case of a one-member and next to the case of an r-member (r-parameter) group G_r of transformations

$$y^i = \varphi^i (x^1, \ldots, x^n; \ a_1, \ldots, a_r)$$

(all parameters are essential) with the infinitesimal operators

$$X_k (f) = X_i^k \frac{\partial f}{\partial x^i}, \ k = 1, 2, \ldots, r.$$

In particular, the necessary and sufficient invariance conditions of integral (2.4) relative to group G_r are

$$X_k (M) + M \sum_{i=1}^{n} \frac{\partial X_i^k}{\partial x^i} = 0, \ k = 1, 2, \ldots, r. \tag{2.5}$$

By analyzing the well-known Bertrand paradoxes in probability theory Poincaré made the remark that in defining an elementary probability (the probability density) it is necessary to take into account the mechanism which realized the random event, which reduces to the requirement of invariance of an integral of form (2.4) relative to a certain transformation group. Deltheil [30] developed Poincaré's remark and derived the invariance conditions (2.5).

In a paper devoted to the construction of the axiomatics of probability theory, equivalent ot Kolmogorov's axiomatics, Hadwiger [44] also discussed the connection between the question of the correct formulation of geometric probability theory problems and the examination of a certain transformation group.

In a group G_r is intransitive, then the number l of linearly independent operators $X_k (f)$ is less than n and the group has $n-l$ absolute invariants, namely, the solutions of the system of equations

$$X_k(f) = 0, \quad k = 1, 2, \ldots, r. \tag{2.6}$$

In this case the integral invariant (2.4) exists, but M is determined to within a factor − an arbitrary function of the solutions of system (2.6).

If the group is simply transitive, then r = n, the operators $X_k(f)$ are linearly independent, the group does not have absolute invariants, and the integral invariant exists and has the property of uniqueness to within a constant factor which can be used for a suitable norming of integral (2.4).

Chebotarev [14] called the integral invariant (2.4) the volume of the group (but this is none other than the invariant measure of a point set!), established the completeness of system (2.5), and, in the case of a simply transitive Abelian group, found the remarkable formula

$$M = \frac{1}{\|X_i^k\|}, \quad \|X_i^k\| = \det(X_i^k).$$

Chebotarev's formula can be generalized to the case of a simply transitive group without assuming it to be Abelian. Namely, if the operators

$$Y_k(f) = Y_i^k \frac{\partial f}{\partial x^i}$$

satisfy the conditions

$$(X_l, Y_k)f \equiv 0, \quad k, l = 1, 2 \ldots, n,$$

and are linearly independent (such operators do exist − the operators of the group reciprocal with G_n are such), then

$$M = \frac{1}{\|Y_i^k\|}$$

In the case of a transitive (but not simply transitive) group, r > n and the system of equations (2.5) can turn out to be inconsistent. Drinfel'd [2] found the consistency conditions. Among the operators of the group there are n linearly independent ones, let them be the operators $X_1(f), \ldots, X_n(f)$ and let

$$X_{n+k}(f) = \lambda_k^j X_j(f), \quad k = 1, 2, \ldots, r - n.$$

The necessary and sufficient conditions for system (2.5) to be consistent are: at least one of the following systems of identities should be fulfilled:

$$\sum_{i=1}^{n} \frac{\partial X_i^k}{\partial x^l} \equiv 0, \quad k = 1, 2, \ldots, r,$$

$$X_s(\lambda_k^s) \equiv 0, \quad k = 1, 2, \ldots, r - n. \tag{2.7}$$

In 1955, apparently without knowing of Chebotarev's work, published in 1937 and also as a monograph [15] in 1940, Stoka [76] called a group G_r measurable if it admits of a unique, to within a constant factor, integral invariant (2.4), and he called this invariant the measure of the point set. The necessary and sufficient conditions for group G_r to the measurable were proved in the following form by Stoka [80]: G_r should be transitive and the coefficients of the infinitesimal operators and the structure constants (C_{uv}^l) of the group should satisfy the equations

$$C_{uv}^l \left[\frac{\partial}{\partial y^i} (X_l^w \overline{X}_w^u \overline{X}_j^v) - \frac{\partial}{\partial y^j} (X_l^w \overline{X}_w^u \overline{X}_i^v) \right] = 0,$$

$$C_{uv}^l X_h^i X_i^w \overline{X}_i^v \overline{X}_w^u - C_{uh}^l X_i^w \overline{X}_w^u = 0,$$

$$i, j, u, v, w = 1, 2, \ldots, n; \quad h, l = 1, 2, \ldots, r, \tag{2.8}$$

where $\overline{X}_w^u$ are the cofactors of the elements X_w^u in the nonzero determinant $\| X_w^u \|$.

Stoka found the measurability conditions for group G_r also in other forms, but always containing C_{uv}^l, and $\overline{X}_w^u$. Ushakova [13] verified that the completeness conditions for system (2.5), together with conditions (2.7), are equivalent to conditions (2.8). Deltheil wrote the group measurability conditions in the form of a system of equations in total differentials. Chern solved the problem of finding the measure of a point set in terms of the structure constants and of the relative components of the group.

Vrănceanu [103] verified that the group measurability conditions of Chern and of Deltheil (equivalent to conditions (2.5) and (2.7)) are equivalent. Stoka's conditions have an intermediate character – they also were used by Vrănceanu.

In the same paper [103] Vrănceanu proved that a group G_r in which a stationary subgroup is perfect is measurable and (as a corollary) that a symmetric Cartan space with a simple stationary subgroup is measurable.

Having in mind applications to integral geometry, Vidal [101, 102] generalized the concept of the Poincaré-Cartan integral invariant to the case of a completely integrable system of Pfaffian equations

$$\omega_i = \sum_{j=1}^{n} \alpha_{1j} du_j = 0, \quad i = 1, \ldots, h, \quad h < n, \tag{2.9}$$

in an n-dimensional space R_n. Equations (2.9) define a certain set of integral manifolds of dimension n−h filling R_n. An arbitrary region $D_0^{(p)}$ of dimension p, under the motion of its points along the integral manifolds of system (2.9), turns into a region $D_1^{(p)}$. If

$$\int\limits_{D_0^{(p)}} \Omega_p = \int\limits_{D_1^{(p)}} \Omega_p,$$

where Ω_p is a pth-order exterior differential form, then $\int \Omega_p$ is called an integral invariant of system (2.9). Vidal proved that the necessary invariance condition is the fulfillment of the equality

$$d\Omega_p = 0 \tag{2.10}$$

on the basis of system (2.9). He proved the sufficiency of this same condition under the stringent assumption

$$\Omega_p = \Omega_n = \omega_1 \wedge \omega_2 \wedge \ldots \wedge \omega_n. \tag{2.11}$$

Vidal left the invariance condition undeciphered and as an example considered only the case of system (2.2) with n = 3.

Drinfel'd and Kyong Kim [7] noted that when system (2.9) has the form

$$dx_j = \sum_{i=1}^{m} X_{ji} dt_i, \quad X_{ji} = X_{ji}(x_1, \ldots, x_n), \tag{2.12}$$

then the integral $\int \Omega_p$ is its integral invariant if and only if it is such for a group with the infinitesimal operators

$$X_i(f) = \sum_{j=1}^{n} X_{ji} \frac{\partial f}{\partial x_j}, \quad i = 1, 2, \ldots, r. \tag{2.13}$$

The condition of complete integrability of system (2.12) signifies that group (2.13) is Abelian, but this, as Kyong Kim [8] has shown, leads to the following results: Vidal's condition (2.10) is necessary and sufficient; integral invariants of orders 1, 2, ..., n exist; there are exactly C_n^p pth-order integral invariants $\int \Omega_p^s$ with linearly unrelated integrand forms and any pth-order integral invariant has the form $\int \lambda_s \Omega_p^s$, where λ_s are functions of the integrals of the system $X_i(f) = 0$.

Since the uniqueness of an nth-order integral invariant holds only $r \geq n$, while even in this case there is no uniqueness for invariants of lower orders, it is doubtful if it is possible to construct correctly an integral geometry in the direction projected by Vidal (Vidal has not obtained even one really meaningful theorem).

However, Vidal's remark that integral invariants of lower orders also should be considered in integral geometry is apparently worthy of attention. It is possible that these invariants are measures in subspaces of lower dimension. A simple example: in a three-dimensional Euclidean space the integral invariants are $\int dxdydz$, $\int dxdy$, $\int dydz$, $\int dzdx$, $\int dx$, $\int dy$, $\int dz$.

§ 3. Measure of a Set of Geometric Elements and Integral Invariants

In § 1 we talked about actual sets of geometric elements and about the measures of such sets which could be discovered by Chern's program. In the present section we propose certain ideas and results obtained by an application of integral invariants.

Suppose we are given a set H of geometric elements in an n-dimensional space. Examples: a point set, a line set, a set of curves of a specified type, a set of geodesics, etc. Suppose that a r-parameter Lie group G_r acts in the space being considered, transitively transforming the elements of H.

Certain familiar formulas, for example, Cartan's formula

$$\iint \frac{dudv}{(u^2 + v^2)^{3/2}}, \tag{3.1}$$

for the measure, invariant relative to the transformations

$$\begin{aligned} x' &= a + x \cos\alpha - y \sin\alpha, \\ y' &= b + x \sin\alpha + y \cos\alpha, \end{aligned} \tag{3.2}$$

of the set of straight lines

$$ux + vy = 1, \tag{3.3}$$

already contain within themselves the idea which in various forms appears in the majority of papers devoted to invariant measures of sets of geometric elements. This idea can be formulated roughly as follows: Let there exist a one-to-one correspondence between the elements of set H and the points of some space W and let the group G_r induce a group Q_s in W.

If in W there exists a measure of a point set, invariant relative to Q_s, then it can be taken as a measure on H. From the probabilistic point of view the measure on a set H_1 of elements of H is the "number" of elements in H_1; if a point set W_1 in W corresponds to the set H_1, then the "number" of these points coincides with the "number" of elements in H_1. Example: to the set of all straight lines (3.3) in a Euclidean plane there corresponds the Plücker plane (u, v) and in a hyperbolic plane, the Beltrami plane. The "number" of straight lines in some subset equals the "number" of points in (u, v), whose coordinates uniquely define the straight line (3.3) and, conversely, the straight line uniquely defines a point in (u, v).

Transformations (3.2) with the infinitesimal operators $\frac{\partial f}{\partial x}$, $\frac{\partial f}{\partial y}$, $y\frac{\partial f}{\partial x} - x\frac{\partial f}{\partial y}$ induce in the Plücker plane the group of transformations.

$$u' = \frac{u \cos \alpha + v \sin \alpha}{1 - au - bv}, \quad v' = \frac{v \cos \alpha - u \sin \alpha}{1 - au - bv}$$

with the infinitesimal operators $v\frac{\partial f}{\partial u} - u\frac{\partial f}{\partial v}$, $u^2\frac{\partial f}{\partial u} + uv\frac{\partial f}{\partial v}$, $uv\frac{\partial f}{\partial u} + v^2\frac{\partial f}{\partial v}$, whose integral invariant (to within a constant factor) is integral (3.1). The transformation group with the infinitesimal operators

$$-y\frac{\partial f}{\partial x} + x\frac{\partial f}{\partial y}, \quad (1 - x^2)\frac{\partial f}{\partial x} - xy\frac{\partial f}{\partial y}, \quad -xy\frac{\partial f}{\partial x} + (1 - y^2)\frac{\partial f}{\partial y}$$

(hyperbolic motion) induces in the Beltrami plane the transformation group with the infinitesimal operators

$$v\frac{\partial f}{\partial u} - u\frac{\partial f}{\partial v}, \quad (u^2 - 1)\frac{\partial f}{\partial u} + uv\frac{\partial f}{\partial v}, \quad (v^2 - 1)\frac{\partial f}{\partial v} + uv\frac{\partial f}{\partial u},$$

having a unique second-order integral invariant

$$\int \frac{\partial u \partial v}{(u^2 + v^2 - 1)^{3/2}}$$

– the measure of the set of straight lines (3.3) in the case of hyperbolic motion. The arguments made and the examples were presented by Drinfel'd in [3] and in this paper, by the same method, he found the measures of the set of planes, of the set of line elements, of the set of Cartesian coordinate systems, etc., but he did not construct a general theory.

In a series of papers published during 1955-67, apparently without knowledge of [3], Stoka developed the same arguments and constructed a sufficiently general theory. In [76] he had already formulated a basic definition.

Let V_p be an s-parameter family of p-dimensional manifolds in the n-dimensional Euclidean space E_n and let W_q be a q-parameter ($q \leq s$) family of manifolds

$$\mathfrak{F}_j(x^1, \ldots, x^n; \alpha^1, \ldots, \alpha^q) = 0, \quad j = 1, 2, \ldots, n-p, \qquad (3.4)$$

from V_p. Let W_q be invariant relative to the group of transformations in E_n,

$$x^i = f^i(y^1, \ldots, y^n; a_1, \ldots, a_r),\qquad (3.5)$$

inducing in the space R_q of parameters $\alpha_1, \ldots, \alpha_q$ a measurable (in the Chebotarev-Stoka sense) group of transformations

$$\alpha^h = \alpha^h(\beta^1, \ldots, \beta^q; a_1, \ldots, a_r).\qquad (3.6)$$

Then as the measure of set W_q Stoka takes the integral invariant

$$\int_{(A_\alpha)} F(\alpha^1, \ldots, \alpha^q)\, d\alpha^1 \wedge d\alpha^2 \wedge \ldots \wedge d\alpha^q$$

of group (3.6), where A_α is a point set in R_q corresponding to the manifolds in W_q.

In this same paper [76] Stoka found in closed form the measures of certain sets of geometric elements. In subsequent papers Stoka developed and refined (often repeating himself) the results obtained by him in [76] and carried out computations in special cases. Perhaps the most clear basic definitions and results of Stoka were presented in [81, 85, 90] and in the monograph [93]. Let us cite Stoka's basic definitions and some results.

Let M_q^p be a family of p-dimensional manifolds defined by the equations

$$\mathfrak{F}_j(x^1, \ldots, x^n; \alpha^1, \ldots, \alpha^q) = 0, \quad j = 1, 2, \ldots, n-p,\qquad (3.7)$$

where all the parameters α^k are essential. Let T be a continuous transformation of the n-dimensional space of points $(x^1, \ldots, x^n)$, leaving the family M_q^p invariant, i.e., $TM = M' \in M_q^p$ for each $M \in M_q^p$. The set of all transformations T possessing the property indicated forms a group Σ. Let $S \in \Sigma$ and $SM = M$ for all $M \in M_q^p$. Transformations S form a group g — an invariant subgroup of group Σ.

The factor group Σ/g is called a maximal invariance group of family M_q^p, and its subgroups are called the invariance groups of the same M_q^p. Let G_r be an invariance group given by the equations

$$y^i = f^i(x^j, a^s), \quad i, j = 1, 2, \ldots, n; \quad s = 1, \ldots, r.\qquad (3.8)$$

Then

$$\mathfrak{F}_j \left(f^i \left(x, a \right), \alpha^k \right) = \mathfrak{F}_j \left(x, \beta^k \right), \tag{3.9}$$

$$\beta^k = g^k \left(a^1, \ldots, a^s; \ \alpha^1, \ldots, \alpha^q \right). \tag{3.10}$$

Transformations (3.10) form a group H_r isomorphic to G_r, which is called a group associated with G_r. The measure of the set $\mathfrak{U}$ of manifolds $M \in M_q^p$ is the integral

$$\mu \left(\mathfrak{U} \right) = \int\limits_{(\mathfrak{U}_\alpha)} F \left(\alpha^1, \ldots, \alpha^q \right) d\alpha^1 \wedge d\alpha^2 \wedge \ldots \wedge d\alpha^q,$$

where $\mathfrak{U}_\alpha$ is a point set of the space of parameters α^k, corresponding to the set $\mathfrak{U}$, if $\mu \left(\mathfrak{U} \right)$ is the unique (to within a constant factor) integral invariant of group H_r.

Among the many theorems of Stoka we note certain ones.

1. If the family M_q^p admits of two invariance groups G_r and G_s $(r > s)$, then G_s is a subgroup of G_r.

2. If $\mathfrak{U}$ is a one-parameter subset, then it has an invariant measure if and only if the group associated with the maximal invariance group is a translation group.

Stoka found some (only sufficient) conditions for the measurability of sets of geometric elements and applied them to the analysis of concrete cases: sets of hyperbolas, parabolas, parallelograms, etc. in a Euclidean plane [91], and sets of p-parametric curves $(1 \leq p \leq 3)$ and surfaces in constant-curvature Riemann spaces V_2 and V_3 [88, 92].

Simple examples suggest that a certain set of geometric elements can be nonmeasurable relative to a given transformation group, but some of its subsets may turn out to be measurable.

For example (Stoka), in the Euclidean plane the family of all circles is nonmeasurable, but this family has three one-parameter and three two-parameter measurable subfamilies:

$$x^2 + y^2 - 2k_1 vx - 2vy + k_2 v^2 = 0 \quad (k_1, k_2 \ \text{are constants})$$

$$x^2 + y^2 - 2ux - 2vy + k(u^2 + v^2) = 0 \quad (k \text{ is a constant})$$

etc.

The question of the existence of measurable subsets of a non-measurable set in general form has been examined by Stoka and Lutsenko.

Stoka [90] proved that M_q^p contains a measurable subset if and only if the group $H(\alpha)$ associated with the maximal invariance group G of the set M_q^p is intransitive or contains an intransitive subgroup. If $H_r(\alpha)$ is such a subgroup (or $H(\alpha)$ itself is) and if its invariant manifold is given by the equations

$$\alpha^\nu = \varphi^\nu(\tau^1, \ldots, \tau^{q_1}), \quad \nu = 1, \ldots, q, \quad (q_1 < q), \tag{3.11}$$

then the set $\mathfrak{F}_{q_1} \subset M_q^p$, obtained from Eqs. (3.9) by replacing α^ν by the expressions (3.11) is measurable if the group $H_r(\tau)$ induced by the group $H_r(\alpha)$ is measurable.

Stoka carried out his analysis of certain special examples right up to the end, but, in general, the application of his theory is hampered by the awkwardness of the measurability conditions (2.8).

Lutsenko [9], starting from the relations established by Stoka,

$$X_k(f) + Y_k(f) = 0, \quad k = 1, 2, \ldots, r, \tag{3.12}$$

between the operators $X_k(f) = \xi_i^k \frac{\partial f}{\partial x^i}$ of group G_r and the operators $Y_k(f) = \eta_i^k \frac{\partial f}{\partial \alpha^l}$ of group H_r, found the following and sufficient condition for the invariance of the set S of geometric elements

$$\mathfrak{F}(x^1, \ldots, x^n; \alpha^1, \ldots, \alpha^q) = 0 \tag{3.13}$$

relative to group G_r: the ranks of the matrices

$$\left\| \frac{\partial^{\nu+1} \mathfrak{F}}{\partial \alpha^1 \partial x^{j_1} \ldots \partial x^{j_\nu}} \cdots \frac{\partial^{\nu+1} \mathfrak{F}}{\partial \alpha^q \partial x^{j_1} \ldots \partial x^{j_\nu}} \frac{\partial^\nu X_k(\mathfrak{F})}{\partial x^{j_1} \ldots \partial x^{j_\nu}} \right\|$$
$$k = 1, 2, \ldots, r, \quad \nu = 0, 1, \ldots.$$

should not exceed q.

In the same paper Lutsenko showed that relations (3.12), differentiated properly and a suitable number of times, allow us to find the operators $Y_k(f)$, after which we can clear up the question of the existence of an invariant measure of the set of manifolds (3.13) and set up the differential equations for the density. He also found a convenient way of determining measurable subsets with the aid of the invariant manifolds of group $H_r(\alpha)$.

As an example Lutsenko found all the subsets of the set of second-order curves, which are measurable relative to Euclidean motions and relative to affine transformations, and computed the measures.

Lutsenko used only the infinitesimal operators of group G_r, which is not a hindrance when the group is given by finite transformations (the finding of the infinitesimal operators is an exercise in differentiation) and is important when only the infinitesimal operators of the group are given.

§ 4. Kinematic Measure

The concept of a kinematic measure (density), introduced by Poincaré from probabilistic considerations, was developed in integral geometry by Santaló and Blaschke and later by other authors. This concept proved to be useful in the applications of integral geometry when the set of intersections of a moving manifold with a fixed one is considered. Precise definitions and important results are collected in Blaschke's "Lectures" and Santaló's "Introduction...". In recent years Chern, Hadwiger, Santaló, Stoka, Fáry, and others have devoted papers to the development, generalization, and applications of the concept of kinematic measure (density).

If X is a k-dimensional Riemann manifold, then we can write the Riemann metric in X in the form

$$ds^2 = \sum_{\alpha=1}^{k} \varphi_\alpha^2 ,$$

where the φ_k are linearly independent Pfaffian forms. Chern [28] defined the functions $S_{\alpha\beta\gamma\delta}$, on the set of orthonormal nets on X, by the relations:

$$\varphi_{\alpha\beta} = -\varphi_{\beta\alpha}, \ d\varphi_\alpha = \sum_{\beta=1}^{k} \varphi_\beta \wedge \varphi_{\beta\alpha},$$

$$d\varphi_{\alpha\beta} = \sum_{\gamma=1}^{k} \psi_{\alpha\gamma} \wedge \varphi_{\gamma\beta} + \frac{1}{2} \sum_{\gamma,\delta} S_{\alpha\beta\gamma\delta} ?_{\gamma} \wedge ?_{\delta}$$

and, next, the functions

$$I_s = \frac{(-1)^{\frac{r}{2}} (k-r)!}{2^{k/2} k!} \sum \delta \begin{pmatrix} \alpha_1, & \dots, & \alpha_s \\ \beta_1, & \dots, & \beta_s \end{pmatrix} S_{\alpha_1 \alpha_2 \beta_1 \beta_2} \cdots S_{\alpha_{s-1} \alpha_s \beta_{s-1} \beta_s},$$

where $0 \le r \le k$, $\delta(\ )$ is the Kronecker symbol, the sum ranges over all values of $\alpha_1, \dots, \alpha_s$, $\beta_1, \dots, \beta_s$, varying independently from 1 to k.

Taking X to be oriented and compact, the author assumes

$$\mu_s(X) = \int_X I_s dv,$$

where dv is a volume element, and proves the remarkable formula

$$\int \mu_s(M^p \cap M^q)\, dg = \sum_{i=0}^{r} c_i \mu_i(M^p) \mu_{r-1}(M^q),$$

where i and r are even, $0 \le r \le p + q - n$, M^p and M^q are compact manifolds, without boundaries, of dimensions p and q, M^p is fixed, M^q moves with a kinematic density dg. The coefficients c_i depend on n, p, q, r, and Chern computed them. Hadwiger [39] announced the following result:

Let K be the smallest ring of sets with respect to a system of convex bodies in a Euclidean space E_n; let $W_i(A)$ be Crofton integrals of the Euler-Poincaré characteristics $\chi(A \cap E_i)$ of the intersection of $A \in K$ with an ith-dimensional plane E_i moving in E_n; let $\varphi(A)$ be a functional, decreasing sufficiently rapidly at infinity and continuous on K, possessing the property: $\varphi(A \cup B) + \varphi(A \cap B) = \varphi(A) + \varphi(B)$. Then the kinematic integral over all permutations of A in E_n is a linear function of $W_i(A)$ with coefficients independent of A.

We do not know if the author has published this in detail.

In [65] Santaló announced a generalization of Blaschke's basic kinematic formula to the case of an n-dimensional space of constant curvature and presented a complete proof in [72].

In an n-dimensional space of constant curvature let Q_0 be a fixed and Q_1 a moving body with boundaries of class C^2. The position of Q_1 is defined by specifying a point x of it and by the mutually orthogonal unit vectors e_1, e_2, ..., e_n starting at x. The kinematic density is given by the formula

$$dQ_1 = dx \wedge dO_1 \wedge dO_2 \wedge \ldots \wedge dO_{n-1},$$

where dx is a volume element, dO_i is an area element of the (i-1)-dimensional sphere described by the vector e_i. Santaló's basic formula expresses the integral of the Euler-Poincaré characteristics $\int \chi(Q_0 \cap Q_1)\, dQ_1$ in terms of the curvature of the space, of the volumes of the spheres O_i and the bodies Q_0 and Q_1, of $\chi(Q_0)$ and $\chi(Q_1)$, and of an integral of the elementary symmetric functions of the curvature of the boundaries of the bodies Q_0 and Q_1.

Certain kinematic formulas can be obtained by an application of the Poincaré-Cartan integral invariants [3, 4, 76].

§ 5.   Abstract Foundations.   Some New Directions

1. Hermann's paper [46] is devoted to the abstract foundations of integral geometry. The author's plan consists in the following. The manifolds, curves, mappings are assumed to be of class C^∞. The space tangent to a manifold M at a point x is denoted by M_x. The mapping $\Phi\colon M \to M'$ induces a linear mapping $\Phi_x\colon M_x \to M'_{\Phi(x)}$ on the tangent vectors, while to the differential form W there corresponds a differential form $\Phi^*(W)$. Let dim M = n, dim X = n, $\Phi\colon X \to M$, and let W be an mth-order differential form on M, nonzero everywhere. We define $\int_M fw$ of the real measurable function f in the usual way. If Θ is an nth-order form on X, then the function $N(x) = \int_{\Phi^{-1}(X)} \Theta$ is defined almost everywhere on M. The equality

$$\int_X \Phi^*(W) \wedge \Theta = \int_M N\Theta$$

is fundamental and so is its generalization to the case of the covering of M by a field F: $x \to F_x \subset X_x$ of tangent subspaces of constant dimension, satisfying the condition of total integrability.

Hermann analyzed one simple problem in uniform spaces, but neither he nor anyone else has as yet examined concretizations of the proposed plan.

2. Gaeta [36] related the problem of determining measure in uniform spaces to the theory of group representation by linear transformations. He showed that in many cases (there is no general proof) we can introduce a point model of the corresponding uniform space in a projective or affine space so that the corresponding group receives an algebraic representation. The invariant measure of the model generates the measure of the uniform space being considered.

In another paper [35] Gaeta, starting from the fact that in integral geometry we consider integrals of the form $\int_G F(\alpha A \cap B) \, d\alpha$,
$\int_{G/H} F(H \cap B) \, dH$, where G is the group of homeomorphisms α,
acting transitively in a topological space of the points a, b, ...;
x, y, ..., interpreted certain formulas of integral geometry in terms of the direct products of spaces and groups. The idea apparently deserves attention, but Gaeta did not obtain any essentially new results.

3. Legrady [48] extended Chern's method and results to symplectic spaces. He proved, in particular, the existence of point density and geodesic density (the formulas are presented), and found Crofton-type formulas.

4. Busemann [24] showed the feasibility of generalizing a number of formulas of integral geometry, including Crofton's formula, to the case of an affine space with an a-area.

The concept of an a-area consists in the following. Suppose that an open convex set D_n is given in an n-dimensional affine space X^n. On each a-dimensional plane A, $1 \le a \le n-1$, $A \cap D_n \neq \emptyset$, we define a nonnegative, completely additive function α (M) of the Borel sets $M \subset A \cap D_n$, vanishing on the $(a-1)$- flats and depending continuously on the vertices of a-simplexes. α (M) also is an a-area.

Bouligand [21] introduced the following generalization of the area of a flat figure. Let the metric be given by the formula

$ds = f(x, y; dx, dy), f(x, y; \rho dx, \rho dy) = \rho f(x, y; dx, dy)$, $\rho > 0$.   If ξ
and η satisfy the equation $f(x, y; \xi, \eta) = 1$, then the locus of the
endpoints of a vector with a start point at (x, y) and with components ξ and η is a closed curve—the indicatrix of the metric. The
quantity $\pi \iint_R \frac{d\sigma}{\alpha}$, is called a generalized area, where $d\sigma$ and α are respectively, an area element of region R and the area bounded by
the indicatrix (the indicatrix is a circle only in the case of the
Euclidean metric).

Bouligand posed the question of transformations which preserve a generalized area. It seems to us that this question (and
others arising from it) relate to integral geometry.

5. Cruceanu [29] proved the following theorems:

1°. In order for a Riemann space to admit of a simply transitive group, preserving volume, with the infinitesimal operators
$\xi_i^j \frac{\partial f}{\partial x^i}$, where the vectors $[\xi_1^j, \xi_2^j, \ldots, \xi_n^j]$ are harmonic, it is necessary and sufficient that there exist a coordinate system such that
the following conditions are fulfilled: $g_{ij} = F_{x^i x^j}, \|g_{ij}\| = \text{const}$.

2°. An infinitesimal transformation (in Riemann space) $\bar{x}^i =$
$x^i + \xi^i dt$, where $[\xi^1, \xi^2, \ldots, \xi^n]$ is a gradient vector, preserves
volume if and only if the orthogonal hypersurfaces form an isother-
mal family.

6. In a two-dimensional Finsler space Ueno [100] introduced
into consideration the density of a vector field,

$$dG = g_{ij} a^i_{\ k} \ b^j g_{lm} b^l dx^k \wedge dx^m,$$

where (g_{ij}) is the fundamental tensor, $b(x)$ is a unit vector field
orthogonal to $a(x)$. This density is invariant relative to variations
of $a(x)$ along a geodesic, under which the support line elements of
infinitely close points are parallel. The kinematic density of
oriented geodesic segments,

$$dK = dG \wedge ds,$$

where s is the distance from the center of the direction element to the
start of the segment, does not depend on the choice of the field.

7. The classical Steiner formula and the numerous analogous
formulas can be characterized in the following manner.

Let M be a compact differentiable manifold of dimension k (k < n) in an n-dimensional space E and let R be a set of points of E, whose distance from M does not exceed r. Then the volume of R is an nth-degree polynomial relative to r with coefficients expressed in terms of the integral of symmetric functions of principal curvatures (the Minkowski integral). Steiner's formulas and the kinematic formulas are closely related.

In the extensive paper [32] by Federer and in the dissertation [22, 23] by Brothers, among other results on integral geometry in Euclidean and Riemann spaces, a considerable place is taken up by a generalization of Steiner's formula and of the fundamental kinematic formula. In both these papers the tool and the scene of action is the modern theory of measure, manifolds, and de Rham flows.

Federer designates as the reach of a set A of points of a Euclidean space E_n the largest number ε such that if $x \in E_n$ and if the distance $d(x, A)$ of x from A is less than ε, then A contains a unique point $\xi_A(x)$ closest to x. The class of sets with a positive reach, considered by the author, contains all convex sets.

Let $P \subset E_n$ be a bounded Borel set, A be a set with a positive reach $\varepsilon(A)$, $0 \le r < \varepsilon(A)$, and V be the n-dimensional volume of the set $E\{x : d(x, A) \le r, \xi_A(x) \in P\}$. Then (a generalization of Steiner's formula)

$$V = \sum_{i=0}^{n} r^{n-i} \alpha(n-i) \Phi_i(A, P),$$

where $\alpha(j)$ is the j-dimensional volume of the unit sphere in E_i, Φ_i is a Minkowski-type integral. The fundamental kinematic formula is also similarly generalized.

Brothers considered an n-dimensional Riemann manifold X with a transitive isometry group G and proper submanifolds A_0, B_0 of dimensions k, l, k + l ≥ n. Let $A \subset A_0$ and $B \subset B_0$ be Borel sets and let C act transitively in the spaces tangent to A and B. Then: if Ψ is a left invariant Haar measure on G, then there exists a constant α, depending only on A_0, B_0, Ψ, such that

$$\int_G H^{k+l-n}(A \cap gB)\, d\Psi_g = \alpha H^k(A) \int_B \Delta \cdot dH^l,$$

where H^r is the r-dimensional Hausdorff measure and Δ is a positive function induced on X by a modular function on G.

Let G act transitively in a set **E** of closed l-dimensional submanifolds from X, $\{g \in G : g(E) = E\}$ be transitive in each $E \in \mathbf{E}$, and **E** have a measure Φ invariant relative to G. Then

$$\int_E H^{k+l-n}(A \cap E)\, d\Phi E = \beta H^k(A) \quad (\beta = \text{const}).$$

If $k + l \le n$, G is a compactum, $\dim G = m + n$, then, with a suitable connection between the metrics of X and G,

$$\int_G H^0(A \cap gB)\, dH_g^{k+l-n} = \gamma H^k(A)\, H^l(B)\,.$$

Brothers obtained generalizations of Federer's results in Euclidean space also.

Chapter II

ACTUAL PROBLEMS AND APPLICATIONS

§ 1. Generalizations of Known Formulas

Masotti Biggiogero [55] obtained the result: let ω be the angle between the tangents PA and PA_1 to an oval; let R and R_1 be the radii of curvature of the oval at the points A and A_1; let T and T_1 be the lengths of the segments PA and PA_1; let L be the length of the oval. If

$$h = \int_0^\pi \frac{f(\omega)}{\sin \omega}\, d\omega, \quad k = \int_0^\pi \frac{\varphi(\omega)}{1 - \cos \omega}\, d\omega,$$

then

$$\int \frac{f(\omega)}{T \cdot T_1}\, dP = 2h\pi, \quad \int \frac{f(\omega)}{T \cdot T_1}(R + R_1)\, dP = 2hL,$$

$$\int \varphi(\omega)\,(T^{-1} + T_1^{-1})\, dP = 2kL, \tag{1.1}$$

where dP is the point density and the integrals are taken over the region external to the oval. When $f(\omega) = \sin \omega$, $f(\omega) = \sin^m \omega$ (m > 0), and $\varphi(\omega) = 1 - \cos \omega$, formulas (1.1) are, respectively, the formulas of Crofton, Lebesgue, and Santaló.

In [54], which is a survey of formulas of integral geometry in the Euclidean spaces E_2 and E_3 (with a bibliography of 286 items), Masotti Biggiogero obtained the formula

$$dR = \frac{\rho_1 \rho_2}{D^3} \, dP$$

for the density of straight lines R intersecting an oval Ω. Here ρ_1 and ρ_2 are the radii of curvature of the oval Ω at the points A_1 and A_2 of its intersection with R, dP is the density of the points P of intersection of the tangents to the oval at A_1 and A_2, D is the diameter of the circle passing through the points A_1, P, A_2.

The same author has found [57] the formula

$$\int\limits_{R \cap L \neq \varnothing} \left\{ \frac{f(r_1)}{\sin \alpha_1} + \frac{f(r_2)}{\sin \alpha_2} \right\} dR = \pi \int\limits_L f \, ds,$$

$$\int\limits_{K \cap L \neq \varnothing} l^m \left(\frac{f(r_1)}{\sin \alpha_1} + \frac{f(r_2)}{\sin \alpha_2} \right) dR = m \int\limits_S dP \int\limits_L f(r) \, \delta^{m-2} ds,$$

where $f(r)$ is an integrable function of the radius of curvature r; α_1 and α_2 are the angles of the intersection of the straight line R with an oval L; l is the length of chord R; dR is the density of the straight lines R; dP is the point density; δ is the distance from P to an arc ds of the oval.

For ovaloids Masotti Biggiogero [54] proved the formula

$$dR = \frac{1}{k_1 k_2} \frac{\sin^4 \omega}{\delta^2} \, dR^1,$$

for the density of straight lines intersecting an ovaloid Σ. Here, dR^1 is the density of straight lines R^1 along which there intersect the planes π_1 and π_2 tangent to Σ at its points of contact with R; k_1 and k_2 are the total curvatures of Σ at those same points; ω is the angle between π_1 and π_2; δ is the length of the chord R in Σ.

She also found [56] formulas which express the area of the surface and the volume of an ovaloid as integrals with respect to the set of straight lines intersecting the ovaloid, and with respect to the set of straight lines which are the lines of intersection of pairs of planes tangent to the ovaloid.

Santaló [68] proved, for any convex figure on a plane, the invariance, relative to a unimodular group of affine transformations, of the integral

$$I_m = \int_0^{2\pi} k^{\frac{m-2}{3}} \Delta^{-m} d\varphi,$$

where $\Delta = \Delta(\varphi)$ is the width of the figure in the direction φ, k is the curvature, and m is any number.

In a three-dimensional space an invariant relative to a unimodular affine group is

$$J_m = \int k^{\frac{m-3}{4}} \Delta^{-m} d\omega,$$

where k is the Gaussian curvature, Δ is the width of the convex body in the direction ω, and the integration is over the unit halfsphere.

Much the same invariants have been obtained by Santaló in [69]. Estimates for I_m and J_m have been obtained in [68, 69], exact from above and hypothetically exact from below.

For the case of a unimodular affine space Santaló obtained [70] a formula of Steiner type

$$V^* = V + \lambda\Omega + \lambda^2 M + \frac{\lambda^3}{3} C,$$

where Ω, M, C, V are, respectively, the (affine) area, total mean curvature, total curvature, and volume, of the surface S and of the body bounded by it, V^* is the volume bounded by the surface S^*: $X^* = X - \lambda I_3$ (I_3 is the affine normal).

In a hyperbolic plane Santaló [74] derived several formulas typical in integral geometry. Let K be an h-convex set (together with points A and B, to it belongs the segment of the horocycle H, defined by these points). Then

$$\int ndH = 4L, \int dH = 2L_1, \int \sigma^3 dH = 6F^2, \int \sigma dH = 2\pi F,$$

where dH is the density of the horocycles (dH = $e^{\pm\rho} d\rho \wedge d\varphi$); the first integral extends onto the set of all horocycles while the remaining ones, onto the set of nonempty $H \cap K$; n is the number of points of intersection of a rectifiable curve (of length L) with the horocycle; L_1 is the length of the boundary of K; σ is the length of the chord $H \cap K$; F is the area of K.

§ 2. The Rashevskii Problem

The importance of the concept of geodesic density and the relation of this concept to the ideas of Rashevskii were ascertained by Yaglom [17]. Tekse [97] generalized the concept of geodesic density by solving the problem:

Let S be some set of curves in an n-dimensional Riemann space V_n with a positive metric, and let X be a 2 (n−1)-parameter set of curves from S, satisfying the conditions: 1°. Through every line element of some neighborhood of a point in V_n there passes exactly one curve from X. 2°. In the neighborhood indicated any two curves are encountered no more than once.

The generalized density of the geodesics is defined as $dx^i \wedge dp_i + f_{ij} dx^i dx^j$.

We are required to find the conditions for the existence of a density of the form indicated.

Tekse found the necessary and sufficient condition in the form of a specific requirement on the structure of the covariant curvature vector of a curve in S and ascertained the physical meaning of this requirement.

Klee, McMinn, and Besicovitch [19] solved the problem of the existence of a linear measure of the set of directions of all vectors lying on a closed convex surface S in a three-dimensional space. The answer is affirmative.

§ 3. Applications of Kinematic Measure

(Lattices and Coverings)

Trandafir [98, 99] carried the concepts and results of the theory propounded in § 10 of [11] over to the case of a three-dimensional Euclidean space E_3 and of a two-dimensional Riemann space V_2 of constant curvature.

The lattice of fundamental domains (cells) is the name given to the sequence of domains $\alpha_0, \alpha_1, ..., \alpha_n, ...$ subject to the conditions: 1. Every point of the space belongs to one and only one region α_i. 2. Every cell α_i can be matched with the fundamental cell α_0 by means of the motion T_i ($T_i\alpha_i = \alpha_0$) which takes the whole lattice into itself. A figure K can be a domain or a system of domains, a curve or a system of curves, a collection of points, etc.

Let K_0 be a fixed figure, K a moving one, f a function integrable on $K_0 \cap K$ ($f = 0$ if $K_0 \cap K = \emptyset$).

In [98, 99] the well known formula ([11], p. 56)

$$\int_{K_0 \cap K \neq \emptyset} f(K_0 \cap K)\, dK = \int_{\dot{a}_0} \left[\sum_{i=0}^{\infty} f(T_i K_0 \cap K) \right] dK,$$

where dK is the Blaschke kinematic density in E_3, was trivially carried over to the spaces E_3 and V_2, while the formula

$$dK = dP \wedge d\omega$$

to V_2 (P is a point rigidly connected with K, ω is the angle between the radius vector of P and the geodesic passing through P and rigidly connected with K).

In E_3 let: C be a simply-connected domain with volume V; $\overline{H}$ and $\overline{h}_0$ be integrals of the mean curvature over the boundaries of domains C and α_0; v_0 and s_0 be the volume and the area of α_0; ζ be the number of parts into which C is divided by the lattice. It turns out that the mean value of ζ equals

$$1 + \frac{v}{v_0} + (s_0 \overline{H} + S \overline{h}_0)\frac{1}{4\pi v_0}. \tag{3.1}$$

Hence: the domain C can be covered by the cells of the lattice so that their number is not greater than the quantity (3.1).

In V_2 let: k be the constant curvature (of the space); the metric be $ds^2 = d\rho^2 + \frac{1}{k}\sin^2\sqrt{k\rho}\,d\theta^2$, where ρ, θ are geodesic coordinates; X_{01}, $X(K_0)$ be, respectively, the Euler-Poincaré characteristics for $K \cap T_i K_0$, K_0; S_0, S, s_0 be, respectively, the area of K_0, K, α_0; L_0, L, l_0 be, respectively, the lengths of the boundaries ∂K_0, ∂K, $\partial \alpha_0$.

It was proved: 1. The mean magnitude of χ_{01} equals

$$\frac{S_0 \chi (K) + S \chi (K_0)}{s_0} + \frac{L_0 L - k S_0 S}{2 \pi s_0}.$$

2. If $K_0 = \alpha_0 \cup \partial \alpha_0$ and η is the number of parts into which the lattice divides K, then the mean magnitude of η equals

$$\frac{1}{2 \pi s_0} \int \eta dK = 1 + \frac{S}{s_0} - \frac{kS}{2\pi} + \frac{l_0 L_0}{2 \pi s_0}. \tag{3.2}$$

3. If K is a curve and n_{01} is the number of intersections of K with the lattice, then the mean value of the quantity n_{01} equals

$$M [n_{01}] = \frac{2 L_0 L}{\pi s_0}. \tag{3.3}$$

From (3.2)-(3.3) follows:

1. Every domain K with an area S, bounded by a curve of length L, can be covered by the cells of the lattice so that their number is not greater than the quantity (3.2).

2. If K is a moving curve of length L and K_0 is a curve of length L_0 lying in a fundamental cell, then we can always find a position of K such that K intersects the lattice no less than $\left[\frac{2 L_0 L}{\pi s_0}\right]$ times.

§ 4. Applications of Kinematic Measure
(Minkowski Integrals), Moments

1. Bodies of Revolution. In an n-dimensional Euclidean space let A be some fixed body, ω_i be the volume of the i-dimensional unit ball, E_i be a moving i-dimensional plane. We call

$$W_i = \frac{1}{c_i} \int \chi (A \cap E_i) \, dE_i, \ 0 \leqslant i \leqslant n, \tag{4.1}$$

the Minkowski integral, where $\chi (A \cap E)$ is the Euler-Poincaré characteristic; dE_i is the density of the moving plane; $c_0 = 1$, $c_i = \binom{n}{i} \omega_{n-1} \cdots \omega_{n-i}/\omega_1 \cdots \omega_i$. If A is convex, then the Minkowski inequalities are valid.

$$\left(\frac{W_k}{\omega_n}\right)^{\frac{1}{n-k}} \geqslant \left(\frac{W_i}{\omega_n}\right)^{\frac{1}{n-i}}, \quad 0 \leqslant i < k \leqslant n-1. \tag{4.2}$$

Hadwiger [41] proved that inequalities (4.2) are true for non-convex bodies of revolution if every hyperplane perpendicular to the axis of revolution intersects the body along a set of points homeomorphic to an (n-1)-dimensional ball. Firey [33] has recently extended some old (1949) inequalities of Hadwiger for convex bodies of revolution which he calls generalized convex bodies of revolution. Among such bodies in a three-dimensional space are the cube, the octahedron, the spindle, the truncated right cylinder and cone. Such bodies are definable in cartesian coordinates as the mapping $\bar{x} = rx/f(x, y)$, $\bar{y} = ry/f(x, y)$, $\bar{z} = z$, $r = \sqrt{x^2+y^2} > 0$ of convex bodies of revolution around the OZ axis. It is assumed that $f(x, y) \geq 0$ and equality holds only when x = 0, y = 0; $f(\lambda x, \lambda y) = \lambda f(x, y)$, $\lambda > 0$; $f(x + x_1, y + y_1) \leqslant f(x, y) + f(x_1, y_1)$.

From the number of theorems obtained by Firey we cite the following:

If K_0 and K_1 are coaxial convex bodies of revolution, then

$$M^2(K_\vartheta) \leqslant (1 - \vartheta) M^2(K_0) + \vartheta M^2(K_1),$$
$$a(K_\vartheta) M(K_\vartheta) \leqslant (1 - \vartheta) a(K_0) M(K_0) + \vartheta a(K_1) M(K_1),$$

where M (K) is the total mean curvature, a (K) is the equatorial radius, K is the Blaschke sum

$$K_\vartheta = (1 - \vartheta) \times K_0 \# \vartheta \times K_1, \quad 0 \leqslant \vartheta \leqslant 1.$$

2. Convex Canal Families. Normal Bodies.

In [43] Hadwiger studied integrals of type (4.1) on convex canal families. The collection of all convex bodies of an n-dimensional Euclidean space with a common projection onto an (n-1)-dimensional plane E is called a convex canal class K if together with two bodies A, B ∈ K the Minkowski sum $\alpha A \times (1-\alpha) B$, $0 \leq \alpha \leq 1$, also belongs to this collection.

A canal family $A(\lambda) \subset K$ (λ is a parameter varying in some interval I) is said to be convex if

$$A(\alpha\xi + \beta\eta) \subset \alpha A(\xi) \times \beta B(\eta), \ \xi, \eta \in I, \ \alpha, \beta \geqslant 0, \ \alpha + \beta = 1.$$

The family is totally convex if its section with a plane of any dimension, containing a straight line perpendicular to E, is convex. The definition of concavity is analogous.

Hadwiger proved that the Minkowski integrals of totally convex (concave) canal families are convex (concave) functions of parameter λ.

Hadwiger [42] examined Crofton-type and Minkowski-type integrals also on the so-called normal bodies. In an n-dimensional Euclidean space a cell associated with a direction $r = [r_1, r_2, ..., r_n]$, where $r_1, ..., r_n$ are linearly independent unit vectors, is a compact set P whose intersection with any subspace E_i parallel to i of the vectors r_s is connected. A set A is called a normal body if it is compact and representable in the form $P_1 \cup P_2 \cup ... \cup P_N$, where P_i are cells. Here it is assumed that the intersection of any number of the P_i is a cell associated with r and, simultaneously, N is bounded for all r.

If A is a normal body, then as the Euler-Poincaré characteristic we take

$$\chi(A) = \sum_{s, v_1, ..., v_s} (-1)^{s-1} \varepsilon (P_{v_1} \cap P_{v_2} \cap ... \cap P_{v_s}),$$

where $\varepsilon(P) = 1$ when $P \neq \emptyset$ and $\varepsilon(P) = 0$ when $P = \emptyset$.

The integrals $\int \chi(A \cap E_i) dE_i$, where dE_i is the kinematic density, exist and can be treated as integrals (4.1). For them:

$$|W_i(A) - W_i(B)| \leqslant \sum_i \frac{1}{c_i} \int |\chi(A \cap E_i) - \chi(B \cap E_i)| dE_i,$$

$$\frac{1}{c_i} \int W_j(A \cap E_i) dE_i = \frac{\omega_i \omega_{n-j}}{\omega_{n-i} \omega_{i-j}} W_i(A), \ 0 \leqslant j \leqslant i \leqslant n - 1. \tag{4.3}$$

The right hand side in (4.3) satisfies the axioms of a metric. Thus, the space of normal bodies is metrizable.

3. Additive Functionals. Hadwiger [38] analyzed additive functionals on convex polyhedra. In an n-dimensional

Euclidean space a convex polyhedron is denoted by P + Q if it is dissected by an (n− 1)-dimensional plane into two convex polyhedra P and Q. The sectioning plane itself, PQ, also is a convex polyhedron. By $\varphi(P)$ we denote a functional defined on the set of all convex polyhedra.

The functional $\varphi(P)$ is called: additive if $\varphi(P) + \varphi(Q) = \varphi(P + Q) + \varphi(PQ)$; simply additive if $\varphi(P) + \varphi(Q) = \varphi(P + Q)$; invariant relative to motion if $\varphi(P) = \varphi(Q)$ for congruent P and Q; bounded if for any cube $T \supseteq P$ there exists a constant c (T) such that $|\varphi(P)| < c(T)$.

The sequence of functionals $\Phi_0(P)$, $\Phi_1(P), ..., \Phi_n(P)$ is called a scale if $\Phi_i(P)$ has been defined for any i-dimensional convex polyhedron P and is a simply additive functional in any i-dimensional Euclidean space.

The additive of a functional is equivalent to its representability as a linear combination of some scale of functionals, while invariance relative to motion is equivalent to the like invariance of the scale.

If $\Phi_i(P) = c_i V_i(P)$, where the c_i are constants, $V_i(P)$ is the i-dimensional volume of an i-dimensional polyhedron, then additive functionals, invariant relative to motion and representable as a linear combination of such a scale, are linearly expressed in terms of the Minkowski integrals $W_0(P)$, $W_1(P), ..., W_n(P)$.

Hadwiger suggested that any bounded additive functional $\varphi(P)$, invariant relative to motion, is a linear combination of Minkowski integrals.

Fáry [31] considered functionals defined on a collection C of all compact convex subsets of an n-dimensional Euclidean space E_n as real continuous functions $\varphi\colon C \to E_n$ subject to the conditions:

$$\varphi(tA) = \varphi(A) \quad (t\colon E_n \to E_n,\ t(x) = x + x_0), \tag{4.4}$$

$$\varphi(A \cup B) + \varphi(A \cap B) = \varphi(A) + \varphi(B),\ (A, B, A \cup B \in C). \tag{4.5}$$

Let U be a fixed proper (containing interior points of E_n) convex body; $\varphi_j(A) = V_j(A, U)$ be mixed volumes; $W_i(A) = U_i(A, B^n)$, where B^n is the unit ball in E_n. Then:

1. The $\varphi_i(A)$ $i = 0, 1,..., $ n, form a basis of a vector space of functionals φ satisfying conditions (4.4), (4.5) and the condition: if $\varphi_i(A) = \varphi_i(B)$ for $i = 1,..., $ n, then $\varphi(A) = \varphi(B)$.

2. $\int \varphi_\nu(A)\, dA = \alpha_\nu^{ij} W_i(A) \cdot W_j(U)$, $\nu = 0, 1,..., n$,   where dA is the kinematic density, α_ν^{ij} are constants.

3. If φ satisfies the hypothesis of Theorem 1, then

$$\int \varphi(A)\, dA = \beta^{ij} W_i(A) W_j(U),$$

where β^{ij} are constants .

   Alemany [18] computed the coefficients α_ν^{ij}, β^{ij}. It turned out that $\alpha_\nu^{ij} \neq 0$ only when $i = \nu$, $j = n - \nu$.

   4. Moments of a Convex Domain and of Its Boundary. Although Müller's paper [60] appeared as far back as 1953, it is not superfluous to recall it here. Let G be a moving convex flat domain and G' a moving one. Let $F[\ ]$, $F_i[\ ]$, $F_{ij}[\ ]$, $U[\ ]$, $U_i[\ ]$, ... denote, respectively, the area, the moment relative to the x_1-axis, the second-order moment, the length, etc., for a domain and its boundary.

   Müller found a number of formulas of the type:

$$\int F_i[G \cap G']\, dG = 2\pi F[G]\, F_i[G'],$$

$$\int F_{ij}[G \cap G']\, dG = 2\pi F[G]\, F_{ij}[G'],$$

$$\int U_i[G \cap G']\, dG = 2\pi \{F[G]\, U_i[G'] + U[G]\, F_i[G']\}.$$

A natural generalization of such formulas to convex domains in a three-dimensional space has apparently not been carried out as yet.

   In the 1955 edition of his "Lectures" Blaschke announced his forthcoming book (co-authored with Müller), "Ebene Kinematik," in which the relation of integral geometry with kinematic geometry was considered. The book was published in 1956.[*] We are not familiar with its contents.

---

[*]Translator's Note: W. Blaschke and H. R. Müller, Ebene Kinematik, Verlag von R. Oldenbourg, Munich (1956), 269 pp.

§ 5.   Certain Affine Invariants

The attention of individual geometers was drawn to the (Kawaguchi) spaces K^n belonging, in Klein's sense, to the group of transformations

$$x^i = a^i_j \bar{x}^j + a^i, \ \det(a^i_j) \neq 0, \ \sum_{j=1}^{n} a^i_j = \det(a^k_j), \ k = 1, 2, \ldots, n. \quad (5.1)$$

For n = 3 Watanabe [104] pointed out a covariant tensor and a contravariant vector, the Gauss-Weingarten formulas, and the Gauss-Codazzi equations. For the curve $x_i = x_i$ (t) he (and before him, Katurada) found the length of the arc, invariant relative to transformations (5.1),

$$s = \int \left\{ \left| \begin{matrix} \dot{x}^1 & \ddot{x}^1 \\ \dot{x}^2 & \ddot{x}^2 \end{matrix} \right| + \left| \begin{matrix} \dot{x}^2 & \ddot{x}^2 \\ \dot{x}^3 & \ddot{x}^3 \end{matrix} \right| + \left| \begin{matrix} \dot{x}^3 & \ddot{x}^3 \\ \dot{x}^1 & \ddot{x}^1 \end{matrix} \right| \right\}^{1/3} dt. \quad (5.2)$$

As is well known [15], an affine arc of a plane curve

$$l = \int \sqrt[3]{y''} dx = \int \left| \begin{matrix} \dot{x} & \ddot{x} \\ \dot{y} & \ddot{y} \end{matrix} \right|^{1/3} dt$$

is an integral invariant of the group obtained by extending an affine group, preserving the area. It seems natural to obtain formula (5.2) also with the aid of direct calculations according to the plan: find the infinitesimal operators of the group, extend the operators (twice), and compute the integral invariant of the extended group. The naturalness of the invariant is automatically explained when calculating according to such a plan. (Watanabe did not touch upon the question of uniqueness.)

Drinfel'd [5] carried out the indicated calculations and also, by extending the infinitesimal operators of group (5.1) under the condition dz = pdx + qdy, found the invariant

$$\iint \left\{ \frac{D(x, y)}{D(u, v)} + \frac{D(y, z)}{D(u, v)} + \frac{D(z, x)}{D(u, v)} \right\} dudv,$$

which, as seems natural, can be called the area of a surface in K^3.

In [105] Watanabe examined the case of any n both for transformations (5.1) as well as for the affine transformations subject to the conditions $\sum\limits_{j=1}^{n} a_j^k (-1)^{j+1} = (-1)^{k+1} \det (a_j^i)$, and found a generalization of formula (5.2).

Onodera [63] generalized Watanabe's results to the case of a space K^n with a given skew-symmetric tensor and with linear connectivity.

The formulas of Watanabe and Onodera for the length of an arc (for any n) can be obtained by a group extension, but the computations become very tedious. Nevertheless we aver that the application of group extensions, which for a long time has demonstrated its value in geometry, can prove to be useful also in integral geometry.

Invariants (measures) in integral geometry have been used by various authors to solve problems not directly related to it. As examples we mention Reshetnyak [10] and Hsiung and Shahin [47].

§ 6. Distribution Functions of Intersections

Sulanke [94], by solving the problem of the estimate of the ratio of the area of nonfertile parts of a field to the area of the whole field, obtained interesting results in a number of questions connected with the consideration of chord lengths of planar and spatial figures in Euclidean spaces of two and three dimensions.

The Lebesgue measure of the set

$$\zeta (g) = \zeta_B (g) = m (g \cap B)$$

is called the chord length of a straight line g in a Borel point set B.

This function of g is measurable for every bounded B and is a random quantity if the straight line is taken at random. The probability $P (\zeta_M (g) \leqslant x \mid g \cap K \neq \varnothing) = F_{M|K} (x)$ is called the conditional distribution function of the chord length on M relative to set K, $M \subseteq K$. The function $F_M (x) = F_{M|M} (x)$ is called the absolute distribution function. It turns out that this function is continuous if M is a nondegenerate convex bounded domain.

Let K_0 be a given bounded domain with a continuous distribution function $F_0 (x)$ of chord length; $\{B_i\}_1^\infty$ be a sequence of domains contained in K_0; $\zeta_i(g) = \zeta_{B_i}(g)$; $\zeta_0(g) = \zeta_{K_0}(g)$; $M ( )$, $D^2 ( )$ be the mean and the variance.

The following theorem is valid: If

$$\lim_{i \to \infty} M\left(\frac{\zeta_i\,(g)}{\zeta_0\,(g)}\right) = k > 0, \quad \lim_{i \to \infty} D^2\left(\frac{\zeta_i\,(g)}{\zeta_0\,(g)}\right) = 0,$$

then

$$\lim_{i \to \infty} F_i\,(x) = F_0\left(\frac{x}{k}\right), \quad \lim_{i \to \infty} \frac{m\,(B_i)}{m\,(K_0)} = k.$$

A set B is called almost convex if for almost all straight lines g the intersection g ∩ B is a segment to within a set of measure zero.

We denote the k-th moment of the chord by

$$S_k\,(B) = \int\limits_{g \cap \overset{\circ}{B} \neq \varnothing} \zeta_B^k\,(g)\,dg.$$

There holds the theorem: For every (flat) bounded Borel set B the inequality $3S_1^2\,(B) \geq \pi^2 S_2\,(B)$ is valid, and equality is attained on almost convex B and only on these.

Thus, among bounded Borel sets it is not the convex but the almost convex sets which are characterized by the properties of the chord moments.

The results cited characterize also convex sets in the class of bounded closed hulls of open sets since the closed hull of an open set is convex if it is almost convex.

Similar Sulanke theorems were proved also for a three-dimensional space. Paper [94] also contains a summary of the known results on the Blaschke moment problem: let there be given a sequence $\{S_k\}_0^\infty$ of positive numbers, then what are the necessary and sufficient conditions for the existence of a convex figure for which the S_k are the k-th chord moments?

To this summary Sulanke appended several of his own results on the generalized problem: what the conditions for the existence of a bounded Borel set B for which $\{S_k\}_0^\infty$ is a sequence of chord moments?

Sulanke takes the probability approach also in [95].

The union G of a finite number m of bounded flat convex arcs which pairwise have no more than endpoints in common is called a net.

Let G be contained in a bounded closed point set E and let P_i denote the probability of the intersection of G with a straight line g $(g \cap E \neq \emptyset)$ at exactly i points.

The collection of probabilities P_i is called the distribution condition for the number of points of intersection of G with the straight line.

The sequence of positive integers $m_1 < m_2 < ... < m_s$ for which $\Phi_{mj} > 0$ is called the distribution type. A number of results on the question of the existence of nets with a given distribution or distribution type were obtained in the paper.

The set of nets distance-metrizable between two nets G_1 and G_2 is defined as the measure of the set of straight lines intersecting these nets at an unequal number of points.

The integral geometry of nets was generalized in this same paper to elliptic and other geometries.

Using the concepts and facts of integral geometry Giger [37] computed the density of particles in a volume by means of counting the particles in a specimen and measuring their cross-sections on the boundaries of the specimen (the problem is of interest to biologists and metallographists).

Let $\{A_k\}$ be a countable set of bounded convex bodies; U be a convex three-dimensional specimen and $\hat{U}$ its boundary; V_k; F_k, M_k, V, F, M, respectively, be the volume, surface area and mean curvature of the body A_k and of U, and $\overline{V}$, $\overline{M}$, $\overline{F}$ be the mean values (in the integral geometric sense) of the quantities V_k, M_k, F_k. Further, let I (U), I ($\hat{U}$) be the number of nonempty $A_k \cap U$, $A_k \cap \hat{U}$; L ($\hat{U}$) be the sum of the perimeters of the nonempty $A_k \cap \hat{U}$; F ($\hat{U}$) be the sum of the areas of the nonempty $A_k \cap \hat{U}$. The formulas

$$D = \frac{1}{V} \left\{ I(U) - \frac{I(\hat{U})}{2} - \frac{L(\hat{U})M}{\pi^2 F} \right\}, \quad \overline{M}D = \frac{2\pi}{F} \left\{ I(\hat{U}) - \frac{2F(\hat{U})}{F} \right\},$$

$$\overline{F}D = \frac{4}{\pi F} L(\hat{U}), \quad \overline{V}D = \frac{1}{F} F(\hat{U}),$$

are valid for the density D of particles A_k in U.

In their own turn, I ($\hat{U}$), L ($\hat{U}$), F ($\hat{U}$) are found by counting and measuring on the boundary and inside a flat specimen.

With the aid of well-known integral geometry formulas Hadwiger [45] computed the probability

$$W\,[G \cap S \neq \varnothing \Rightarrow G \cap S \subset P \cup Q,\; G \cap P \neq \varnothing,\; G \cap Q \neq \varnothing],$$

where P and G are measurable sets on the unit sphere S of the Euclidean space E_k, G is a random curve all of whose positions are equally probable.

§ 7. Integral Geometry and Pattern Recognition

Such is the title of a report [62] by Novikoff, a member of one of the engineering research centers in the U.S.A., presented at a symposium on self-organization held at the University of Illinois. A large part of the report is devoted to an exposition of the generally known elementary concepts and facts of integral geometry. Novikoff assumes that integral geometry can become the theoretical basis for the construction of pattern recognition devices. The author's statement: "Let there exist a preassigned alphabet of patterns which need to be recognized and suppose that the designer of the machine knows this alphabet very well; a complete geometrical description of it is at his disposal. However, he knows nothing whatsoever about the disposition and orientation of the pattern on the recognition machine's grid. The experimenter wants to be able to distinguish the patterns, i.e., to be able to say which representation of the alphabet is to be found on the grid."

As a first example, making clear the question's connection with integral geometry, Novikoff uses the classical needle problem: "It is necessary to recognize two patterns which are lattices of an infinite number of parallel lines of infinite length, differing in the distances between the lines: in one pattern this distance equals d, in the other, d'.

"Suppose that d < d'. In what way can we recognize which of the two patterns is present if we do not know the orientation of the lines?" "The designer of the recognition machine simply selects a needle of length equal to the smallest distance between the parallel lines, throws it many times onto the plane covered by the parallel lines with spacing d, and averages the number of cases when the needle intersected the lines. If the number obtained is close to $2/\pi$, then this was the net of parallel lines with the smaller distance between them..."

It is possible that Novikoff has overestimated the role of integral geometry in the design of pattern recognition machines. However, we must say that his arguments may serve as a stimulus for posing new problems and for finding new interpretations of the integral geometry problems already solved.

BIBLIOGRAPHY

1.   S. A. Demidova, "Application of the theory of integral invariants of Lie groups to certain aspects of integral geometry," Dissertation, Kharkov (1956).

2.   G. I. Drinfel'd, "On measure in Lie groups," Khar'kov. Gos. Univ. Uch. Zap., 29 (1949); Zap. Mat. Otd. Fiz.-Mat. Fak. i Khar'kov. Mat. Obshch., 21:47-57 (1949).

3.   G. I. Drinfel'd, "Some basic formulas of integral geometry. II," Khar'kov. Gos. Univ. Univ. Uch. Zap., 40 (1952); Zap. Mat. Otd. Fiz.-Mat. Fak. i Khar'kov. Mat. Obshch., 23:61-71 (1952).

4.   G. I. Drinfel'd, "Theory of integral invariants and integral geometry," Proc. Third All-Union Math. Congress, Vol. I [in Russian], Akad. Nauk SSSR, Moscow (1956), p. 151.

5.   G. I. Drinfel'd, "Computation of an affine arc and of an affine area," Ukr. Geometr. Sb., (1968).

6.   G. I. Drinfel'd and Kyong Kim, "Integral invariants of completely integrable Pfaffian systems of differential equations," Dopovidi Akad. Nauk Ukrain. RSR, No. 6, 713-716 (1963).

7.   G. I. Drinfel'd and A. V. Lutsenko, "The measure of the set of second-order curves," Dopovidi Akad. Nauk Ukrain.RSR, No. 1, 14-17 (1965).

8.   Kyong Kim, "Integral invariants of a completely integrable system of equations in terms of total differentials," Vestnik Khar'kov. Gos. Univ., Ser. Mekh.-Mat., No. 3, 51-64 (1965).

9.   A. V. Lutsenko, "The measure of sets of geometric elements and of their subsets," Ukr. Geometr. Sb., No. 1, 39-57 (1965).

10. Yu. G. Reshetnyak, "Integral-geometric methods in the theory of curves," Proc. Third All-Union Math. Congress, Vol. I [in Russian], Akad. Nauk SSSR, Moscow (1956), p. 164.

11. L. Santaló, Introduction to Integral Geometry [Russian translation], Izd. Inostr.
     Lit., Moscow (1956), 183 pp.

12. M. Stoka, "The measure of families of manifolds in the space E_3," Rev. Math.
     Pures et Appl. (RPR), 4(2):305-316 (1959).

13. E. G. Ushakova, "Conditions for the existence of an integral invariant of a
     transitive group," Khar'kov. Gos. Univ. Uch. Zap., 115 (1961); Zap. Mat. Otd.
     Fiz.-Mat. Fak. i Khar'kov. Mat. Obshch., 27:149-152 (1961).

14. N. G. Chebotarev, "The determination of volume in Lie groups," Zap. Mat. Otd.
     Fiz.-Mat. Fak. i Khar'kov. Mat. Obshch., 14:3-20 (1937).

15. N. G. Cebotarev, Lie Groups [in Russian], Moscow-Leningrad (1940).

16. L. M. Yurtova and A. V. Lutsenko, "Measures of sets of pairs," Ukr. Geometr.
     Sb., (1968).

17. I. M. Yaglom, "Integral geometry in the set of line elements," Appendix to the
     book: L. Santaló, Introduction to Integral Geometry [Russian translation], Izd.
     Inostr. Lit., Moscow (1956), 183 pp.

18. R. E. Alemany, Valores numericos de ciertas constantes relacionadas con volu-
     menes mixtos de cuerpos convexos. Rev. Unión mat. argent. y Asoc. fis. argent.,
     21(3):113-118 (1963).

19. A. S. Besicovitch, "On the set of directions of linear segments on a convex sur-
     face," in: Convexity, Amer. Math. Soc. (1963), pp. 24-25.

20. W. Blaschke, Vorlesungen über Integralgeometrie, 3 ed., VEB Dtsch. Verl. Wiss.
     (1955), Vol. 8, 130 pp.

21. G. Bouligand, Sur les transformations ponctuelles conservant les aires. Gaz. mat.,
     14(56):1-4 (1953).

22. J. E. Brothers, "Integral geometry in homogeneous spaces," Doctoral dissertation,
     Brown Univ. (1964), 87 pp.; Dissert. Abstr., 25(8):4717-4718 (1965).

23. J. E. Brothers, Integral geometry in homogeneous spaces. Trans. Amer. Math.
     Soc., 124(3):480-517 (1966).

24. H. Busemann, Areas in affine spaces. III. The integral geometry of affine area.
     Rend. Circolo mat. Palermo, 9(2):226-240 (1960).

25. G. Chakerian, Integral geometry in the Minkowski plane. Duke math. J., 29(3):
     375-381 (1962).

26. Shiing-shen Chern, On integral geometry in the Klein spaces. Ann. math., 43:
     178-189 (1942).

27. Shiing-shen Chern, Differential Geometry and Integral Geometry, Proc. Internat.
     Congr. Mathematicians, 1958. Cambridge Univ. Press, Mass. (1960), pp. 440-449.

28. Shiing-shen Chern, On the kinematic formula in integral geometry. J. Math.
     and Mech., 16(1):101-118 (1966).

29. V. Cruceanu, Asupra transformărilor infinitezimale ale unui spatiu Riemann cu
     păstrarea volumului. Studii și cercetári științ. Acad. RPR. Fil. Iași Mat., 9(2):
     181-188 (1958).

30. R. Deltheil, Probabilités géometriques, Paris, (1926), Traité du calcul des pro-
     babil. et de ses applic., tome II, fasc. II.

31. I. Fáry, Functionals related to mixed volumes. Illinois J. Math., 5(3):425-430
     (1961).

32. H. Federer, Curvature measures. Trans. Amer. Math. Soc., 93(5):418-491 (1959).

33. W. J. Firey, Generalized convex bodies of revolution. Canad. J. Math., 19(5):
     972-996 (1967).

34. F. Gaeta, Algunas observaciones sobre la realiracion efectiva del programa de
     Chern geometria integral. Math. notae., 17(1-2):1-29 (1959).

35. F. Gaeta, Sobre un proceso de linearizacion aplicable en problemas de geome-
     tria integral. Publs Fac. cienc, fisicomat. Univ. nac. La Plata, No. 224, pp. 17-
     32 (1960).

36. F. Gaeta, Sobre la subordinacion de la geometria integral a la teoria de la re-
     presentacion de grupos mediante transformaciones lineales. Contribs. cient.
     Univ. Buenos Aires. Fac. cienc. exactas y natur. Mat., 2(2):27-87 (1960).

37. H. Giger, Ermittlung der mittleren Masszahlen von Partikeln eines Körpersystems
     durch Messungen auf dem Rand eines Schnittbereichs. Z. angew. Math. und Phys.,
     18(6):883-888 (1967).

38. H. Hadwiger, Über additive Funktionale k-dimensionaler Eipolyeder. Publs.
     mathematical, 3(1-2):87-94 (1953).

39. H. Hadwiger, Zur kinematischen Hauptformel der integralgeometrie. Proc. In-
     ternat. Congr. Math., Amsterdam, 2:225-226 (1954).

40. H. Hadwiger, Altes und Neues über konvexe Körper, Birkhäuser, Basel (1955),
     116 pp.

41. H. Hadwiger, Minkowskis Ungleichungen und nichtconvexe Rotationskörper.
     Math. Nachr., 14(4-6):377-383 (1955).

42. H. Hadwiger, Körper im euklidischen Raum und ihre topologischen und metris-
     chen Eigenschaften. Math. Z., 71(2):124-140 (1959).

43. H. Hadwiger, Über konkave und konvexe Eikörperscharen. Publs. math. Debre-
     can, Nos. 1-2, pp. 97-101 (1957).

44. H. Hadwiger, Zur Axiomatik der innermathematischen Wahrscheinlichkeits-
     theorie. Mitt. Verein. schweiz. Versicherungsmathematiker, 58(2):151-165
     (1958).

45. H. Hadwiger, Geometrische Wahrscheinlichkeiten bei Durchstichen von Geraden
     durch Kugelflächen. Mitt. Verein. schweiz. Versicherungsmathematiker, 68(1):
     27-35 (1968).

46. R. Hermann, Remarks on the foundations of integral geometry. Rend. Circolo
     mat. Palermo, 2(9):91-96 (1960).

47. Chuan-Chih Hsiung and J. K. Shahin, Affine differential geometry of closed
     hypersurfaces. Proc. London Math. Soc., 17(4):715-735 (1967).

48. K. Legrady, Symplektische Integralgeometrie. Ann. mat. pura ed appl., 41:139-
     159 (1956).

49. R. E. Luccioni, Sobre la existencia de medida para hipercuadricas singulares en
     espacios projéctivos. Rev. Univ. nac. Tucumán, A14(1-2):269-276 (1962).

50. R. E. Luccioni, Geometria integral en espacios projectivos. Rev. Univ. nac.
     Tucumán, A15(1-2):53-80 (1964).

51. R. E. Luccioni, Sobre la existencia de medida para conjuntos de subespacios
     lineales en espacios afines. Rev. Univ. nac. Tucumán, A16(1-2):219-227 (1966).

52. G. Lüko, On the mean length of the chords of a closed curve. Israel J. Math.,
     4(1):23-32 (1966).

53. C. D. Maraval, El problema de la aguja de Buffon en el espacio de n dimensio-nes. Gac. mat., 11(3-4):74-75 (1959).

54. G. Masotti Biggiogero, La geometria integrale. Rend. Seminar mat. e fis. Milano, 1953-1954; 25:164-231 (1955).

55. G. Masotti Biggiogero, Sulla geometria integrale: generalizzazione di formule di Crofton, Lebesgue e Santaló. Rev. Union mat. argent. y Asoc fis. argent., 17:125-134 (1955).

56. G. Masotti Biggiogero, Sulla geometria integrale: nuove formule relative agli ovaloidi. Scritti mat. onore Filippo Sibirani. Bologna, (1957), pp. 173-179.

57. G. Masotti Biggiogero, Nuove formule di geometria integrale relative agli ovali. Ann. mat. pura ed appl., 58:85-108 (1962).

58. G. Masotti Biggiogero, Nuove formule di geometria integrale relative agli ovaloidi. Rend. Ist. lombardo sci. e lettere. Sci. mat., fis., chim. e geol, A69(3):666-685 (1962).

59. M. Masuyama, On a fundamental formula in bulk sampling from the viewpoint of integral geometry. Repts Statist. Applic. Res., Union Japan Scientists and Engrs, 4(3):85-89 (1956).

60. H. R. Müller, Über Momente ersten und zweiten Grades in der Integral-geometrie. Rend. Circolo mat. Palermo, 2(1):119-139 (1953).

61. Z. Nádenik, O Integralni geometri. Pokroky mat., fiz. a astron., 7(2):75-79 (1962).

62. A. Novikoff, "Integral geometry as a tool in pattern perception," in: H. von Foerster and G. W. Zopf, Jr. (Eds.), Principles of Self-Organization, Pergamon Press, Oxford (1962), pp. 347-368.

63. T. Onodera, On hypersurfaces in certain n-dimensional spaces. Tensor, 19(1): 55-63 (1968).

64. M. de Resmini, Un ramo relativamente nuevo della matematica: la geometria integrale. Archimede, 13(3):134-144 (1961).

65. L. A. Santaló, "On the kinematic formula in spaces of constant curvature," Proc. Internat. Congr. Math., 1954, Amsterdam (1954), pp. 251-252.

66. L. A. Santaló, "Sur la mesure des espaces linéaires qui coupent un corps convex et problèmes qui s'y rattachent," Colloq. quest. réalité géom., Liège, 1955, Liège (1956) pp. 177-190.

67. L. A. Santaló, On the mean curvatures of a flattened convex body. Istanbul univ. fen. fac. mecm., A21(3-4):189-194 (1956).

68. L. A. Santaló, Un nuevo invariante afin para las figuras convexes del plano y del espacio. Math. notae, 16(3-4):78-97 (1958).

69. L. A. Santaló, Two applications of the integral geometry in affine and projective spaces. Publs math. Debrecen, 7(1-4):226-237 (1960).

70. L. A. Santaló, La fórmula de Steiner para superficies paralelas en geometria afin. Rev. Univ. nac. Tucumán, A13(1-2):194-208 (1960).

71. L. A. Santaló, On the measure of sets of parallel linear subspaces in affine space. Canad. J. Math., 14(2):313-319 (1962).

72. L. A. Santaló, Sobre la formula fundamental cinematica de la geometria integral en espacios de curvatura constant. Math. notas, 18(1):79-94 (1962).

73. L. A. Santaló, "Integral geometry of the projective groups of the plane depending

on more than three parameters," An. ştiinţ, Univ. Iaşi, Sec. Ia, 11b:307-335
(1965).

74. L. A. Santaló, Horocycles and convex sets in hyperbolic plane. Arch. Math.,
     18(5):529-533 (1967).

75. L. A. Santaló, Grupos del plano respecto de los cuales los conjuntos de puntos
     y de rectas admiten una medida invariante. Rev. Unión mat. argent. y Asoc. fis.
     argent., 23(3):119-148 (1967).

76. M. Stoka, Măsura unei mulţimi de varietăţi dintr-un spatiu R_n. Bul ştinţ. Acad.
     R. P. Rumîne. Sec. mat. şi fiz., 7(4):903-937 (1955).

77. M. Stoka, Asupra măsurii mulţimii cercurilor din plan. Gaz. mat. şi fiz., A7(10):
     556-559 (1955).

78. M. Stoka, Asupra subgrupurilor unui grup G_r măsurabil. Comun. Acad. RPR, 6(3):
     393-394 (1956).

79. M. Stoka, Asupra grupurilor G_r măsurabile din plan. Bul. ştiinţ. Acad. RPR. Sec.
     mat. şi fiz., 9(2):341-380 (1957).

80. M. Stoka, Asupra grupurilor G_r măsurabile dintr-un spaţiu R_n. Comun. Acad.
     RPR, No. 6, pp. 581-585 (1957).

81. M. Stoka, Asupra măsurii mulţimilor de varietăţi dintr-un spaţiu euclidian E_n.
     Comun. Acad. RPR, No. 3, pp. 313-317 (1957).

82. M. Stoka, Geometria integrale in uno spazio euclideo E_n. Boll. Unione mat.
     ital., 13(4):170-485 (1958).

83. M. Stoka, Invariantii integrali ad unui grup Lie de Transformări. An. Univ.
     "C. I. Parchon". Ser. ştiinţ. natur, No. 20, pp. 33-35 (1958).

84. M. Stoka, Asupra grupurilor G_r măsurabile dintr-un spaţiu E_n. Comun. Acad.
     RPR, 9(1):5-10 (1959).

85. M. Stoka, Géométrie intégrale dans un espace E_n. Rev. math. pures et appl.,
     (RPR), 4(1):123-156 (1959),

86. M. Stoka, Congruences de variétés mesurables dans un espace E_n. Rev. math.
     pures et appl. (RPR), 4(3):431-439 (1959).

87. M. Stoka, Famiglie di varietà misurabili in uno spazio E_n. Rend. Circolo mat.
     palermo, 8(2):192-205 (1959).

88. M. Stoka, Integralgeometrie in einem Riemannschen Raum V_n. Rev. math. pures
     et appl. (RPR), 5(1):107-120 (1960).

89. M. Stoka, Geometrie integrală intr-un spatiu riemannian V_n. Studii şi cercetări
     mat. Acad. RPR. Fil. Cluj. 11(2):381-395 (1960).

90. M. Stoka, Das Mass der Untersysteme von Mannigfaltigkeiten in einem Raum X_n.
     Rev. math. pures et appl. (RPR), 5(2):275-286 (1960).

91. M. Stoka, Geometrie integrală într-un spatiu euklidian E_n. Lucrările conf. geo-
     metrie şi topol., 1958, Bucureşti, Acad. RPR (1962) pp. 109-116.

92. M. Stoka, Familii de varietăti măsurabile intr-un spatiu riemannian V_3 cu curbura
     constantă negativă. Studii şi cercetari mat. Acad. RPR, 14(3):365-376 (1963).

93. M. Stoka, Geometria integrală, Bucuresti, Acad. RPR (1967), 238 pp.

94. R. Sulanke, Die Verteilung der Sohnenlängen an ebenen und räumlichen Figuren.
     Math. Nachr., 23(1):51-74 (1961).

95. R. Sulanke, Integralgeometrie ebener Kurvennetze. Acta math. Acad. scient.
     hung., 17(3-4):233-261 (1966).

96. R. Sulanke, Croftonsche Formeln in Kleinschen Räumen. Math. Nachr., 32(3-4): 217-241 (1966).

97. K. Tekse, Riemann-térz intégralgeometriá jának néhény problémájáról. Magyar tud. acad. Mat. és. fiz. tud. aszt. közl., 11(3):289-304 (1961).

98. R. Trandafir, Problems of integral geometry of lattices in an Euclidean space E_3. Boll. Unione math. ital., 22(2):228-235 (1967).

99. R. Trandafir, Problems of integral geometry of lattices in a riemannian space V_2 with constant curvature. Boll. Unione math. ital., 1(2):244-248 (1968).

100. S. Ueno, On the densities in a two-dimensional generalized space. Mem. Fac. Sci. Kyûsyû Univ., A9(1):65-77 (1955).

101. A. E. Vidal, A generalization of integral invariants. Proc. Amer. Math. Soc., 10(5):721-727 (1959).

102. A. E. Vidal, Generalización de los invariantes integrales y aplicación a la geometria integral en los espacios de Klein y de Riemann. Collect. math., 12(2): 71-102 (1960).

103. G. Vrănceanu, The measurability of Lie groups. Ann. polon. math., 15(2):179-188 (1964).

104. S. Watanabe, On hypersurfaces of spaces belonging to certain transformation group. Acta math. Acad. scient. hung., 17(1-2):137-145 (1966).

105. S. Watanabe, On hypersurfaces of spaces belonging to a certain transformation group. II. Tensor, 18(1):90-96 (1967).

Differential-Geometric Methods
in the Calculus of Variations

N. I. Kabanov

Introduction

The so-called metric theories, based on the concept of a
metric, occupy an important place in differential geometry. Most
of all the giving of these metrics can be looked upon as the giving
of functionals of various variational problems on appropriate spaces,
which leads to the interconnection of differential geometry and the
calculus of variations. From this connection differential geometry
obtains the possibility of the development of new theories and of
the posing of new problems, while the application of intrinsic
geometrical methods in the calculus of variations, in addition to
clarifying the essence of the results by means of a geometrical interpretation, can prove to be useful for the creation of new research
methods.

In this survey we consider only those geometric theories
which are connected with the classical problems of the calculus
of variations and we use local differential-geometric methods of
investigation. The majority of these theories were developed
using Riemannian geometry as a model, which makes it possible to
construct a theory of differential invariants needed for solving the
problems of equivalence and for classifying the corresponding
variational problems. However, the approach indicated frequently
requires constraints, connected with the positive definiteness and
completeness of the metric, which are unnatural from the calculus
of variations point of view and are barely related with the problems
of seeking an extremum. Another approach, free from the defi-
ciencies indicated, is based on the concept of an indicatrix introduced as far back as the beginning of this century by Carathéodory

[77] for the simplest problem in the calculus of variations. This approach allows us to specify all the classical variational problems by a single purely geometric method and to obtain in geometric form the basic extremum conditions in these problems. Here appears the feasibility of applying differential-geometric methods to the study of an indicatrix as a surface in some Klein space. The fundamentals of this approach have been set forth by Vagner [147].

In view of the breadth of the area being considered the survey does not touch upon the theories which do not have a direct bearing on the geometry or the calculus of variations or do not use differential-geometric methods (for example, the theories of Busemann G-spaces, of spaces of support elements, in particular, of line elements, of extensors, of higher-order connections, etc.). Here a significant place has been given to those branches which were not sufficiently completely dealt with in Bliznikas' survey [1] on Finsler spaces and their generalizations (for example, the indicatrix concept, the geometry of the Lagrange problem, the geometric interpretation of the extremum conditions).

Papers which appeared after 1953, as well as the fundamental papers of an earlier period, are considered in more detail. The same holds for the Bibliography.

I am deeply grateful to my teacher Professor V. V. Vagner for reading the manuscript and for making a number of suggestions, and also to my colleague M. V. Losik for help in the writing of the third and fourth sections and for compiling the Bibliography.

§ 1.   Geometrization of the Simplest
n-Dimensional Variational Problem.
Finsler Space

The variational problem stated in the heading is the problem of seeking the extremum of the line function

$$I\,[\gamma] = \int_{t_1}^{t_2} F(x^i,\,\dot{x}^i)\,dt \quad (i,j=1,2,\ldots,n),$$ (1.1)

defined on   a set of oriented regular curves γ:

$$x^i = x^i\,(t),\quad t\in[t_1,\,t_2]$$ (1.2)

of a certain differentiable manifold X_n, passing through two distinct fixed points. The collection of n quantities $x^i = \frac{dx^i}{dt}$ is to be interpreted as a tangent vector to the curve γ, belonging to a vector tangent space $T_n(P)$, at the point P $(x^i(t)) \in X_n$. The collection of 2nd numbers $(x^i, \dot{x}^i)$ defines a tangent line element of some point of curve γ.

The line element $(x, \dot{x})$ is called admissible relative to the variational problem (1.1) if the function F $(x, \dot{x})$ is defined for it. If the tangent line elements of curve γ are admissible, then the value of integral (1.1) is called the arc length of this curve. If the direction of the vector $v^i \in T_n$ is admissible, i.e., $v^i = \lambda \dot{x}^i$, $\lambda > 0$, then the number v defined by the equality

$$v = F(x^i, v^i) \tag{1.3}$$

is called its length (in particular, its zero length or even its negative length). In this case the vector v^i is termed measurable.

On the basis of the function F, a symmetric tensor

$$g_{ij}(x, \dot{x}) = \frac{1}{2} \frac{\partial^2 F^2}{\partial \dot{x}^i \partial \dot{x}^j}, \tag{1.4}$$

called a metric, is associated with each line element.

In the classical definition [122] X_n is called a Finsler space F_n at each tangent T_n of which the metric (1.3) is introduced, if the function F $(x, \dot{x})$ satisfies the following three conditions: 1. F $(x, \dot{x})$ is positive homogeneous in the first degree relative to $\dot{x}$; 2. F $(x, \dot{x}) > 0$ for $\dot{x} \neq 0$; 3. the quadratic form $g_{ij}(x, \dot{x})\xi^i\xi^i$ in the variables $\xi^i \neq 0$ is positive definite.

The Finsler metric (1.3) is called regular if the rank of the matrix $\left(\frac{\partial^2 F}{\partial x^i \partial \dot{x}^j}\right)$ equals n−1. If this rank equals n−r−1, then the metric is called a singular metric of singularity class r. In correspondence with this definition the Finsler space also is called regular or singular of the same singularity class.

If any direction in T_n is measurable, then the metric is called complete. Otherwise the metric (1.3) is called incomplete. Vagner

[10] calls an F_n with an incomplete metric a generalized Finsler space.

Of important significance in the geometrization of variational problem (1.1) is the concept of its indicatrix, first introduced by Carathéodory [77]. The equation

$$F(x^i, v^i) = 1 \tag{1.5}$$

defines a (2n-1)-dimensional cutting hypersurface $X_{n+(n-1)}$ called the global indicatrix of variational problem (1.1) (the Finsler metric). Its intersection with the tangent T_n at the point P (x^i) defines a hypersurface which is called the local indicatrix at this point. If with each point P $\in$ X_n we associate the (n + 1)-dimensional space $T_n(P) \times R$, where R is a one-dimensional arithmetic space, then the equation

$$v^{n+1} = F(x^i, v^i) \tag{1.6}$$

defines a global semiconical indicatrix since together with the point $v_0^\alpha (\alpha = 1, ..., n + 1)$ belonging to surface (1.6), the whole ray ρv_0^α, $\rho > 0$, also belongs to it by virtue of the positive homogeneity of the function F. Every local indicatrix (1.5) is a section of the semiconical indicatrix (1.6) by the plane $v^{n+1} = 1$. The introduction of the semiconical indicatrix is appropriate in view of the possibility of waiving the requirement that the fundamental function F be positive, unnatural from the calculus of variations point of view.

In the adjoint vector space $T_n'(P)$ corresponding to $T_n(P)$ we can analogously construct for a regular F_n a hypersurface (a hyperfamily of hypersurfaces) by means of the equation

$$H(x^i, y_j) = 1, \tag{1.7}$$

called a figuratrix [74]. Here, $y_j \in T_n'(P)$ is a covariant vector defining the tangent hyperplane to the indicatrix. Equations (1.5) and (1.7) dually define a contravariant and a covariant [10] vector metric (the Minkowski metrics). We require of the function H that it possess the properties of the function F.

In the consideration of the generalized geometry F_n (in the geometrization of the simplest variational problem (1.1)) it is appropriate to write the equations for the indicatrix and the figuratrix in parametric form [10]:

$$v^i = l^i (x^j, \eta_i^a),$$
$$(a = 1, \ldots, n-1) \qquad (1.8)$$
$$y_i = l_i (x^j \, \eta_i^a),$$

where the η^a are parameters defining the positions of the point and of the hyperplane on the surfaces indicated.

With each point P $(x, \dot{x}_0)$ of the indicatrix there is associated a hyperquadric

$$g_{ij} (x, \dot{x}_0) \, x^i x^j = 1, \qquad (1.9)$$

with the help of which a space R^n with Euclidean metric is associated with each line element $(x, \dot{x}_0)$ having a common center x. This, a point of view of Cartan, signifies that a Finsler space can be looked upon as a fiber space $R^n(X_{n+(n-1)})$ whose base is a $(2n-1)$-dimensional cutting hypersurface $X_{n+(n-1)}$ the hypersections by which are the local indicatrices, while the layers over the points are Euclidean metric spaces R^n.

Another point of view, more suitable for the geometrization of the calculus of variations itself, is due to Winternitz [155], Vagner [10], and others, and is that F_n is treated as a fiber space whose base is a differentiable manifold X_n, while the layers are the tangent spaces T_n of this X_n with Minkowski metrics given in them. If the metric is defined by means of the indicatrix, then F_n can be considered as a fiber space $X_{n+(n-1)}$ whose base is X_n, while the layers are the local indicatrices of the tangent spaces associated with the points of the base space itself [10]. The latter signifies that the geometric theory of variational problem (1.1) is reduced to the analysis of a general fiber space (of a composite manifold, in Vagner's early terminology [12]).

Since we can treat a local indicatrix as a hypersurface of a central-affine space, it is appropriate to start the study of a global indicatrix with the study of local indicatrices, and then to go on to the study of the field of local hypersurfaces (of the corresponding fiber space). This makes it possible to apply central-affine geometry to the calculus of variations. The geometry of a local indicatrix as a hypersurface of the tangent T_n was studied by Vagner [10], Laugwitz [114], Kawaguchi [107], and Varga [144].

The obtaining of the differential invariants associated with variational problem (1.1) and needed for answering the question about the equivalence of two such problems relative to diffeomorphisms of the underlying differentiable manifolds, is based on the concept of connection in the corresponding fiber space. Various versions of connection in F_n have been found by many authors (see Bliznikas' survey [1]). It is inappropriate to consider all of them here and so we shall dwell briefly on Vagner's connection as the one most convenient in applications to the calculus of variations.

Vagner [10] dually defines the parallel displacement of contravariant (X^i) and covariant (Y_i) vectors by means of the vanishing of the absolute differentials

$$\overset{1}{\delta} X^i = dX^i + \overset{1}{\Gamma}{}^i_j(x^k,\ X^k)\, dx^j = 0,$$

$$\tag{1.10}$$

$$\overset{2}{\delta} Y_i = dY_i - \overset{2}{\Gamma}{}_{ij}(x^k,\ Y_k)\, dx^j = 0,$$

where $\overset{1}{\Gamma}$ and $\overset{2}{\Gamma}$ depend nonlinearly on X^k and Y_k and are called the coefficients of nonlinear partial connection, respectively, contravariant and covariant. For the case of a regular F_n Vagner, using certain conditions [10], found an explicit expression for the connection coefficients in terms of the metric function F. In this same paper, using the nonlinear connection in (1.10), he gave an expression for the second variation of the length of an arc in F_n, derived the Hamilton-Jacobi equation, examined a new form of the equations of the extremals (the geodesics) in correspondence with the parametric description (1.8) of the indicatrix equations, and geometrically interpreted the various necessary and sufficient conditions for the extremum of integral (1.1). He found the conditions for the global indicatrix to be constant (Eq. (1.8) does not contain the variables x^i in a suitable coordinate system), i.e., essentially he found the conditions for reducing the variational problem (1.1) to the case when the integrand depends only on the derivatives. In this paper Cartan's connection is distinctively highlighted from the point of view of the theory of fiber spaces with the aid of a central-affine rolling of a hypersurface.

The singular variational problem (1.1) was studied from the geometric point of view in [8]. In this case the local indicatrices are singular hypersurfaces of the first singularity class, i.e., every hypersurface is the envelope of an $(n-2)$-parameter family of hyperplanes. In another paper [3] he analyzed the geometric theory of Carathéodory transformations in a regular F_n, i.e., homological transformations of the form

$$'v^i = \frac{\varepsilon v^i}{1 - \sigma_j v^j}, \quad 'w_i = \varepsilon (w_i - \sigma_i) \ (\varepsilon = \pm 1) \tag{1.11}$$

in point and in hyperplane coordinates, respectively, where σ_j is the gradient of some scalar function $\sigma(x^k)$ and $\varepsilon(1 - \sigma_j v^j) > 0$ (as is well known, the Carathéodory transformations preserve the extremals of variational problem (1.1). Having introduced the connection in the corresponding problem (1.1) by means of a fiber space invariant relative to transformations (1.11), Vagner found the conditions for reducing a global indicatrix to a constant one by means of the Carathéodory transformations.

Kabanov [38], on the basis of these investigations, found the connection in a singular F_n of the first singularity class, invariant relative to the Carathéodory transformations, found the conditions for reducing with the aid of these transformations the variational problem (1.1) to the case when the integrand does not depend on the coordinates of the points X_n, and proved the reduction theorem. (Here it is pertinent to remark that the geometric theory of the Carathéodory transformations in variational problems for ordinary and multiple integrals requires the use of central-projective and affine geometrices of surfaces, which thus find an unexpected application in the geometry of the calculus of variations.)

Losik [49] gave a geometric interpretation of the torsion of the nonlinear connection in (1.10) as the deviation of this connection from the connection conjugate to it.

Zhotikov [28, 29] constructed a geometrical theory of a singular variational problem (1.1) of an arbitrary singularity class.

The differential-geometric formalism of investigation in the simplest variational problem with an algebraic function F has been constructed by Liber [44], Tonooka [142, 143], and Ermakov [23, 24].

The differential-geometric apparatus developed by Vagner [12] under the former name "Composite manifold" is appropriate in the analysis of the geometric theory of various variational problems. In their own papers the authors listed essentially make use of this research tool.

Other aspects of Finsler geometry, in particular, various classes of Finsler spaces, are omitted here in the interest of brevity, all the more so because they are presented with sufficient completeness in the survey [1] already mentioned.

§ 2.   Variational Problem for Functionals

Containing Higher Derivatives.

Kawaguchi Spaces

We consider the problem of finding the extremum of a function of the oriented line

$$I[\gamma] = \int_{t_1}^{t_2} F(x^i,\ x^{(\alpha)i})\, dt, \tag{2.1}$$

defined on the set of curves (1.2) of the differentiable manifold X_n. Here the ordered system of nM numbers $x^{(\alpha)i}$ ($\alpha = 1, 2, \ldots, M$), where $x^{(\alpha)i} = \left(\dfrac{d^\alpha x^i}{dt^\alpha}\right)_{t=t_0}$, is called a line element of order M at the point $x_0^i = x^i(t_0)$. The integrand F should satisfy the Zermelo conditions

$$\sum_{\alpha=1}^{M} \alpha x^{(\alpha)i} F_{(\alpha)i} = F,$$

$$\sum_{\alpha=p}^{M} \binom{\alpha}{p} x^{(\alpha-p+1)i} F_{(\alpha)i} = 0, \quad (p = 2, 3, \ldots, M),$$

where $F_{(\alpha)i} = \dfrac{\partial F}{\partial x^{(\alpha)i}}$. The set of all line elements of order M at the point x_0^i is called a tangent space of order M at this point. Every quantity depending on $x^{(\alpha)i}$ and not varying under a transformation of parameter t along a curve is called intrinsic.

The manifold X_n in which integral (2.1) is given, invariant relative to an admissible parametrization along the curves (1.2), is called a Kawaguchi space. Thus, the theory of Kawaguchi space is a geometric theory of the variational problem (2.1) in parametric form. The value of integral (2.1) along an admissible curve is called the arc length of this curve. The arc length of a curve, expressible by an integral with higher derivatives, is encountered in the analysis of Klein spaces (affine, projective arc length, etc.). A Kawaguchi space is a natural generalization of Finsler space. In spite of this the geometry of Kawaguchi space is considerably more complicated than the geometry of Finsler spaces because it relates to a differential theory of spaces of higher order.

The original construction of a geometric theory for the variational problem (2.1) with $n = M = 2$ is due to Cartan. However, his approach is unfit for higher dimensions. Here we should at once say that subsequent results in Kawaguchi geometry have been obtained in a purely formal way, based on a number of relations in the theory of higher-order tangent spaces and in the theory of extensors whose geometric meaning has not been ascertained, which makes it difficult to clarify the relation of the results obtained to the calculus of variations.

The foundations of the theory of spaces with metric (2.1) were laid by Kawaguchi in the Thirties [101, 102, 104], in whose honor they were named Kawaguchi spaces. By generalizing Cartan's method, applied by him in the construction of Finsler geometry, Kawaguchi examined a certain fiber space whose base is the space of line element X_n of order N, while the layers are n-dimensional vector spaces which can be identified with the tangents T_n to X_n at the corresponding points. Connection in this space is defined by specifying a covariant differential of the intrinsic contravariant vector $V^i(x^i, x^{(\alpha)i})$ $(\alpha = 1, 2, ..., N)$

$$\delta V^i = dV^i + \sum_{\alpha=1}^{N} \overset{(\alpha)}{\Gamma}{}^i_{kj} V^k dx^{(\alpha)j}, \tag{2.2}$$

where $\overset{(\alpha)}{\Gamma}{}^i_{kj}$ are functions of x^i and $x^{(\alpha)i}$. This is too little to be able to construct an apparatus for covariant differentiation, therefore, Kawaguchi additionally introduced the so-called base connection which is defined by the equality:

$$\delta x^{(\alpha)i} = M^{\alpha i}_j \left(dx^{(\alpha)i} + \sum_{\beta=0}^{\alpha-1} \Lambda^{\alpha j}_{\beta k} dx^{(\beta)k} \right), \quad (\alpha = 1, ..., N), \tag{2.3}$$

where $M_j^{\alpha i}$ are tensors for all α and $M_j^{\alpha i}$ and $\Lambda_{\beta k}^{\alpha j}$ depend on a line element of order N, and, moreover, $M_k^{0j} = \delta_k^j$, while $\Lambda_{\beta k}^{0j} = 0$. Now (2.2) can be represented in the form:

$$\delta V^i = \sum_{\alpha=0}^{N} \nabla_j^{(\alpha)} V^i \delta x^{(\alpha)j},$$

where $\nabla_j^{(\alpha)} V^i$ are various types of covariant derivatives of V^i.

Using Synge's vectors (133] Kawaguchi constructed from the metric function F an intrinsic symmetric tensor g_{ij} depending on a line element of order $2M-1$, which in the general case is non-degenerate and can be chosen as the metric tensor in a Kawaguchi space. With the aid of this tensor Kawaguchi succeeded in uniquely determining the connections (2.2) and (2.3) for $N = 2M-1$, and this emerged as the foundation of the theory of Kawaguchi spaces of general form. This theory was developed during the Thirties and Forties by Kawaguchi himself and by others, mainly, Japanese geometers.

Because of the reasons mentioned above the theory of Kawaguchi spaces of general form was developed but little during the subsequent years, although certain aspects of it were generalized in various directions (the theory of higher-order connections, the generalization of the theory of extensors, the theory of higher-order paths, etc.). All this has no direct bearing on the geometry of the calculus of variations and is not considered here.

Among the papers of the Fifties and Sixties we note the following: Pen-wang Sun [55], by generalizing Cartan's method [79], obtained the basic differential invariants of a two-dimensional Kawaguchi space of arbitrary order; Ide [91-94] derived a new higher-order connection which allowed him to obtain a new connection in Kawaguchi space and to find its relation to the known connection.

Losik [46, 48] showed that in any Klein space there exists an invariant Kawaguchi metric whose automorphism group coincides with the fundamental group of the given space, i.e., any Klein space is completely characterized by some lattice of a Kawaguchi space. In another paper [47], on the basis of the geometry of the local indicatrix, i.e., the surface

$$F(x^i, x^{(a)i}) = 1$$

of the tangent space of order M at the point x , he found the geometric interpretation of a number of extremum conditions of the corresponding variational problem (2.1): transversability, the necessary and sufficient Weierstrass conditions, and the necessary conditions for discontinuous solutions (the Erdman-Weierstrass-Zimmermann conditions), which are analogous to the corresponding interpretations for problem (1.1) [10].

In addition to the general Kawaguchi space whose metric is defined by integral (2.1), Kawaguchi spaces of special types have also been studied. An extensive theory has been constructed for a Kawaguchi space with the metric

$$\int (A_i x''^i + B)^{\frac{1}{p}} dt,$$

where A_i and B are functions of x^i and x'^i (the notation is the usual one: $x'^i = \frac{dx^i}{dt}, x''^i = \frac{d^2x^i}{dt^2}$). The general theory is not applicable in this case because the metric tensor g_{ij} is degenerate. Therefore, Kawaguchi [105] chose

$$G_{ik} = 2A_{i(k)} - A_{k(i)},$$

where $A_{i(k)} = \frac{\partial A_i}{\partial x'^k}$, as the fundamental tensor of the given space. This tensor is not symmetric, is homogeneous of degree p−3 relative to the x' , and, in the general case, where p ≠ 3/2, is nondegenerate. The theory of such a special Kawaguchi space is simplified at the expense that it is constructed, as also the Finsler geometry, on the basis of the space of first-order line elements.

Kawaguchi [105] introduced two connections in the given space. The first is defined by specifying the covariant differential of a vector $v^i (x^j, x'^j)$ which is homogeneous of degree zero relative to the x'^j:

$$\delta V^i = dV^i + \Gamma^i_{(j)k} V^j dx^k,$$

and the second, called the base connection, is defined by the equality

$$\delta x'^i = dx'^i + \Gamma^i_{(l)}dx^j,$$

where $\Gamma^i = \frac{1}{2}G^{ij}(2A_{jk}x'^\kappa - B_{(j)})$, $A_{jk} = \frac{\partial A_j}{\partial x^k}$, and $G_{ik}G^{il} = \delta^l_k$. The covariant differential relative to the base connection has the form:

$$\delta V^i = dV^i + \Gamma^i_{(j)k}V^j dx^k + C^i_{jk}V^j\delta x'^k,$$

where the torsion tensor C^i_{jk} is expressed in definite form in terms of the metric function and is a function of the zeroth degree of homogeneity relative to the x^i.

In a short survey it is not possible to dwell on all aspects of the theory of this special Kawaguchi space. Those interested may address themselves to the papers [115, 100, 90, 148, 113, 98, 99, 91, 93, 94, 127, 124-126, 119, 111, 112, 156, 123].

Other particular types of Kawaguchi spaces [87-89] also have been studied, in addition to the special one above. At the end of the Fifties and the beginning of the Sixties there appeared several papers by Watanabe [149-151], in which he introduced a generalized affine space, namely, a Kawaguchi space with the metric

$$\int \left(A_i x^{(M)i} + B\right)^{\frac{1}{p}} dt,$$

where M = n, A_i and B are functions of a line element of order M−1, and moreover, the following conditions are fulfilled:

$$A_{i(M-1)l_1(M-1)j_1} = 0,$$

$$A_{i(M-1)l_1(M-2)l_2\ldots(M-\alpha)l_\alpha(M-\alpha-1)l_{\alpha+1}(M-\alpha-1)j_{\alpha+1}} = 0,$$

$$(\alpha = 1, 2, \ldots, M-2).$$

This metric is a generalization of the invariant arc length in Euclidean space and, when n = 2, is a particular case of the special Kawaguchi space. Mainly, a three-dimensional generalized affine space was studied. In [151] Watanabe found the conditions under which the extremal curves of a generalized affine space possess certain properties which are inherent to them in an affine space. Other particular cases of Kawaguchi spaces were considered in [152–154].

In [45, 48] Losik found invariant Kawaguchi metrics for spaces of affine, equiaffine, and almost-symplectic torsion-free connections, which completely characterize these spaces. In particular, this yields the characteristic Kawaguchi metrics of an affine, an equiaffine, and a symplectic space, which coincide with the known invariant arc lengths in these spaces.

§ 3. Variational Problem for

Multiple Integrals. Spaces with Areal Metric

The problem mentioned here consists in seeking the extremum of the m-fold integral

$$\int\limits_{(m)} F\left(x^i, p_a^i\right) du^1 \ldots du^m \quad (a, b = 1, \ldots, m) \tag{3.1}$$

as a function of an oriented m-dimensional surface

$$x^i = x^i \left(u^a\right) \tag{3.2}$$

with a fixed or a moving boundary. The value of the integral (3.1) when $p_a^i = \dfrac{\partial x^i}{\partial u_a}$ is called the area of the surface (3.2). The metric function F satisfies the conditions:

a) Under the transformations $\widetilde{p}_a^i = A_a^b \, p_b^i$, $\mathrm{Det}\left|A_b^a\right| > 0$, F is transformed to $\widetilde{F} = \mathrm{Det}\left|A_b^a\right| F$

b) If the vectors p_a^i $(a = 1, \ldots, m)$ are linearly dependent, then $F\left(x^i, p_a^i\right) = 0$. If, moreover, the condition $F\left(x^i, p_a^i\right) > 0$ is fulfilled for linearly independent p_a^i, then F is said to be positive definite (then the variational problem (3.1) itself is said to be likewise). A space X_n in which integral (3.1) is given, which is invariant relative to an admissible parametrization of surface (3.2), is called a space with an m-areal metric. The properties of F show that on any m-dimensional surface it induces a scalar density of weight -1 which can be used for finding the area of the surface.

It is well known that F can be represented as a function of the point x^i and of the simple contravariant m-vector $p^{i[m]} = m! p_{[1}^{i_1} p_2^{i_2} \ldots p_{m]}^{i_m}$, $F(x^i, p^{i[m]})$, and is positive homogeneous of first degree relative to $p^{i[m]}$, which signifies that F is defined on the fiber space of simple tangent m-vectors of the manifold X_n. The layer of this space is the set of simple m-vectors of an n-dimensional vector space, called the Grassmann cone, serving as a geometrical interpretation of the condition that the m-vector be simple. If F is defined not on the whole fiber space but only on a certain region of it, then the m-areal metric is called incomplete.

When m - 1 the areal space becomes a Finsler space and when m = n − 1 it is called a space with a hyperareal metric or a Cartan space. The name proper areal space is usually reserved only for the case 1 < m < n − 1.

For tying up with the calculus of variations, the concept of the indicatrix of the areal metric, introduced by Vagner [14], is of importance. Applying the terminology used in [147] we designate as a global indicatrix the cutting hypersurface in the fiber space indicated, defined by the equation

$$F(x^i,\ p^{i[m]}) = 1, \tag{3.3}$$

whose intersections with the layers − the Grassmann cones − are hypersurfaces on these cones and are called the local indicatrices. A global indicatrix completely defines an areal metric.

1. Space with a Hyperareal Metric. When m = n − 1 the Grassmann cone introduced above becomes an n-dimensional linear space which is a considerably simple matter in comparison with the general case. Therefore, analysis of spaces with a hyperareal metric began earlier and they have been studied more fully.

A one-to-one relation holds between contravariant (n − 1)-vectors and covariant vector densities of weight −1, therefore, in our case the metric function can be represented as a function $F(x^i, p_i)$ which is positive homogeneous of degree 1 relative to p_i, where the p_i are the components of a covariant vector density of weight −1. The regularity and the singularity of the corresponding variational problem are defined just as in the simplest case. The local indicatrix (3.3) at each point x^i will now be a hypersurface in the space of the densities indicated.

Cartan [78] established the foundations of the geometry of a space with a hyperareal metric. Essentially, as also in the case of a Finsler space, he considered a certain fiber space with a base, the space of tangent hyperflat elements to X_n, defined by p_i, and with a layer, the n-dimensional vector space with the symmetric metric tensor

$$g^{ij} = (\mathrm{Det}\,|\,a^{kl}\,|\,)^{-\frac{1}{n-1}}a^{ij},$$

where $a^{ij} = \frac{1}{2}\frac{\partial^2 F^2}{\partial p_i \partial p_j}$. If the variational problem for an $(n-1)$-fold integral is regular, then the metric tensor is nondegenerate and, together with the one dual to it, it can be used to raise or lower indices.

Cartan defined a metric connection with the aid of the absolute differential of a contravariant vector $V^i(x^k, p_k)$, positive homogeneous of degree 0 relative to p :

$$\delta V^i = dV^i + (\Gamma^i_{jk}dx^k + C^{ik}_j dp_k)\,V^j,$$

where Γ^i_{jk} and C^{ik}_j are functions of x^i and p_i, positive homogeneous of zeroth degree relative to p_i. If the Cartan tensor $H^{ij} = g^{ij} + A_k A^{kji}$, where $A^{..k}_{ij} = \frac{F}{\mathrm{Det}\,|\,g^{ij}\,|}C^{..k}_{ij}$, $A_k = A^{..j}_{kj}$, is nondegenerate, then the coefficients Γ and C can be expressed, by a series of invariant conditions, in terms of the metric function F. We note that the vanishing of the tensor $A^{..k}_{ij}$ characterizes the Cartan spaces defined by a certain Riemannian metric. By associating with a curve the set of hyperflat elements normal to it, Cartan introduced the arc length of a curve, curves of extremal length, and autoparallel curves characterized by the condition $\delta l_i = 0$ where l_i is the unit tangent vector.

With the hypersurface (3.2) (a, b = 1, ..., n−1) of a Cartan space there is associated the set of its tangent hyperflat elements and on it there are defined two quadratic forms: $g_{ab}du^a du^b$, where $g_{ab} = g_{ij}p^i_a p^i_b$, and $b_{ab}du^a du^b = l_i\,\delta^2 x^i = dx^i\,\delta l_i$. The minimality

condition for the hypersurface (i.e., for the extremality of its hy-
perarea) is the vanishing of its mean curvature $H = g^{ab}b_{ab}$.

Certain properties of curves and of hypersurfaces in a regular Cartan space were studied by Berwald [71, 72] and Alardin
[66]. In [71] Berwald found an invariant form of the second variation of integral (3.1) ($a = 1, ..., n-1$) for the minimality of the hypersurface – the so-called Koschmieder's normal form – and found
a geometric meaning for the Koschmieder invariant occurring in
this variation. Later, Buchin Su [128] obtained by another method
the associated differential equation of the minimal surface and the
Koschmieder invariant. Using the Lie derivative in the space of
hyperflat elements Khmelevskii [58, 59] found an expression in invariant form for the first and the second variations of the integral
indicated.

Moór [118] solved the problem of the existence of a Riemann
space osculating to a Cartan space along a one-parameter family
(x^i (t), p_i (t)) of hyperflat elements, by means of which a connection
is defined in the Cartan space, coinciding with the corresponding
Riemannian connection along the curve x^j (t).

Berwald [72] also noted that a Cartan space with $A_i = 0$ is
dual in a certain sense to the analogous class of Finsler spaces,
while Vagner [14] found the necessary and sufficient conditions for
a similar duality between m-dimensional metrics and (n- m)-
dimensional areal metrics. Duality between Finsler and Cartan
spaces has been investigated by Moór [116, 117], showing that under a correspondence defined by him many of the quantities in the
dual spaces coincide. As it turned out these results are of interest
only for nonpositive definite Cartan spaces since positive Cartan
spaces of such type are Riemann space according to a theorem of
Deicke [85].

Barthel [68] found the conditions to be imposed on the Cartan
connection in order that the autoparallels and the "hyperplanes"
(hypersurfaces on which the second quadratic form vanishes) would
simultaneously be the extremals and the minimal surfaces, respectively. The Alardin connection [73] is obtained from this as a
special case. For n = 3 Bettingen [73] defined the geodesic curvature of a curve on a surface and derived a formula, analogous to

the Gauss-Bonnet formula, which relates the geodesic curvature with the total curvature of the surface.

Vasil'eva [17, 18] obtained the fundamental objects of a Cartan space by Laptev's method of extending and enveloping differential-geometric objects.

A global definition of a Cartan space, of connection in it, etc., has been given by Debever [83], Reeb [120], and Tashiro [139].

It seems that the approach to the geometry of an $(n-1)$-fold integral, most convenient from the viewpoint of applications to the calculus of variations, is the one proposed by Vagner [13]. Analogously to that which he did in the construction of Finsler geometry [10], he studied a space with (in the general case) an incomplete hyperareal metric on the basis of a systematic exploitation of the indicatrix concept. As we have already remarked, in this case the local indicatrices are hypersurfaces in the space of covariant vector densities of weight -1 and they can be interpreted as hyperfamilies of hyperplanes in a linear space of contravariant vector densities of weight $+1$. Just as in the case of Finsler geometry, this permits us to make use of the theory of hypersurfaces in a central-affine space for the construction of connection in a fiber space, which is a global indicatrix, and for obtaining the geometric interpretation of a number of extremum conditions of the corresponding variational problem: the necessary conditions of Weierstrass and Legendre, the transversality conditions, the sufficient condition of Weierstrass, obtained on the basis of Carathéodory's method.

The connection applicable here is the general linear connection in a fiber space, which we have already referred to [12, 147].

Kabanov [36, 41], using these general ideas, analyzed a Cartan space given to within the Carathéodory transformations which now are specified to be point translations in the corresponding spaces of densities, defined by a contravariant vector density of weight $+1$ which preserves the measurable $(n-1)$-paths and forms a solenoidal field in X_n. As is well known, the extremals of a hypersurface are carried into extremals under these transformations. He has constructed a connection which is invariant relative to the

transformations indicated, has proven a reduction theorem, and has
obtained the necessary and sufficient conditions for the reducibility
of the global indicatrix of a hyperareal space to a constant with the
aid of the Carathéodory transformations.

Vagner's approach permits the geometric theory of a singular variational problem in this case to be studied in a manner
analogous to that used in the case of the singular variational problem (1.1). The case of a minimal singularity class ($r = 1$) was
examined by Kabanov [39, 41]. He constructed a connection in the
corresponding fiber space, proved a reduction theorem, and obtained the necessary and sufficient conditions for the constancy of
the global indicatrix. For problems with an arbitrary singularity
class he obtained analogous results [40], however, under certain
constraints imposed on the dimension of the local indicatrices. On
the basis of the theory of local hyperstrips, developed by Vagner
[11] and applied to the calculus of variations, Kaganov [43] constructed a geometric theory of this variational problem under
rather general assumptions.

Tonooka [140, 141] analyzed the case of a hyperareal metric
depending on second-order derivatives and, in particular, found
generalizations of the Synge vectors and of the metric tensor, obtained a connection, and for it constructed a theory of curvature.
Using Laptev's method, an $(n-1)$-fold integral preserving second
derivatives was studied by Evtushik [20, 21] who defined certain
objects generalizing the well-known objects of Cartan space.

2. Proper Areal Spaces. The differential-geometric
theory of proper areal spaces ($1 < m < n-1$) was first constructed
by Kawaguchi [42, 103] who introduced the notion of connection in
these spaces, which, essentially, is a connection in a fiber space
who base is the space of m-dimensional flat elements tangent to X_n
and whose layer is an n-dimensional vector space. He defined the
connection under the condition that there exists a nondegenerate
metric tensor $g_{ij}(x^i, p^{i[m]})$ which is positive homogeneous of zeroth
degree relative to $p^{i[m]}$. Subsequent development of this theory
has dealt mainly with the problem of seeking the connection defined
invariantly by the metric function F and of the geometrization of
the various extremum conditions of the corresponding variational
problem.

The construction of the differential invariants of the metric function F was complicated by the fact that it is impossible to differentiate F directly with respect to the components of $p^{i\,[m]}$ when $1 < m < n-1$. This difficulty was overcome by Debever [82], Iwamoto [95-97], and Kawaguchi [106] by the introduction of differential operations possessing the properties of first and second order partial derivatives. Then these operations were systematically studied by Barthel.

A purely geometric approach to this problem was taken somewhat earlier by Vagner [14]. At each point of indicatrix (3.3) he constructed a specific tangent hyperplane in the case of m-vectors; the envelope of the obtained family of hyperplanes is a hypersurface in the space of m-vectors and determines a continuation of the metric function onto some region in this space, which then becomes differentiable with respect to its arguments, i.e., with respect to the components of the m-vector. Using the operations indicated he constructed the 2m-index metric tensor of Iwamoto, $g_{i\,[m],\,j\,[m]}$ $(x^i,\, p^{i\,[m]})$, analogous to the metric tensor in Finsler geometry. It is positive homogeneous of zeroth degree relative to $p^{i\,[m]}$, can be applied to measuring the lengths of the contravariant m-vectors, and, in particular, satisfies the identity

$$g_{i[m],\,j[m]}p^{i[m]}p^{j[m]} = (m!)^2 F^2.$$

If the determinant of order $\binom{n}{m}$, constructed from $g_{i\,[m],\,j\,[m]}$, is not equal to zero, then the areal space is called regular.

However, the whole problem of constructing even a two-index metric tensor g_{ij} from a metric function has not been solved in the general case, and, therefore, classes of spaces have been considered where such a tensor already exists. The first to be studied were areal spaces of metric class, introduced by Debever [82] (also see Bliznikas' survay [1]), which are a natural generalization of Finsler and Cartan spaces and of areal spaces generated by a Riemannian metric. The metric tensor g_{ij} is uniquely defined in these spaces under certain conditions. Spaces of metric class were studied by Debever [82], Brickell [75], Iwamoto [95-97], Kawaguchi [106], and Varga [145], who found, in particular, a number of properties leading to the proof of Tandai's important theorem

[134] on the fact that any regular areal space of metric class, with $1 < m < n-1$, is a Riemann space.

A generalization of areal spaces of metric class are areal spaces of submetric class introduced by Kawaguchi [106]. In these spaces the metric tensor g_{ij} is constructed algebraically from the metric function and its first and second order derivatives with respect to $p^{i[m]}$, but the fundamental conditions imposed on the corresponding tensors of the spaces of metric class may not be fulfilled. From the results of Kawaguchi and Hokari [109] it follows that any regular areal space with mutually prime n and m is a space of submetric class. The theory of areal spaces of submetric class was simplified by Kawaguchi and Tandai [110] by means of introducing the so-called normalized metric tensor $g_{ij}{}^*$.

An interesting subclass of spaces of submetric class was examined by Tandai [138]. This subclass is defined by the condition of factorization of the symmetric part of the Legendre form $L_{ij}{}^{(ab)} = g_{ij}{}'' g^{ab}$ and is called the semimetric class of areal spaces. In another paper [135] Tandai attempted to construct a theory of areal spaces of a general form. The set of solutions (not always real) of a certain system of algebraic equations in the components of the metric tensor g_{ij}, in which the number of equations coincides with the number of unknowns, defining nondegenerate tensors, falls into the classes of solutions which are closed relative to multiplication by the mth-degree primitive root of unity and to passing to the adjoint. Each such class, called the class of the metric tensor, defines a certain geometry in the original space. In this geometry there is uniquely defined, with the aid of a number of conditions, a real connection, etc., in particular, there is found an invariant expression for the equations of the minimal surfaces.

Since in the above-mentioned papers of Kawaguchi and Tandai the connection coefficients were defined as the solutions of some system of linear equations, while their expressions directly in terms of the metric function were not given, Davies [80, 81] obtained these expressions for the coefficients of $\Gamma_{jk}{}^i$ with the help of a number of new conditions in the case of a space of submetric class by the method of osculating Riemann spaces. This connection was used by him for seeking the second variation of the integral of the corresponding variational problem and for simplifying the equations of the minimal surfaces. Kawaguchi [108] found expressions for the connection coefficients in the geometry of a special class of the metric tensor.

Certain aspects of the geometry of areal spaces of submetric class, in particular, the conditions for the affiness of the Davies connection [80] and for the independence of the metric function from the x^i in some coordinate system, have been considered by Gama [86].

Barthel [70] tried to approach the study of an areal space of general form from a new point of view, by taking as the fundamental object in the theory a simple contravariant m-vector. He defined the metric in the set of m-vectors as the metric tensor $g_{i[m],j[m]}$, while as the device for absolute differentiation he specified the giving of the absolute differential of the m-vector $V^{i[m]}(x^i, p^{i[m]})$:

$$\delta V^{i[m]} = dV^{i[m]} + (\Gamma^{i[m]}_{k[m],j}dx^j + C^{i[m]}_{k[m],j[m]}dp^{j[m]})\, V^{k[m]},$$

which is related with the metric function by certain conditions which do not uniquely define $\Gamma^{i[m]}_{k[m],j}$. The absolute differentiation can be extended to covariant (n−m)-vector densities of weight −1, which makes it possible to obtain an invariant form for the first variation of the corresponding integral.

Ch'ao-hao Ku and Buchin Su [19] have found the first and second variations of integral (3.1) for an areal space with a connection satisfying certain conditions, while in subsequent papers [129-132] Buchin Su reworked these conditions so that the indicated variations took on a simpler form. These conditions are satisfied by the Cartan connection in Finsler geometry and in a space with a hyperareal metric. However, they admit of much arbitrariness in the choice of the connection.

We should take note of Evtushik's paper [22] in which there is found with the aid of Lie differentiation the invariant forms of the first and second variations of the integral of an areal metric, depending on the first and second order derivatives.

Vagner [7, 14] studies an areal space, in the general case, with an incomplete metric, on the basis of the concepts of a global and a local indicatrix (3.3) and of the theory of composite manifolds (general fiber spaces) [12]. In his long fundamental work [14] he

constructed a geometric theory of a local indicateix as a hypersurface on a Grassmann cone, with the help of which he obtained a
geometric interpretation of the tranversality conditions and of the
Weierstrass and Legendre necessary conditions; the Weierstrass
sufficient conditions has been analyzed on the basis of the geometric
interpretation of Carathéodory's method. In another paper [7] he
introduced a linear partial connection (in Vagner's sense [147]) in
a fiber space associated with a global indicatrix, with the use of
the Kawaguchi-Hokari metric tensor [109] for the case when n and
m are relatively prime. This made it possible to construct the
geometry of a measurable m-surface and to find invariant expressions for the first and second variations and to find the Jacobi
equations of the ordinary variational problem for multiple integrals.

The substance of Zhotikov's papers [32, 33] border on these
researches of Vagner, partially repeating and somewhat supplementing his results. Also relevant here is Chernyi's paper [64]
which examines the geometric theory of a singular variational
problem with a double integral in X_4 of singularity class r = 3, i.e.,
when the set of simple tangent hyperplanes of a local indicatrix is
one-dimensional. In the fiber space associated with this problem
a partial linear connection was found and necessary and sufficient
conditions for the constancy of the gloval indicatrix were obtained.

During the Fifties and Sixties attempts were made at a global
analysis of the theory of areal spaces with the aid of the theory of
connections in fiber spaces. Brickell [76] based such an approach
on the Legendre form $L_{ij}{}^{ab}\delta p_a{}^i \delta p_b{}^j$, which is defined on the layers
of some fiber space, constructed a mapping ρ associated with every
areal space, and obtained by a new way a number of the results of
the Japanese school of Kawaguchi. Tandai [136, 137] examined the
tangent m-frame bundle X_n in which an areal metric is defined by
a scalar function of special form and he analyzed the various fibra-
tions of this space. With the aid of general linear connections he
found an invariant affine connection in these spaces.

Dedecker [84] analyzed globally a **variational** problem for a
multiple integral with the help of a form on a differentiable manifold. In particular, he introduced the concepts of an indicatrix, of
a figuratrix, of transversality, examined the Hamilton-Jacobi
equation, and determined its total integral.

§ 4. Intrinsic Lagrange Problem

for Ordinary Integrals

The traditional definition of this Lagrange problem in geometric form consists in the finding of the extremum of the line
function (1.1) on a set of oriented curves satisfying the system of
differential equations

$$\overset{p}{f}(x^i, \dot{x}^i) = 0, \ (p = m + 1, \ldots, n), \tag{4.1}$$

where $\overset{p}{f}$ are functions, positive homogeneous of first degree relative to $\dot{x}^i$, called admissible curves. The system of equations

$$\overset{p}{f}(x^i, v^i) = 0, \tag{4.2}$$

where v^i are the coordinates of the points of the tangents $T_n(X_n)$,

defines (if the rank of the matrix $\left[\dfrac{\partial \overset{p}{f}}{\partial v^i}\right] = n - m$) a semiconical

m-cutting surface $S_{n+(m)}$, while at each T_n it defines an m-dimensional semicone generated by halflines issuing from its center.
Then the curves satisfying system (4.1) will be tangent at each
point to some generating halflines of the corresponding semicone.

The reduction of the Lagrange problem on a conditional extremum to a problem on an unconditional extremum, when integral
(1.1) has been given right from the start on a set of admissible
curves, is effected in the following manner. We define $S_{n+(m)}$ by
the parametric equations

$$v^i = L^i(x^j, \eta^a), \ (a = 1, \ldots, m), \tag{4.3}$$

where L^i is positive homogeneous of first degree relative to η^a.

Obviously, the admissible curves satisfy the equations

$$\dot{x}^i = L^i(x^j, \eta^a),$$

from which we find, along each admissible curve (1.2), the values
of the curvilinear coordinates η^a defining the tangent vector to the

curve as a function of a parameter t, $\eta^a = \eta^a$ (t). The function of a curve, given only on a set of lines satisfying equations (4.3), can be defined analytically by means of the integral

$$\int_{t_1}^{t_2} F(x^i, \eta^a)\, dt, \tag{4.4}$$

where F is a function of the variables η^a, positive homogeneous of first degree. Such a variational problem is called an intrinsic Lagrange problem [147].

If the function (4.4) of an oriented line is positive definite, then to specify it in the intrinsic Lagrange problem is equivalent to specifying $(m-1)$-dimensional surfaces in T_n, called the local indicatrices of the Lagrange problem, which are direction surfaces for the semicones (4.2) and are given on them by the equation

$$F(x^i, \eta^a) = 1, \tag{4.5}$$

defining the global indicatrix in this problem.

Equation (4.5) for the global indicatrix in the intrinsic Lagrange problem is conveniently written in the parametric form:

$$v^i = l^i(x^j, \chi^a) \quad (a = 1, \ldots, m-1).$$

From the geometric point of view the study of the intrinsic Lagrange problem reduces to the study of its global indicatrix [147] as some fiber space of general form. It is clear that same can be said also relative to the theory of the differential invariants of this problem. Thus, the basic problem here is to define in intrinsic fashion the connection in an appropriate fiber space, which is given by a system of Pfaffian differential equations of the form

$$\delta \chi^a = d\chi^a + \Gamma_i^a(x^j, \chi^a)\, dx^i = 0, \tag{4.6}$$

where the Γ_i^a depend nonlinearly on χ^a and are called the coefficients of a nonlinear partial connection.

When looking at the geometric theory of extremum conditions in the Lagrange problem it is convenient to reduce it to a Mayer problem in a space one higher in dimension, and then to treat the latter problem as a theory of semicones in tangent spaces.

Concerning the determination of sufficient conditions in the Lagrange problem, it is natural here to make use of the Carathéodory conditions, which preserve the extremal properties of the curves and which are geometrically reduced to projective transformations in the tangent spaces of form (1.11). Therefore, it is of interest to construct a theory of the differential invariants of this problem given to within arbitrary Carathéodory transformations, i.e., to find the connection in an appropriate fiber space of form (4.6), already invariant relative to these transformations, etc.

Finally, we should remark that the study of the Mayer problem (as is well known, it is equivalent to the Lagrange problem) requires, from the geometric point of view, the construction of a theory of the differential invariants of the field of central semicones in tangent spaces, which, thus, also is of interest.

The geometrization of the Lagrange problem, leading to a generalization of Finsler geometry, was given by Vagner in the Forties [9, 2, 11, 146]. In subsequent papers [15, 16] he introduced, from the geometric point of view, the concept of an anormal curve, of a variation vector along an admissible curve, of a system of differential equations of anormal curves, of pseudoarclengths, etc. It turns out that the Lagrange problem can be treated as the problem of finding curves of extremal pseudolengths in X_n with a differential pseudometric induced by the field of central semicones, which corresponds to the reduction of this problem to the Mayer problem in a space one higher in dimension. Finally, in an extensive article [147], almost a monograph, Vagner pointed out the way to the geometrization of almost all the known problems of the calculus of variations with first derivatives, by means of the concept of a semiconical indicatrix.

An expression for the second variation and the Jacobi equations in invariant form have been obtained by Tokarev [56] for the Lagrange problem. Rzhekhina [53, 54] has examined a particular case of this problem, when the number of equations in (4.1) equals $n-2$. He found a partial connection in the corresponding fiber space, the complete system of differential invariants, and the conditions for the constancy of the global indicatrix.

The geometric theory of the Carathéodory transformations in the Lagrange problem with one additional condition (4.1) has been considered by Kabanov [37]. He constructed the connection in the space associated with the global indicatrix, invariant relative to the Carathéodory transformations, proved the reduction theorem, and found the necessary and sufficient conditions for the reducibility of the global indicatrix to a constant by means of the Carathéodory transformations.

On the basis of the general ideals of Vagner, Zhotikov has examined, in a series of papers [25, 31, 38], the geometry of the local and the global indicatrices of the Lagrange problem under rather general assumptions, and, moreover, has given in [25] an expression for the first variation and a geometric interpretation of the transversality conditions.

Pershin [50-52] considered a singular Lagrange problem in X_4 and the geometric theory of the Carathéodory transformations in this problem. In particular, he gave a geometric interpretation of the transversality conditions and of the Weierstrass sufficient condition and found the conditions for the constancy of the global indicatrix.

In the last 15-20 years foreign mathematicians have been very little occupied with the geometric theory of the Lagrange problem for an ordinary integral. Here we should note only Rund's paper [121] in which there is introduced for the study of this problem in the corresponding fiber space nonholonomic connection coefficients used by him to construct the covariant δ-derivative [122] and to construct the differential invariants of the problem itself.

The geometric theory of the central semicone field in X_n, needed, as already mentioned, in the geometry of the Mayer problem, was considered by Vagner [9, 16], Chashechnikov [60], Zhotikov [26, 27], and Frolova [57].

§ 5.   Intrinsic Lagrange Problem

for Multiple Integrals

Usually the Lagrange variational problem for multiple integrals is formulated as the problem of seeking the extremum of a scalar function of an oriented m-surface (3.2) in X_n, expressible by means of integral (3.1), and, moreover, this extremum is sought on the set of oriented integral surfaces of a system of first-order partial differential equations

$$\overset{q}{f}\left(x^i, \frac{\partial x^i}{\partial u^a}\right) = 0 \quad (q = 1, \ldots, m(n-m) - r), \tag{5.1}$$

where the functions $\overset{q}{f}$ satisfy conditions analogous to those satisfied by the integral F in (3.1). These integral surfaces are called admissible surfaces. If we consider the $\binom{n}{m}$- dimensional space $M_{\binom{n}{m}}$ of all n-dimensional m-vectors $v^{i_1,\ldots,i_m}$, then the conditions for the simplicity of the m-vector can be interpreted as the equation of an $(m(n-m) + 1)$-dimensional cone in $M_{\binom{n}{m}}$, called, as is well known, the Grassmann cone. A simple m-vector can be represented as an alternating product of m vectors v^i_a, specified to within an arbitrary linear unimodular combination, i.e., the $m \cdot n$ quantities v^i_a can be looked upon as some redundant coordinates of the points on the Grassmann cone. This signifies that the equation

$$\overset{q}{f}(x^i, v^i_a) = 0 \tag{5.2}$$

defines an $(r + 1)$-dimensional semicone at each $M_{\binom{n}{m}}$, associated with some point of the space X_n, and, together with the equation

$$F(x^i, v^i_a) = 1,$$

defines an r-dimensional surface on the Grassmann cone, called the local indicatrix in the Lagrange problem for multiple integrals. The integral surfaces of system (5.1) possess the property that at each point of such a surface the tangent m-path will correspond to the path of some generating halfline of the semicone (5.2).

The reduction of the Lagrange problem on a conditional extremum to the problem on an unconditional extremum, when the function of surface (3.1) is given right from the start only on a set of admissible surfaces, is effected in the following way [147].

The equations of the semicones (5.2) are considered in the parametric form

$$v^{i_1 \ldots i_m} = L^{i_1 \ldots i_m}(x^i, \zeta^k), \quad (k = 1, \ldots, r+1), \qquad (5.3)$$

where the functions L are positive homogeneous of first degree relative to the parameters ζ^k. The equalities

$$m! \, x_1^{i_1} \ldots x_m^{i_m} = L^{i_1 \ldots i_m}(x^i, \zeta^k), \quad \left(x_a^i = \frac{\partial x^i}{\partial u^a} \right)$$

should be satisfied for the admissible surfaces, from which we can find, along each admissible surface of form (3.2), the values of the curvilinear coordinates defining its tangent m-vector as a function of the parameters u^a: $\zeta^k = \zeta^k(u^a)$.

Then the function of an oriented surface, specified only on a set of admissible surfaces, is given analytically in the form of the integral

$$\int_{(m)} F(x^i, \zeta^k) \, du^1 \ldots du^m, \qquad (5.4)$$

where F is a function positive homogeneous of first degree relative to the variables ζ^k. Such a variational problem with multiple integrals is called an intrinsic Lagrange problem. When F > 0, the analysis of the intrinsic Lagrange problem is equivalent to the study of the field of r-dimensional surfaces (a fiber space) lying on the Grassmann cones in the spaces $M_{\binom{n}{m}}$, associated with the points of X_n, which are direction surfaces for the cones (5.2) and are defined on them by the equation

$$F(x^i, \zeta^k) = 1. \qquad (5.5)$$

called the global indicatrix equation in this variational problem. The global indicatrix equation (5.5) can also be written in the parametric form

$$v^{i_1 \ldots i_m} = l^{i_1 \ldots i_m}(x^i, \eta^a) \quad (\alpha, \beta = 1, \ldots, r) \qquad (5.6)$$

or

$$v_a^i = l_a^i(x^j, \eta^a),$$

if we make use of the redundant coordinates of the points of the Grassmann cone. Thus, the study of the intrinsic Lagrange problem for multiple integrals, as also the analogous problem for an ordinary integral, is reduced to the study of its global indicatrix as one r-cutting surface or as a fiber space whose base is X_n and whose layers over points are the local indicatrices lying on the corresponding Grassmann cones.

The necessity of the Euler condition in this problem, expressible by means of the usual Lagrange multiplier rule, has not yet been proved, it seems, for arbitrary n and m. In 1945 Barker succeeded in proving the necessity of the Euler condition only for n = 4 and m = 2 by a method which does not admit of an immediate generalization to the general case. However, as far as looking for the sufficient conditions is concerned, we can apply, as above, Carathéodory's method, consisting now of transforming the Lagrange variational problem by means of translations of the form

$$'v^{\{i_1 \ldots i_m\}} = \frac{\varepsilon v^{\{i_1 \ldots i_m\}}}{1 - \sigma_{\{i_1 \ldots i_m\}} v^{\{i_1 \ldots i_m\}}}, \qquad (\varepsilon = \pm 1) \qquad (5.7)$$

$$'w_{\{i_1 \ldots i_m\}} = \varepsilon (w_{\{i_1 \ldots i_m\}} - \sigma_{\{i_1 \ldots i_m\}}),$$

at the point and the hyperflat coordinates, respectively, taking the hyperplane $\sigma_{\{i_1 \ldots i_m\}}$ into an infinitely distant hyperplane and preserving the Grassmann cone (the m-vector σ is assumed to be a gradient vector).

In this way Vagner [5] was the first to succeed in obtaining by purely geometric methods the sufficient extremum conditions in the Lagrange problem for multiple integrals, while in another paper [6] he gave a geometric interpretation of extremal surfaces in this problem, i.e., the integral surfaces of the formally constructed Euler equation.

Chernyi [61, 62] examined the geometric theory of the Lagrange problem for a double integral in X_4 under the condition that the number of additional differential equations (5.1) equals three. In this case the global indicatrix is a 1-cutting surface. He constructed a partial connection in the corresponding fiber space and found the complete system of differential invariants for this problem.

The beginnings of a geometric theory of the differential invariants of the Lagrange variational problem with multiple integrals
for the general case have been given in Zhotikov's recent papers
[31-35]. He found a partial connection of type (4.6) in Vagner's
sense [12] in the fiber space $X_r(X_n)$ $(1 \leq r \leq m\,(n-m))$ corresponding to the global indicatrix (5.6), under the condition that certain
complex determinants are nonzero [31]. In this same sense the
connections in the fiber spaces $E_r(X_n(X_r))$, $(E_m(X_r(X_n))$, and
$E_{n-m}(X_r(X_n))$ associated with the fiber space $X_r(X_n)$ were sought
with the aid of a system of Pfaffian equations. The connection objects occurring in these equations define an invariant research tool
in the intrinsic Lagrange problem for multiple integrals. Identities
needed in the search for the complete system of differential invariants and for the problem of the equivalence of global indicatrices were found, in particular, the conditions for the constancy
of the global indicatrix were obtained. An invariant theory of the
first and second variations of a multiple integral and the Jacobi
equations in the Lagrange problem were constructed. A regular
variational problem on an unconditional extremum for multiple integrals was analyzed in [32, 33] as a particular case of the Lagrange problem with $r = m\,(n-m)$. He also constructed [35] a
geometric theory of the Carathéodory transformations (5.7) in this
general problem with $r \geq (n-m)/m$. At first there was developed
the central-projective geometry of an r-surface in $M\,\binom{n}{m}$ on a
Grassmann cone, and next there was sought a partial connection of
the corresponding r-cutting surface, invariant relative to translations (5.7). The conditions for the reducibility of the global indicatrix (r-cutting surface) to a constant one were found with the
aid of the Carathéodory transformations and, finally, expressions
were derived for the first and second variations of the area of a
region of an admissible surface with a fixed boundary and for the
Jacobi equations, invariant relative to the Carathéodory transformations.

BIBLIOGRAPHY

1.   V. I. Bliznikas, "Finsler spaces and their generalizations," (in press).
2.   V. V. Vagner, "Geometry of the field of central plane curves in X_3," Dokl.
     Akad. Nauk SSSR, 48:382-384 (1945).

3. V. V. Vagner, "Homological transformation of a Finsler metric," Dokl. Akad. Nauk SSSR, 46:263-265 (1945).

4. V. V. Vagner, "Geometry of the field of local central plane curves in X_3 and the simplest case of the Lagrange problem in the calculus of variations," Dokl. Akad. Nauk SSSR, 48:229-232 (1945).

5. V. V. Vagner, "A sufficient condition in the Lagrange problem for multiple integrals," Dokl. Akad. Nauk SSSR, 54:479-482 (1946).

6. V. V. Vagner, "Geometric interpretation of extremal surface in the Lagrange problem for multiple integrals," Dokl. Akad. Nauk SSSR, 55:87-90 (1947).

7. V. V. Vagner, "Geometry of an n-dimensional space with an m-dimensional Riemannian metric and its application to the calculus of variations," Mat. Sb., 20(62):3-26 (1947).

8. V. V. Vagner, "Geometric theory of the simplest n-dimensional singular problem of the calculus of variations," Mat. Sb., 21(63):321-364 (1947).

9. V. V. Vagner, "Theory of the field of local curves and local conical surfaces in X_3 and its applications to the calculus of variations and to the theory of partial differential equations," Tr. Seminara Vectorn. Tenzorn. Analizu, No. 6, 257-364 (1948).

10. V. V. Vagner, "Finsler geometry as a theory of the field of local hypersurfaces in X_n," Tr. Seminara Vectorn. Tenzorn. Analizu, No. 7, 65-166 (1949).

11. V. V. Vagner, "Theory of the field of local hyperstrips," Tr. Seminara Vectorn. Tenzorn. Analizu, No. 8, 197-272 (1950).

12. V. V. Vagner, "Theory of composite manifolds," Tr. Seminara Vectorn. Tenzorn. Analizu, No. 8, 11-72 (1950).

13. V. V. Vagner, "Geometry of a space with a hyperareal metric as a theory of the field of local hypersurfaces in a composite manifold," Tr. Seminara Vectorn. Tenzorn. Analizu, No. 8, 144-196 (1950).

14. V. V. Vagner, "Geometry of a space with an areal metric and its applications to the calculus of variations," Mat. Sb., 19(61):341-406 (1946).

15. V. V. Vagner, "Differential-geometric methods in the calculus of variations," Uch. Zap. Kazansk. Univ., 115(10):4-7 (1955).

16. V. V. Vagner, "The calculus of variations as the theory of the field of central semicones," Nauchn. Ezhegodnik. Saratovsk. Univ., Mekh.-Mat. Fak., 1955, Saratov, 1959, pp. 27-34.

17. M. V. Vasil'eva, "The geometry of an integral," Mat. Sb., 36(1):57-92 (1955).

18. M. V. Vasil'eva, "Invariant description of the Cartan geometry of an integral," Uch. Zap. Mosk. Gos. Ped. Inst. im. V. I. Lenina, No. 208, 76-85 (1963).

19. Ch'ao-hao Ku and Buchin Su, "First and second variations of a multiple integral in a space with a multiple areal metric," Shusyue Syuebao, 2(4):231-245 (1953).

20. L. E. Evtushik, "On the geometry of a double integral," Mat. Sb., 37(1):197-208 (1955).

21. L. E. Evtushik, "The geometry of the integral $\int F(x^\alpha, x^n, x^n_\alpha, x^n_{\alpha\beta})[dx^1, \ldots, dx^{n-1}]$ " Nauchn. Dokl. Vysshch. Shkoly, Fiz.-Mat. Nauk, No. 6, 114-118 (1958).

22. L. E. Evtushik, "The Lie derivative and the differential field equations of a geometrical object," Dokl. Akad. Nauk SSSR, 132(5):998-1001 (1960).

23. Yu. I. Ermakov, "Three-dimensional space with cubic semimetric," Dokl.
     Akad. Nauk SSSR, 118(6):1070-1073 (1958).

24. Yu. I. Ermakov, "Spaces X_n with an algebraic metric and semimetric," Dokl.
     Akad. Nauk SSSR, 128(3):460-463 (1959).

25. G. I. Zhotikov, "The Lagrange variational problem specified in intrinsic fashion,"
     Tr. Fiz.-Mat. Fak. Kirgizsk. Gos. Univ., No. 3, 36-49 (1956).

26. G. I. Zhotikov, "On the theory of the field of local conical surfaces in a first-
     order tangent composite manifold $E_n(X_n)$. I," Izv. Vysshch. Uchebn. Zavedenii,
     Matematika, No. 3, 53-64 (1959).

27. G. I. Zhotikov, "On the theory of the field of local conical surfaces in a first-
     order tangent composite manifold $E_n(X_n)$. II," Izv. Vysshch. Uchebn. Zavedenii,
     Matematika, No. 4, 64-69 (1959).

28. G. I. Zhotikov, "On the theory of the field of local surfaces in a first-order
     tangent composite manifold $E_n(X_n)$," Tr. Seminara Vectorn. Tenzorn. Analizu,
     No. 11, 189-218 (1961).

29. G. I. Zhotikov, "Differential singular Finsler metric, definable in X_n by a field
     of local singular hypersurfaces of singularity class n-m-1," Uch. Zap. Bashkirsk.
     Univ., No. 20, 32-45 (1965).

30. G. I. Zhotikov, "Differential Lagrangian (m + 1)-metric, definable in X_n by a
     field of local m-surfaces," Uch. Zap. Bashkirstk. Univ., No. 20, 9-31 (1965).

31. G. I. Zhotikov, "Geometric theory of the Lagrange variational problem for
     multiple integrals," Uch. Zap. Bashkirsk. Univ., No. 31, 19-66 (1968).

32. G. I. Zhotikov, "Regular variational problem on an unconditional extremum for
     multiple integrals," Uch. Zap. Bashkirsk. Univ., No. 31, 67-84 (1968).

33. G. I. Zhotikov, "Geometry of X_n with an m-dimensional Riemannian metric and
     its applications to the calculus of variations for multiple integrals," Uch. Zap.
     Bashkirsk. Univ., No. 31, 85-109 (1968).

34. G. I. Zhotikov, "Introduction to the geometry of the calculus of variations for
     multiple integrals," Uch. Zap. Bashkirst. Univ., No. 31, 3-18 (1968).

35. G. I. Zhotikov, "Certain aspects of the geometric theory of Carathéodory trans-
     formations in the calculus of variations for multiple integrals and a sufficient
     condition for an extremum in the Lagrange problem" Uch. Zap. Bashkirsk.
     Univ., No. 31, 110-214 (1968).

36. N. I. Kabanov, "Cartan space, defined to within Carathéodory transformations,"
     Uch. Zap. Bashkirstk. Gos. Ped. Inst., 3:47-77 (1958).

37. N. I. Kabanov, "Geometric theory of the Carathéodory transformations in the
     Lagrange problem," Tr. Seminara Vectorn. Tenzorn. Analizu, No. 11, 219-240
     (1961).

38. N. I. Kabanov, "Singular Finsler space, defined correctly within Carathéodory
     transformations," Sibirsk. Mat. Zh., 2(5):655-671 (1961).

39. N. I. Kabanov, "On the geometric theory of the simplest singular variational
     problem for an (n-1)-fold integral," Tr. Seminara Vectorn. Tenzorn. Analizu,
     No. 12, 239-268 (1963).

40. N. I. Kabanov, "Geometry of a singular variational problem for an (n-1)-fold
     integral," Report Abstracts Second All-Union Geom. Conf. Kharkov, Sept. 17-23,
     1964 [in Russian], Kharkov (1964), pp. 98-99.

41. N. I. Kabanov, "On the geometric theory of the Carathéodory transformations in the simplest singular variational problem for an (n-1)-fold integral," Tr. Seminara Vectorn. Tenzorn. Analizu, No. 14, 174-199 (1968).

42. A. Kawaguchi, "An n-dimensional space with a connection depending on m-dimensional flat elements," Tr. Seminara Vectorn. Tenzorn. Analizu, No. 5, 290-300 (1941).

43. S. A. Kaganov, "Geometry of a space with a singular hyperareal metric," Mat. Sb., 42(4):497-512 (1957).

44. A. E. Liber, "Two-dimensional spaces with an algebraic metric," Tr. Seminara Vectorn. Tenzorn. Analizu, No. 9, 319-350 (1952).

45. M. V. Losik, "A certain class of Kawaguchi spaces," Dokl. Akad. Nauk SSSR, 134(6):1299-1302 (1960).

46. M. V. Losik, "Klein spaces as Kawaguchi spaces," Dokl. Akad. Nauk SSSR, 139(6):1299-1301 (1961).

47. M. V. Losik, "Geometric interpretation of certain conditions in an ordinary variational problem with higher derivatives," Sibirsk. Mat. Zh., 2(4):556-566 (1961).

48. M. V. Losik, "Kawaguchi spaces related with Klein spaces," Tr. Seminara Vectorn. Tenzorn. Analizu, No. 12, 213-237 (1963).

49. M. V. Losik, "Infinitesimal connections in tangent fiber spaces," Izv. Vysshch. Uchebn. Zavedenii, Matematika, No. 5, 54-60 (1964).

50. A. I. Pershin, "On the geometric theory of a singular Lagrange variational problem in X_4," Investigations on Integro-Differential Equations at Kirgiz, No. 2 [in Russian], Akad. Nauk Kirg.SSR, Frunze (1962), pp. 335-342.

51. A. I. Pershin, "On the geometric theory of the Radon case of a singular Lagrange variational problem in X_4," Izv. Vysshch. Uchebn. Zavedenii, Matematika, No. 5, 95-99 (1964).

52. A. I. Pershin, "Geometric theory of the Carathéodory transformations of a singular Lagrange variational problem in X_4," Tr. Molodykh Uchenykh Saratovsk. Univ., Vyp. Mat., Saratov (1964), pp. 73-87.

53. N. F. Rzhekhina, "On the theory of the field of local curves in X_n," Dokl. Akad. Nauk SSSR, 72:461-464 (1950).

54. N. F. Rzhekhina, "Theory of the field of local hypertorses in X_n," Tr. Seminara Vectorn. Tenzorn. Analizu, No. 9, 411-430 (1952).

55. Pen-wang Sun, "Equivalence problem for the integral $\int F(x, y, y', \ldots, y^{(n)})dx$." Shyusyue Syuebao, 4(2):223-224 (1954).

56. P. I. Tokarev, "Geometric theory of the second variation in the Lagrange variational problem," Tr. Seminara Vectorn. Tenzorn. Analizu, No. 9, 431-455 (1952).

57. I. V. Frolova, "On the theory of the field of local rigged m-cones in $E_n (X_n)$," Uch. Zap. Bashkirsk. Univ., No. 31, 303-315 (1968).

58. É. I. Khmielevskii, "Application of the Lie derivative to seeking extremals in spaces of hyperflat elements," Sb. Aspirantsk. Rabot. Kazansk. Univ., Mat., Mekh., Fiz., Kazan (1964), pp. 94-96.

59. É. I. Khmielevskii, "Invariant form of the second variation of an (n-1)-fold integral, obtained by Lie differentiation in Cartan space," Sb. Aspirantsk. Rabot. Kazansk. Univ., Mat., Mekh., Fiz., Kazan (1964), pp. 61-68.

60. S. M. Chashechnikov, "Theory of the field of local hypercones," Dokl. Akad. Nauk SSSR, 117:765-768 (1957).

61. D. E. Chernyi, "Geometry of the Lagrange variational problem with a double integral in X_4," Izv. Vysshch. Uchbn. Zavedenii, Matematika, No. 4, 153-165 (1964).

62. D. E. Chernyi, "Geometric theory of the differential invariants of the Lagrange problem with a double integral in the space X_4," Tr. Molodykh Uchenykh Saratovsk. Univ., Vyp. Mat., Saratov (1964), pp. 108-119.

63. D. E. Chernyi, "A sufficient condition in the Lagrange problem with higher-order derivatives," Volzhsk. Mat. Sb., No. 5, 374-379 (1966).

64. D. E. Chernyi, "Geometric theory of a singular variational problem with a double integral in the space X_4," Volzhsk. Mat. Sb., No. 3, 332-338 (1965).

65. D. E. Chernyi, "Geometry of 1-cutting and semiconical 2-cutting surfaces in the fiber space $M_{4+\binom{4}{2}}$ and its applications to the calculus of variations," Tr. Seminara Vectorn. Tenzorn. Analizu, No. 14, 200-228 (1968).

66. F. Alardin, L'autoparallélisme des courbes extrémales dans les espaces métriques fondés sur la notion d'aire. J. Math. pures appl., 27:255-336 (1948).

67. C. B. Barker, The Lagrange multiplier rule for two dependent and two independent variables. Amer. Journ., Mathem., 67:256-276 (1945).

68. W. Barthel, Über das Verhältnis der Vectorübertragung zu den Variations-problemen in Cartanschen Räumen. Rend. Circolo mat. Palermo, 3(2):270-281 (1954).

69. W. Barthel, Über homogene Funktionen auf dem Grassman-Kegel. Arch. Math., 9:262-274 (1958).

70. W. Barthel, Über metrische Differentialgeometrie, begründet auf dem Begriff eines p-dimensionalen Areals. Math. Ann., 137(1):42-63 (1959).

71. L. Berwald, Über die n-dimensionale Cartanschen Räume und eine Normalform der zweiten eines (n−1)-fachen Oberflächenintegrale. Acta Math., 71:191-198 (1939).

72. L. Berwals, Über Finslersche und Cartansche Geometrie. II. Invarianten bei der Variation vielfacher Integral und Parallelhyperflächen in Cartanschen Räumen. Comp. Math., 7:141-176 (1939).

73. W. Bettingen, Zum Satz von Gauss−Bonnet in der dreidimensionalen metrischen Differentialgeometrie. Rend. Circolo mat Palermo, 9(3):347-359 (1960).

74. W. Blaschke, Über die Figuratrix in der Variationsrechnung. Arch. Math. Phys., 3(20):28-44 (1912).

75. F. Brickell, On the existence of metric differential geometries based on the notion of area. Proc. Cambridge. Phil. Soc., 46:67-72 (1950).

76. F. Brickell, On areal spaces. Tensor, 13:19-30 (1963).

77. C. Carathéodory, Über die diskontinuierlichen Lösungen in der Variationsrechnung," Dissertation, Göttingen (1904), 71 pp.

78. E. Cartan, "Les espaces metriques fondés sur la notion d'aire," Acualites scientifiques et industrielles 72, Hermann, Paris (1933).

79. E. Cartan, La geometrie de l'integrale $\int F(x, v, y', y'')dx$. J. Math. pures. appl., 15:42-69 (1936).

80. E. T. Davies, Areal spaces. Ann. Mat. pura ed appl., 55:63-76 (1961).

81. E. T. Davies, On the use of osculating spaces. Tensor, 14:86-98 (1963).

82. R. Debever, "Sur une classe d'espaces a connexion euclidienne," Dissertation, Brussels (1947).

83. R. Debever, "Sur une structure infinitésimale régulière associée aux intégrales d'hypersurfaces de calcul de variations," Convegno internationale di geometria differenziale, Italy, 1953, ed. Cremonese, Rome (1954), pp. 214-221.

84. P. Dedecker, "Calcul des variations formes differentielle et champs géodési- ques," Colloq: internat. centre nat. rech. scient. Strasbourg 52, Paris (1953), pp. 17-34.

85. A. Deicke, Über die Finsler-Räume met $A_i = 0$. Arch. Math., 4(1):45-51 (1953).

86. M. Gama, On areal spaces of the submetric class. I. II. III. Tensor 16(3):262- 268 (1965); 16(3):291-293 (1965); 17(1):79-85 (1966).

87. S. Hokari, Die Geometrie des Integrals $\int (a_i a_j x''^l x''^j + 2b a_i x''^i + c)^{\frac{1}{p}} dt$. Proc. Acad. Tokyo, 12:209-212 (1936).

88. S. Hokari, Die Theorie des Kawaguchischen Raumes mit der Massbestimmung von einer bestimmten Gestalt. J. Fac. Sci Hokkaido Univ., Ser. I. Math., 8:63- 78 (1940).

89. H. Hombu, Geometrie des Integrals $\int (Ly''' + M) dx$. Proc. Imp. Acad., Tokyo, 12:159-161 (1936).

90. S. Ide, On the theory of curves in n-dimensional space with the metrics
$$s = \int (A_i(x, x') x'^i + B)^{\frac{1}{p}}$$ I. II. Tensor, 9:25-29 (1949); 2:89-98 (1952).

91. S. Ide, On the connections in higher order spaces. Tensor, 3:84-90 (1954).

92. S. Ide, On the connections in higher order spaces. II. Tensor, 4(3):135-140 (1955).

93. S. Ide, On the Wirtinger's connections in higher order spaces. J. Fac. Sci. Hokkaido Univ., Ser. I, 13:75-119 (1956).

94. S. Ide, On the geometrical meanings of Wirtinger's connections based on Kawa- guchi's. Tensor, 14:216-218 (1963).

95. H. Iwamoto, Über eine geometrische Theorie der mehrfachen Integrale. Japan. J. Math., 19:479-512 (1948).

96. H. Iwamoto, On geometries associated with multiple integrals. Math. Japonicae, 1:74-91 (1948).

97. H. Iwamoto, On the geometry in a space based on the notion of area. I., Ten- sor, 9:7-12 (1949).

98. C. Kano, Conformal geometry in an n-dimensional space with the arc length
$$s = \int (A_i(x, x') x''^i + B(x, x'))^{\frac{1}{p}} dt.$$ Tensor, 5(3):187-196 (1956).

99. C. Kano, Conformal geometry in an n-dimensional space with the arc length
$$s = \int (A_i(x, x') x''^i + B(x, x'))^{\frac{1}{p}} dt.$$ II. Tensor, 10(3):210-217 (1960).

100. Y. Katsurada, On the theory of curves in a higher order space with some special metrics. Tensor, 7:58-64 (1944).

101. A. Kawaguchi, Theory of connections in the generalized Finsler manifold. Proc. Imp. Acad. Tokyo, 7:211-214 (1937).

102. A. Kawaguchi, Die Differentialgeometrie in der verallgemeinerten Mannigfaltig- keit. Rend. Circolo mat. Palermo, 56:245-276 (1932).

103. A. Kawaguchi, Ein metrischer Raum, der eine Verallgemeinerung des Finslerschen Raumes ist. J. Math. und Phys., 43:289-297 (1936).

104. A. Kawaguchi, Theory of connections in a Kawaguchi space of higher order. Proc. Imp. Acad. Tokyo, 13:237-240 (1937).

105. A. Kawaguchi, Geometry in an n-dimensional space with the arc length $s=$ $\int (A_i(x, \; x')x''^i + B(x, \; x'))^{\frac{1}{p}} \; dt.$ Trans. Amer. Math. Soc., 44:153-167 (1938).

106. A. Kawaguchi, On areal spaces. I. II. III. Tensor, 14-45 (1950); 1:67-88 (1951); 1:89-101 (1951).

107. A. Kawaguchi, On the theory of non-linear connections. II. Theory of Minkowski space and of non-linear connections in a Finsler space. Tensor, 6(3):165-199 (1956).

108. A. Kawaguchi, On the theory of areal spaces. Bull. Calcutta Math. Soc., 56(2-3): 91-104 (1964).

109. A. Kawaguchi and S. Hokari, Grundlegung der Geometrie der n-dimensionalen metrischen Räume auf Grund des Begriffs des k-dimensionalen Flächeninhalts. Proc. Imp. Acad. Tokyo, 16:320-325 (1940).

110. A. Kawaguchi and K. Tandai, On areal spaces. V. Normalized metric tensor and connection parameters in a space of the submetric class. Tensor, 2:47-58 (1952).

111. S. Kawaguchi, On some properties of projective curvature tensor in a special Kawaguchi space. Tensor, 13:83-88 (1963).

112. S. Kawaguchi, On a special Kawaguchi space of recurrent curvature. Tensor, 15(2):145-158 (1964).

113. S. Kawaguchi and T. Nobuhara, On extremal curves in a special Kawaguchi space. Tensor, 5(3):197-200 (1956).

114. D. Laugwitz, Eine Beziehung zwischen affiner und Minkowskischer Differentialgeometrie. Publs Math., 5(1-2):72-76 (1957).

115. T. Michihiro, Theory of curves in a two-dimensional space with arc-length $s=\int (A_i x''^i \; +B)^{\frac{1}{p}} \; dt,$ Tensor, 4:63-66 (1944).

116. A. Moór, Über die Dualität von Finslerschen und Cartanschen Räumen. Acta Math., 88:347-370 (1952).

117. A. Moór, Ergänzungen zu meiner Arbeit: "Über die Dualität von Finslerschen und Cartanschen Räumen.", Acta Math., 91(3-4):187-188 (1954).

118. A. Moór, Die osculierenden Riemannschen Räume regulärer Cartanscher Räume. Acta Math. Acad. sci hung., 5(1-2):59-72 (1954).

119. M. Okumura, On some remarks of special Kawaguchi spaces. Tensor, 11(2):154-160 (1961).

120. G. Reeb, "Sur les espaces de Finsler et les espaces de Cartan," Colloq. internat. centre nat. rech. scient. Strasbourg 52, Paris (1953), pp. 35-40

121. H. Rund, Über nicht-holonome allgemeine metrische Geometrie. Math. Nachr., 11(1-2):61-80 (1954).

122. H. Rund, The Differential Geometry of Finsler Spaces, Springer, Berlin (1959), Vol. 8, p. 283.

123. N. K. Sharma and R. Behari, Some properties of spaces with the arc-length $s=\int (A_i(x^i, \; \dot{x}^i) \ddot{x}^i)^{\frac{1}{3}} dt.$ Ganita, 15(1):1-8 (1964).

124. T. N. Srivastava, Generalized Bianchi and Veblen identities in a special Kawaguchi space. Tensor, 15(2):87-102 (1964).

125. T. N. Srivastava, Les identités généralisées de Bianchi et de Veblen dans les espaces de Kawaguchi spéciaux. C. r. Acad. sci., 254(15):2706-2708 (1962).

126. T. N. Srivastava, A few remarks on special Kawaguchi spaces. Tensor, 15(1): 12-19 (1964).

127. T. N. Srivastava and R. S. Mishra, Bianchi identies in special Kawaguchi space. Tensor, 11(1):43-50 (1961).

128. Buchin Su, Koschmieder invariant and the associate differential equation of minimal hypersurface in a regular Cartan space. Math. Nachr., 16(2):117-129 (1957).

129. Buchin Su, On the theory of affine connections in an areal space. Bull. Math. Soc. sci. math. et phys. RPR, 2(2):185-190 (1958).

130. Buchin Su, The geometry of spaces with areal metrics. Math. Nachr., 16(5-6): 281-287 (1957).

131. Buchin Su, On the determination of certain affine connections in an areal space. Sci. Rec., 1(4):195-198 (1957).

132. Buchin Su, Certain affinely connected spaces with areal metrics. Scientia Sinica, 6:967-975 (1957).

133. J. L. Synge, Some intrinsic and derived vectors in a Kawaguchi space. Amer. J. Math., 57:679-691 (1935).

134. K. Tandai, On areal spaces VI. On the characterization of metric areal spaces. Tensor, 3(1):40-45 (1953).

135. K. Tandai, On areal spaces. VII. The theory of the canonical connection an m-dimensional subspaces. Tensor, 4(2):78-90 (1954).

136. K. Tandai, On general connections in an areal space. II. General Connections on the tangent m-frame bundle. Tensor, 19:26-46 (1963).

137. K. Tandai, On general connections in an areal space. I. General connection on a fibre of the tangent m-frame bundle. Tensor, 13:277-291 (1963).

138. K. Tandai, On areal space. VIII. Theory of a space of the semi-metric class. Tensor, 10:161-166 (1960).

139. Y. Tashiro, A theory of transformation groups on generalized spaces and its application to Finsler and Cartan spaces. J. Math. Soc. Japan, 11(1):42-71 (1959).

140. K. Tonooka, On the geometry of an $(n-1)$-ple integral of order two. Tensor, 1:53-59 (1951).

141. K. Tonooka, Theory of subspaces in a geometry based on a multiple integral I. Metric tensor and theory of connections. Tensor, 3:75-83 (1954).

142. K. Tonooka, On a geometry of three-dimensional space with an algebraic metric. Tensor, 6(1):60-68 (1956).

143. K. Tonooka, On three and four dimensional Finsler spaces with the fundamental form $\sqrt[3]{a_{\alpha\beta\gamma}x'^{\alpha}x'^{\beta}x'^{\gamma}}$. Tensor, 9(3):209-216 (1959).

144. O. Varga, Die Krümmung der Eichfläche der Minkowskischen Raumes und die geometrische Deutung des eigen Krümmungtensors des Finslerschen Raumes. Abhandl. Math. Seminar Univ. Hamburg, 20(1-2):41-51 (1955).

145. O. Varga, Eine Charakterisierung der Kawaguchischen Räume metrischer Klasse mittels eines Satzes über derivierte Matrizen. Publs. math., 4(3-4):418-430 (1956).

146. V. V. Vagner [Wagner], Theory of a field of local (n−2)-dimensional surfaces in X_n and its application to the problem of Lagrange in the calculus of variations. Ann. Math., 49:141-188 (1948).

147. V. V. Vagner [Wagner], Geometria del calcolo delle variazioni, Vol. II, Centro Internationale Matematico estivo (CEME), Ed. Cremonese, Rome (1965), 172 pp.

148. S. Watanabe, On special Kawaguchi spaces. Tensor, 7(2):130-136 (1957).

149. S. Watanabe, On special Kawaguchi spaces. II. A generalizations of affine Spaces. Tensor, 8(3):169-176 (1958).

150. S. Watanabe, On special Kawaguchi spaces. III. Generalizations of affine spaces and Finsler spaces. Tensor, 11(2):144-153 (1961).

151. S. Watanabe, On special Kawaguchi spaces. IV. Extremal curves in the generalized affine spaces. Tensor, 11(3):254-262 (1961).

152. S. Watanabe, On special Kawaguchi spaces. V. Some remarks on the special Kawaguchi spaces. Tensor, 11(3):279-284 (1961).

153. S. Watanabe, On special Kawaguchi spaces. VI. Some transformations in certain special Kawaguchi spaces. Tensor, 12(3):244-253 (1962).

154. S. Watanabe and M. Yoshida, On special Kawaguchi spaces. VII. Some transformations in certain special Kawaguchi spaces II. Tensor, 13:31-41 (1963).

155. A. Winternitz, Über die affine Grundlage der Metrik eines Variationsproblems. S.−B. preuss. Akad. Wiss., 26:457-469 (1930).

156. M. Yoshida, On the connections in a subspace of the special Kawaguchi space. Tensor, 17(1):49-52 (1966).